How Democracies Lose Small Wars

State, Society, and the Failures of France in Algeria, Israel in Lebanon, and the United States in Vietnam

In *How Democracies Lose Small Wars*, Gil Merom argues that modern democracies fail in wars of insurgency because they are unable to find a winning balance between expedient and moral tolerance of the costs of war. Small wars, he argues, are lost at home when a critical minority shifts the center of gravity from the battlefield to the marketplace of ideas.

This minority, from among the educated middle-class, abhors the brutality involved in effective counterinsurgency, but also refuses to sustain the level of casualties that successfully combatting counterinsurgency requires. Government and state institutions further contribute to failure as they resort to despotic patterns of behavior in a bid to overcome their domestic predicament.

Merom proceeds by analyzing the role of brutality in counterinsurgency, the historical foundations of moral and expedient opposition to war, and the actions states traditionally took in order to preserve foreign policy autonomy. He then discusses the elements of the process that led to the failure of France in Algeria and Israel in Lebanon.

In the Conclusion, Merom considers the Vietnam War and the influence that failed small wars has had on Western war-making and military intervention.

Gil Merom is a Research Fellow at the University of Sydney and an Assistant Professor of Political Science at Tel-Aviv University. He has published articles on international relations theory, international security, Israeli security and strategy, intelligence estimates, low-intensity conflict, the Algerian war, and other topics in edited volumes and professional journals, including *Political Science Quarterly*, *Armed Forces and Society*, and the *Journal of Strategic Studies*.

How Democracies Lose Small Wars

State, Society, and the Failures of
France in Algeria, Israel in Lebanon,
and the United States in Vietnam

GIL MEROM
Tel-Aviv University
University of Sydney

CAMBRIDGE
UNIVERSITY PRESS

CAMBRIDGE UNIVERSITY PRESS
Cambridge, New York, Melbourne, Madrid, Cape Town, Singapore, São Paulo

Cambridge University Press
40 West 20th Street, New York, NY 10011-4211, USA

www.cambridge.org
Information on this title: www.cambridge.org/9780521804035

© Gil Merom 2003

First published 2003

Printed in the United States of America

A catalog record for this publication is available from the British Library.

Library of Congress Cataloging in Publication Data

Merom, Gil, 1956–
How democracies lose small wars : state, society, and the failures of France in Algeria,
Israel in Lebanon, and the United States in Vietnam / Gil Merom.
    p.   cm.
Includes bibliographical references and index.
ISBN 0-521-80403-5 – ISBN 0-521-00877-8 (pbk.)
1. Low intensity conflicts (Military science) – France   2. Low intensity conflicts
(Military science) – United States.   3. Low intensity conflicts (Military science) – Israel.
4. Military doctrine – France.   5. Military doctrine – United States.   6. Military doctrine –
Israel.   7. Counterinsurgency.   I. Title.
U241.M47 2003
355.02 – dc21   2002041490

ISBN-13   978-0-521-80403-5 hardback
ISBN-10   0-521-80403-5 hardback

ISBN-13   978-0-521-00877-8 paperback
ISBN-10   0-521-00877-8 paperback

To the professors of the department of government,
Cornell University, 1987–1991

Contents

List of Figures and Tables

Figures

Tables

Acknowledgments

Peter Katzenstein, a scholar and friend of a rare breed, supervised my doctoral dissertation, out of which this book grew. I owe him, as many others do, a debt he would never bother to collect, nor would I be able to pay. No one so wisely and generously supports others' research and at the same time fiercely protects their intellectual independence.

Alan Dowty was instrumental in my decision to study for an advanced degree. He probably does not realize how much I learned from him throughout the years. Martin Shefter, Lawrence Scheinman, and Jim Goldgeier improved my work with wise suggestions. Jim, in particular, put his own work aside simply because I needed his advice. John Garofano, Dafna Merom, and Amir Weiner followed and supported my research from its inception. John put at my service his critical mind and literary talent; Dafna forced me to rethink and distill ideas; and Amir, a compulsive historian, contributed his analysis, insight, and advice.

I am indebted to Andrew Mack and Samuel Huntington, who must be unaware of their contributions. Mack's seminal article in *World Politics* about small wars ignited my interest in the topic and inspired many of my ideas. Huntington's essays on political order in changing societies shaped much of my thinking about how the state, as an institution, approaches society.

Many others have also supported my work. The anonymous reviewers of Cambridge University Press invested much time in carefully reading the manuscript and offering superb and detailed suggestions for improvement. Richard Ned Lebow was instrumental in bringing me to Cornell, and Judith Reppy helped me thereafter. I still treasure her good advice. Professor Jean Pierre Rioux helped me in Paris with bibliographical advice. Other peers discussed various parts of my work with me, commented on drafts, corrected me, and contributed exciting ideas. In particular, I thank Gad Barzilai, Michal Ben-Josef-Hirsh, Avi Ben-Zvi, Assif Efrat, Yair Evron, John Fousek, Azar Gat, Benjamin Ginsberg, Tami Hermann, Stanley Hoffmann, Steve Jackson, Mary Katzenstein, Aharon Klieman, Walter Lafeber, Theodore Lowi, Roger

Marcwick, John Mearsheimer, Joe Nye, David Patton, Bruce Porter, Barry Posen, Jeremy Rabkin, Yossi Shain, Jim Sickmeier, and David Vital. For expressing confidence in my work and showing me the generous face of academic institutions, I am indebted to Avi Ben-Zvi, Deborah Brennan, Azar Gat, Graeme Gill, Samuel Huntington, Alfred James, Eti C. Lenman, Anne Madden, Stephen Nicholas, Martin Painter, Bruce Porter, M. Ramesh, Merrilee Robb, Yossi Shain, David Siddle, and Rod Tiffen.

Several programs and foundations supported my research. The Peace Studies Program became my home within Cornell. The Mellon Fellowship Program at Cornell and the Institute for the Study of World Politics in Washington, D.C., granted me generous fellowships. The Cornell Center for International Studies awarded me a Mario Einaudi Research Travel Grant, and the university's Western Societies Program granted me a SICCA Research Fellowship in Europe. The Center for International Affairs and the John M. Olin Institute for Strategic Studies in Harvard provided me with various privileges, including a distinguished Olin Postdoctoral Fellowship. The Israeli Council on Higher Education extended me its prestigious Alon Fellowship, and the University of Sydney granted me a U2000 Research Fellowship. The Department of Government and International Relations at the University of Sydney has also helped by covering some of the production costs of this book.

This book could not have been written without the help of the devoted staff of the following institutions: the libraries of Haifa and Tel-Aviv Universities, the Archive of the *Ha'aretz* daily; the libraries of the Sorbonne, University Paris VIII (St. Denis), and University of Paris X (Nanterre); Fondation Nationale de la Science Politique in Paris; the Olin library at Cornell; the Weidner and other libraries at Harvard; the library of the Army War College in Carlisle, PA; and the Fisher Library of the University of Sydney. I also thank Frank Cass for permission to use my article in *Small Wars and Insurgencies*, as a basis for discussion, particularly in Chapter 2, and to photographer James Nachtwey and Photo Agency VII of Paris, France, for permission to use his photo from the book *Deeds of War* on the cover of this book.

Many friends have helped me in various ways, often by offering me critically needed logistic and other assistance that made my research easier. In particular, I thank Vera and Lucio Acioli, Anne and Mayor Admon, Nurit and Noga Alon, Julia Erwin-Weiner, Nurit and Avi Israeli, Darryl Jarvis, John and Monica Garofano, Sarah Garofano, José Fernández, Marnie Goodbody, Mary and Peter Katzenstein, Paula and Bertrand Lazard, Carmela and Aryeh Merom, Shaul Mishal, Jim T. Pokorny, Maria Robertson, and Gideon–"Gidi the Bear"–Schaller.

Finally, I am extremely grateful to my editors, Aryeh Merom and Susan Park, to my copy editor Ronald Cohen, and to Lewis Bateman from Cambridge University Press for patiently guiding me throughout the laborious process of turning my book-draft into an acceptable manuscript. I am also grateful to Robert Swanson for preparing the index.

In large measure, my work was pleasant and rewarding because of the outstanding individuals I have mentioned. There is something of each of them in the book and, happily, by now, in myself.

PART I

I

Introduction

Put narrowly, this book explains how democracies fail in small wars in spite of their military superiority. Yet this book is not so much about military matters and the interaction between unequal parties on the battlefield. Rather, it is largely about societal processes within democracies that are engaged in counterinsurgency and about how these processes affect world politics. Indeed, the explanatory power of the argument and the implications of this study transcend the phenomenon of small wars, and, as I explain in the Conclusion, are relevant to a number of important issues of political science, foreign policy, and international relations.

The Biased Study of War and the Neglect of Small Wars

Paradoxically, the study of war in political science and international relations may have narrowed the understanding of war. By and large, war was subjected to mechanistic models that considered variables such as industrial capacity, military hardware, levels of forces deployed, and organizational structures and routines that lend themselves to easy operationalization and measurement. Society was hardly ever the focal point of the research on war. Indeed, rarely were social friction, cultural attributes, prevailing values, and norms taken into serious consideration as determinants of the outcomes of war. Rather, the amorphous collective of society was by and large considered important only in relation to its potential as a source for the men and material needed for war. Human beings were the subjects of study of war, mostly when they were dwellers of the upper level of the political order: politicians, bureaucrats, and generals. Ironically, conventional wars were studied by political scientists mostly from the state perspective, precisely when the relations between state, society, and war became ever more complex.

Within the field of international relations, the study of war was biased in yet another way. Major systemic wars attracted most of the attention, whereas small wars – those that pitted powerful states against insurgent communities – were relegated to the margins of intellectual discourse. Two unjustified assumptions seem to have underlain this unfortunate development: the assumption that the asymmetric nature of the power-relations between the protagonists left little to be studied, and the assumption that, irrespective of the outcomes, small wars were irrelevant to world politics.[1]

Why Study Small Wars?

A small war has the following distinct characteristics: It involves sharp military asymmetry, an insurgent that fights guerrilla war, and an incumbent that uses ground forces for counterinsurgency warfare. The incumbent can be an indigenous government that fights on its own or with external participation, or a foreign power that imposes itself on the population.[2]

Small wars are important for several reasons. First, they date to antiquity and are relatively widespread. Indeed, the quantitative study *Resort to Arms* suggests that 43 percent of a total of 118 international violent conflicts that occurred between 1816 and 1980 were small wars.[3] Second, the study of small wars is important because their recent outcomes seem to involve a phenomenological novelty. History is not short of cases of communities that decided to challenge powerful conquerors despite desperate military inferiority and bleak prospects of success. However, in many of these cases, the daring underdog lost, at times to the point that it could not even recover. In the twentieth century, and particularly after 1945, underdogs seem to have done rather well in small wars, particularly when their enemies were democratic.[4] Indeed, they have succeeded against Britain (in Palestine), France

[1] See, for, example Kenneth N. Waltz, *Theory of International Politics* (Reading, MA: Addison-Wesley Publishing, 1979), 190–91.
[2] This definition is closest to that of Eliot A. Cohen's in "Constraints on America's Conduct of Small Wars," *International Security*, 9:2 (1984), 151. See also the definition in Charles E. Callwell's seminal manual on guerrilla warfare (first published in 1896), *Small Wars: Their Principles and Practice* (University of Nebraska Press, 1996), 21. For other definitions and discussions, including of *Low Intensity Conflict* (LIC), see Loren B. Thompson, "Low-Intensity Conflict: An Overview," in Loren B. Thompson (ed.), *Low-Intensity Conflict* (Lexington, MA: Lexington Books, 1989), 2–6; Andrew F. Krepinevich, Jr., *The Army and Vietnam* (Baltimore, MD: Johns Hopkins University Press, 1986), 7; and see Walter Laqueur, *Guerrilla* (Boston: Little, Brown and Co., 1976), 29.
[3] Data from Melvin Small and J. David Singer, *Resort to Arms* (Beverly Hills: Sage Publications, 1982), 52, 59–60, table 2.2.
[4] Global fluctuation in territory acquisition and loss and the demise of European colonialism and imperialism seem also to support this temporal contention. See David Strang, "Global Patterns of Decolonization, 1500–1987," *International Studies Quarterly*, 35 (1991), 429–54, particularly p. 435, figure 1. For data on the changing number of independent political units, see also Charles Tilly, *The Formation of National States in Western Europe* (Princeton: Princeton University Press, 1975), 23; William Eckhardt, *Civilizations, Empires and Wars: A Quantitative History of War* (Jefferson, NC: McFarland and Company, 1992), 147. While the proliferation of independent states was not caused by one factor, it is worthwhile to recall that (a) Western states were the most powerful actors during this period of proliferation; (b) domination over

(in Algeria), the United States (in Vietnam), and Israel (in Lebanon) in spite of clear battlefield inferiority. These successes seem all the more spectacular and puzzling as the democratic protagonists were among the most experienced, successful (except for France), and resilient states to have fought conventional wars in modern times.

Third, small wars are important because their study can produce new insight concerning international relations. Above all, the legacy of small wars can be shown to have a lasting effect on matters that concern the military intervention of democracies. The recent efforts of the Russian state to monopolize the media in the context of the war in Chechnya, and the recent events of the Israeli-Palestinian conflict, are good examples. Moreover, this legacy may be of even greater importance because situations that call for intervention, and have the potential of regressing into small wars, seem to multiply (as we have witnessed in the post-Cold War world, in the Balkans, the Caucusus, Africa, and South-East Asia). Finally, the states that are most likely to consider military intervention in these situations – Western democracies – are the subjects of this study.

International Relations Theory and Small Wars

The most compelling explanations of the outcomes of small wars originate in either realist or motivational theories. Both theoretical schools offer valuable insights that I shall take into consideration after discussing the elements of my thesis. However, none is able to adequately explain the changing pattern of the outcomes of small wars.

Realism, the Balance-of-Power, and Small Wars
The idea that militarily inferior protagonists can prevail in a violent conflict poses a paradox for realism. In a conceptual world that emphasizes the idea that the most important difference between actors is the respective amount of power they possess, and maintains that the ultimate arbiter of international conflict is power defined in terms of military capabilities, the victory of weaker protagonists is hardly conceivable. Indeed, realists address arguments about the futility of military power as alleged rather than proven. Their predisposition – the belief that outcomes of war necessarily reflect the relative power of the protagonists – dictates their research question. Therefore, realists do not really ask how underdogs win wars (as this question is inherently answered by their deductive logic, which contends that if underdogs won, then somehow they must have overcome their military inferiority and ceased being underdogs). Rather, they shift their focus to the question

many territories was gained and then lost in small wars; (c) failures in small wars encouraged Western powers to relinquish more territories; (d) European empires were not replaced in the age of democratic supremacy.

of how the balance of power changed in favor of those who were previously weaker.

When the question is framed in this way, realists venture several explanations for the outcomes of small wars. Robert Gilpin composed a general argument, which is also echoed in Paul Kennedy's writing, about shifts in the balance of power.[5] Gilpin did not discuss small wars, but his reasoning can be easily adapted to explain their outcomes in realist terms. According to Gilpin, hegemonic powers expand to the point that the cost of maintaining their empires exceeds the benefits they yield. Over time, investment in the means of destruction, the cost of upholding the order in the imperial periphery, and domestic decay rob hegemons of their prowess. Rising contenders, unhampered by decadence and the burden of empire maintenance, innovate, invest more wisely and efficiently in production, and send hegemonic empires into the dustbin of history.

The implications of this argument in the context of small wars are fairly clear. The decline of a hegemonic power, particularly when it has widespread territorial commitments and is facing multiple or major challenges, permits underdogs to create a favorable balance of power in their vicinity. James Lee Ray and Ayse Vural use this logic in order to explain the failure of strong powers in small wars.[6] In fact, they were ready to carry the "economic strain" and "cost-of-empire" arguments to the extreme, suggesting that the war in Vietnam was limited (and therefore lost) because the United States had "other foreign policy goals, competing for resources, attention, and effort...."[7]

Zeev Maoz submits a different realist explanation for the puzzle of failed small wars.[8] Simplified somewhat, Maoz's argument is that display of "excessive" power provokes third parties, be they enemies of the powerful protagonist or previously indifferent actors, to join the struggle against the rising menace. This argument does not pin the cause of failure on overdispersion of resources, gradual weakening of imperial powers, or the opportunity both create for underdogs. Rather, it deals with the fear that potential victims develop when they consider the consequences of the aggressive behavior of an ambitious actor, and with the feeling of the former that something has to be done before it is too late.

Another realist way to explain the puzzle is to consider the impact of macro-developments on the balance of power between the West and the

[5] Robert Gilpin, *War and Change in World Politics* (Cambridge: Cambridge University Press, 1981); and Paul Kennedy, *The Rise and Fall of Great Powers: Economic Change and Military Conflict from 1500 to 2000* (NY: Random House, 1987).

[6] James Lee Ray and Ayse Vural, "Power Disparities and Paradoxical Conflict Outcomes," *International Interactions*, 12:4 (1986), 315–42.

[7] Ibid., 323.

[8] Zeev Maoz, "Power, Capabilities, and Paradoxical Conflict Outcomes," *World Politics*, 41:2 (1989), 239–66.

Third World. For example, Hans Morgenthau argues that the failure of Western colonial powers in small wars should be attributed to the disappearance of the "technological, economic, and military differential between the white man of Europe and the colored man of Africa and Asia" – a differential that Morgenthau believes "allowed Europe to acquire and keep its domination over the world."[9]

Finally, students of strategy and war who rely on realist logic attribute the failure of strong powers in small wars to various malfunctions in the process of converting superior resources into effective military preponderance. More specifically, these arguments contend that small wars were lost because the stronger party did not develop adequate military doctrine, adopted a failed strategy, suffered from poor military leadership, or failed in other organizational or operational ways to convert its resource-advantage into combat dominance.[10] Douglas Blaufarb, Andrew Krepinevich, Stephen Rosen, and Robert Pape have made such arguments, in the context of the Vietnam War.[11] More recently, Ivan Arreguín-Toft has added to this family of realist "battlefield" arguments an ambitious quantitative study that contends that "strong actors will lose asymmetric conflicts when they use the wrong strategy vis-à-vis their opponents' strategy."[12]

The Strength of Realism, and its Weaknesses

Realist arguments explain well much of the history of wars, including small ones. The historical relations between undisputed military superiority and the political outcomes of war were almost always straightforward. When protagonists of marked unequal capabilities were about to meet on the battlefield, the underdog was often presented only with a cruel choice between benign submission and catastrophic defeat. Failure to recognize the meaning of inferiority, or a decision to valiantly defy the iron rule of power, often ended, as the fate of vanishing peoples such as the American Indians reminds us, in national calamities of untold proportions.

Occasionally, underdogs did manage to win conflicts with rivals whose general resources and overall forces were far superior to theirs. Still, realist logic accounted for these cases, which often involved empires that

---

[9] Hans J. Morgenthau, *Politics among Nations* (NY: Knopf, 1973), 351.
[10] See, for example, Cohen, "Constraints on America's Conduct of Small Wars," 151–81, particularly p. 177.
[11] Blaufarb, *The Counterinsurgency Era: U.S. Doctrine and Performance* (NY: The Free Press, 1977), 252–55, 298–300; Krepinevich, *The Army and Vietnam*, 164, 259; Rosen, "Vietnam and the American Theory of Limited War," *International Security*, 7:2 (1982), 83, 98–103; and Pape, "Coercive Air Power in the Vietnam War," *International Security*, 15:2 (1990), 107–08.
[12] Arreguín-Toft, "How the Weak Win Wars: A Theory of Asymmetric Conflict," *International Security*, 26:1 (2001), 95.

overextended, decayed, and were under pressure from strong competitors.[13] Indeed, as realists maintain, the powerful actor in such cases failed to bring to bear, in the battlefield, its basic superiority. The British crown gave up the American colonies because of troubles elsewhere, and because of French pressure at sea and in Canada. Napoleon's forces faced vicious guerrilla warfare in Spain, but they could not defeat it because, among other things, they faced a powerful British contingent and a very talented British commander in Iberia, Wellington. Moreover, Napoleon finally gave up Spain because he needed his forces for the 1812 invasion of Russia. A century and a quarter later, Hitler faced a similar dilemma in the Balkans. He was never able to concentrate enough power in order to have a real chance of pacifying Yugoslavia because he needed every military unit he could muster for the vast Soviet battlefield.

Still, beyond a certain point in time, realist explanations become increasingly inadequate. Specifically, the details and fate of a significant number of small wars, in particular after 1945, defy realist logic. In cases such as the British struggle in Palestine (1946–48), the French war in Algeria (1954–62), Israel's invasion of Lebanon (1982–85) and its struggle there thereafter, and even in the case of the American war in Vietnam, the powers that lost enjoyed unquestionable military superiority all along. All in all, then, the outcomes of these conflicts cannot be attributed to inferiority in the battlefield.

Indeed, each of the realist explanations I reviewed suffers from detrimental weaknesses. Morgenthau's argument about the vanishing power differential between Western powers and the people of the Third World is simply at odds with history. A leading military historian such as Michael Howard, for example, concluded that the Western "technological superiority was at its most absolute" in the twentieth century.[14] Indeed, he was mystified not by the alleged technology spillover but rather by the fact that underdogs were successful in the twentieth century *in spite of* the unprecedented military capabilities and resources that their Western enemies brought to bear in war.[15]

Ray and Vural's resource-strain argument is problematic because it is not clear at all whether wars in general, and small ones in particular, necessarily hurt the economy of powerful protagonists. Scholars have demonstrated – most notably Miles Kahler – that expensive expansionist wars, and even a vast allocation of resources to security – as indicated by Japan's remarkable

[13] See Michael Doyle's analysis of the decline of empires in *Empires* (Ithaca: Cornell University Press, 1986), particularly pp. 92–103, 116–22.
[14] Michael Howard, "The Military Factor in European Expansion," in Hedley Bull and Adam Watson, *The Expansion of International Society* (Oxford: Clarendon Press, 1984), 41.
[15] See Howard on Vietnam, in "The Military Factor," 41. The relative rate of casualties in modern small wars strongly suggests that Howard is correct. Small and Singer did not find any consistent pattern between the level of casualties imperial and colonial powers suffered in small wars and the outcome of those wars. See Small and Singer, *Resort to Arms*, 188.

economic development in the 1930s – do not necessarily drain the economy.[16] Moreover, even if small wars require considerable investment, it is not necessarily the case that powerful protagonists pay the bills. The latter can, in principle, finance the war without over-straining their current economies. Most obviously, they can borrow on international or domestic money markets, or shift the cost of war, in full or in part, to the vanquished (as Napoleon often did) or to their allies (as the French did in the case of Indochina).

The argument of Maoz deserves only brief attention. While it is big on promise, it is short on delivery. On the one hand, its theoretical contentions against realism are unconvincing, and on the other hand, it rests on wobbly empirical legs.[17] Maoz's "paradox" presumably suggests that "some of the major premises of the realist paradigm must be reexamined."[18] However, paradoxically it does so, after he explained the outcome of the Lebanon war in terms of a dialectic balancing process, that even by his own (vacillating) admission, is perfectly consistent with realism.[19] More importantly, at least one prominent realist seems to be in one mind with Maoz's "heretic" half. Thus, Kenneth Waltz makes the following points concerning the outcomes of the Vietnam War: (1) He argues that *the causality of power should not be overstated* and that power is only a means, whose outcomes are necessarily uncertain.[20] (2) He argues that military force is not sufficient for the specific purpose of pacification, particularly *if the country* using military power for such *purposes has a faction of people politically engaged and active.*[21] (3) He concludes that "such a case as Vietnam ... [is] a clear illustration of the *limits* of military force in the world of *the present as always.*"[22] Now, it is not for me to speculate why Waltz was ready to compromise the realist idea that military power is the ultimate arbiter in international relations. I would merely like to reiterate his two critical departures from realism (which I accept only in a qualified manner). First, he argues that in a recent major conflict, the outcomes depended on the domestic structure of the powerful party rather than on the relative military power of the protagonists. Second, he suggests that military power was futile not only in this particular case, but rather in similar cases throughout history.

Finally, the major weakness of "battlefield" studies is the result of their being doubly narrow. Most draw conclusions from a research of only one

[16] Kahler, "External Ambition and Economic Performance," *World Politics*, 40:4 (1988), 419–51.
[17] Maoz separates himself from realism in "Power, Capabilities, and Paradoxical Conflict Outcomes," 447. For his questionable description of war encounters and his own doubts concerning the value of his "historical" interpretation, see ibid., 247, 249.
[18] Ibid., 264.
[19] Ibid., 261–63.
[20] Kenneth Waltz, *Theory of International Politics*, 192. See also the discussion in ibid., 189–92.
[21] Ibid., 189.
[22] Ibid, 189–90 (italics added).

dimension – the military – of one case study – Vietnam. Thus, although
their insights are interesting, their theoretical value, as far as explaining failure in small wars, is a priori limited.[23] In this sense, Arreguín-Toft's work
is different and potentially more significant. He backs his ambitious statement with quantitative research, and adds a brief discussion of Vietnam that
supposedly supports his causal argument, and illustrates how its elements
operate.[24] Unfortunately, however, underlying his arguments are assumptions and statements that are inconsistent at times and that suggest that the
fundamental causality of failure in small wars (at least those involving modern democracies) should be looked for elsewhere than on the battlefield. As
noted, Arreguín-Toft's thesis contends that incumbents lose insurgency wars
if and when they choose the wrong strategy vis-à-vis the strategy of the insurgents. Conversely, of course, incumbents are expected to win when they
choose the correct strategy of barbarism, defined as "the systematic violation of the laws of war ... [in which the] most important element is depre-
dations against noncombatants (viz., rape, murder, and torture) ... used to
destroy an adversary's will and capacity to fight."[25] Indeed, he hypothe-
sizes that "when strong actors employ barbarism to attack weak actors [that
fight guerrilla warfare] ... all other things being equal, strong actors should
win."[26] Oddly however, Arreguín-Toft also suggests that "strong actors also
lose asymmetric wars when, in attempting to avoid increasing costs ... they
yield to the temptation to employ barbarism ... [because] even when militarily effective, it is risky ... [as it] carries the possibility of domestic political
discovery (and opposition) as well as external intervention."[27] Barbarism,
he sums up, "sacrifices victory in peace for victory in war – a poor policy at
best."[28] But surely Arreguín-Toft would not suggest that great powers drew
further external intervention because they behaved barbarically in the Third
World, or that despotic incumbent regimes feared "domestic discovery" of
their "lawless" counterinsurgency actions? In other words, if his hypothesis is accepted (ignoring the operationally odd definition of barbarism and
the inconsistency in what is attributed to this "strategy"), one is inclined to
suspect that the key to understanding failure in guerrilla war must lie in "all
[those] other things" that are *not* "equal." That is, we should look for the
sources of failure (as his intuitive account suggests) somewhere else than in
the strategic choice or on the battlefield.

[23] See such honest self-criticism (in a somewhat different context) in John Ellis, *From the Barrel of a Gun: A History of Guerrilla, Revolutionary, and Counter-Insurgency Warfare, from the Romans to the Present* (London: Greenhill, 1995), 12.
[24] As Arreguín-Toft observes, his quantitative design can produce correlations but cannot explain causation. See "How the Weak Win Wars," 112.
[25] Ibid., 101.
[26] Ibid., 109.
[27] Ibid., 105–06.
[28] Ibid., 123.

Motivation, The Balance of Will and Interests, and Small Wars

Non-realist scholars have long argued over occasional disjunctions between the power-base of states and their ability to influence other international actors. Still, the failure of realist arguments to explain certain international outcomes by reference to material capabilities does not necessarily negate the utility of the concept of power. At a minimum, however, it casts doubt over the realist definition of power.[29] Indeed, the concept of power was amended, and instead of the rigid approach – which emphasized material and tangible dimensions such as technological capabilities, manpower, and economic strength – scholars introduced "softer" definitions of power that included intangible elements such as motivation, will, national cohesion, and the readiness to sacrifice.[30] Different schools used these amended definitions for the study of various issue areas. For our purposes, however, the discussion of coercive diplomacy is the most relevant.[31]

According to scholars who studied coercive diplomacy, what counts during its exercise (which may occasionally include the use of military force) is the relative motivation of the protagonists. Motivation, the argument goes, derives from what is at stake for the parties to a conflict, or from their relative interests. The argument concludes logically that one of the sides would have the bargaining-advantage when both realize that each has a clear grasp of the objective balance-of-interests and will.[32]

For those thinking in these terms, and in particular for those who witnessed how America failed in Vietnam in spite of spectacular material advantage, it seemed clear that indigenous people won insurgency wars because the balance of interests and the balance of will favored them. Insurgents had more at stake, therefore they also had greater motivation and readiness to sacrifice, and consequently they won.[33] Balance-of-will scholars, then,

[29] For discussions of what precisely constitutes power see David A. Baldwin, "Power Analysis and World Politics: New Trends Versus Old Tendencies," *World Politics*, 31:2 (1979), 161–94; Jeffrey Hart, "Three Approaches to the Measurement of Power in International Relations," *International Organization*, 30:2 (1976), 289–305; K. J. Holsti, "The Concept of Power in the Study of International Relations," *Background*, 7:4 (1964), 179–94; and Herbert A. Simon, "Notes on the Observation and Measurement of Political Power," *Journal of Politics*, 15:4 (1953), 500–16.

[30] Realists such as Morgenthau did consider some "soft" elements of power. See also Steven Rosen, "War Powers and the Willingness to Suffer," in Bruce Russett (ed.), *Peace, War, and Numbers* (Beverly Hills: Sage, 1972), 167–83.

[31] See, for example, Robert Jervis, "Bargaining and Bargaining Tactics," in Roland J. Pennock and John W. Chapman (eds.), *Coercion* (NY: Atherton, 1972), 272–88; Paul Gordon Lauren, "Theories of Bargaining with Threat of Force: Deterrence and Coercive Diplomacy," in Paul Gordon Lauren, *Diplomacy: New Approaches in History, Theory, and Policy* (NY: The Free Press, 1979), 183–211; and Gordon A. Craig and Alexander L. George, *Force and Statecraft* (NY: Oxford University Press, 1990), 197–246.

[32] Jervis, "Bargaining and Bargaining Tactics," 282.

[33] See, for example, Richard Betts, "Comment on Mueller," *International Studies Quarterly*, 24:4 (1980), 520–24.

introduced a formula that specifies direct relationship between what is at
stake in conflict, the motivation of the opponents, and the outcomes of the
confrontation.

Two arguments that were developed independently of the coercive diplomacy literature also merit consideration because they employ an essentially
similar logic. I refer to the arguments of Sir Robert Thompson – who was
chief architect of the British counterinsurgency campaign in Malaya – and
of Michael Howard.[34] Thompson, writing about the American policy in
Vietnam, argued that in cases of popular insurgency the crucial variables of
time, space, and costs have different effects on the protagonists. Howard
argued that strategy must be understood in four dimensions (operational,
logistical, technological, and social) of which the social had become increasingly important since the demise of absolute monarchies, particularly in cases
of "revolutionary wars." Both Howard and Thompson, it is important to
note, emphasized the social strength of the underdog, by and large ignoring
the possible social "weakness" of the powerful protagonist.[35]

The Strength of Motivational Explanations, and their Weaknesses

The idea that war had important non-material dimensions and that it involved a complex process that happens within, as much as between, states,
is not novel. Clausewitz had already defined war as an affair of will, morale,
and motivation, as much as of matter,[36] and as a phenomenon involving the
"people" (society) no less than the army, command, and government (the
state).[37] Irrespective of originality, however, the departure of motivational
theories from Newtonian realist definitions of power – that focused on the
state, material capabilities, and tangible variables – and the adoption of a
view that considers less tangible social attributes is of significance in the
context of research in international security.[38] That noted, it remains unfortunate that motivational arguments do not prove superior to realist theories
as far as small wars are concerned. If anything, they generate new problems.
How is one supposed to measure the relative intensity of will independently,
and without falling into the tautological trap of inferring it from the results

---

[34] Thompson, "Squaring the Error," *Foreign Affairs*, 46:3 (1968), 442–53; and Howard, "The
Forgotten Dimensions of Strategy," *Foreign Affairs*, 57:5 (1979), 975–86.

[35] Joel Migdal seems to be of one mind with Howard and Thompson. See *Strong Societies and
Weak States* (Princeton: Princeton University Press, 1988), 3.

[36] Carl von Clausewitz (ed., trans. by Michael Howard and Peter Paret), *On War* (Princeton:
Princeton University Press, 1984), 77, 97.

[37] Ibid., 89.

[38] Jacek Kugler and William Domke observed that "the foundation of power in the global
system is the relationship between state and society ... [and] this linkage is usually overlooked ... because power politics and the system structure perspective seldom deal with
changes in the domestic structures and their impact in the global system." See Kugler and
Domke, "Comparing the Strength of Nations," *Comparative Political Studies*, 19:1 (1986), 40.

of war? Are "will," "motivation," and the "intensity of interests" truly independent variables? And, if not, how do we study the basic attributes that define them, and do so independently of other characteristics of the confrontation? How does one measure the "strength" of the "feelings" of the warring parties?[39] Whose feelings, strength, motivation, and interests define the amount of will of the protagonists? Are they those of the leadership, different elite circles, or the masses? Are there grounds to argue that successful insurgents had higher motivation and fought for more critical interests than unfortunate ones?

At the heart of all of these questions lie the problems of whether motivation, will, and interests are independently formed and directly related to conflict outcomes. If they are not, as I argue next, then the value of the motivational approach is sharply depreciated. Let us start with the simple proposition that had interest defined motivation and had the latter defined the outcomes, underdogs would have been predisposed to win small wars because of the built-in asymmetry of the parties' stakes and threats. Considered in this way, defeat for the powerful protagonist, painful as it may be, rarely grows to existential proportions, while victory is mostly of additive value. Defeat for the underdog, however, tends to involve potential catastrophic consequences, while victory promises an ultimate reward: independence. In terms of the stakes then, the balance of interests inherently favors the underdog, and therefore so does the balance of motivation. With both balances favoring the underdog, should not the chances of victory also favor it?

On the face of it, this logical sequence seems appealing. However, it cannot be accepted as reliable unless it is in conformity with historical facts. The point is that this logic and history are inconsistent, mainly because the actual interest and motivation of the underdog in small wars (as opposed to abstract aspirations) rarely reflect this abstract and simple formulation. Rather, the formation of interests, goals, and motivation is usually embedded in the particular context, and often dependent on the underdog's perception of the resolve of his mighty enemy, the possible cost of confrontation, and the chances of success.[40] The variable of interest, then, is not independent, because it is inseparable from utility calculations, including those of the possible conflict outcomes. Moreover, even if the interests of underdogs (independent or not) are dearer to them than those of powerful parties to themselves, I still argue that the evidence does not support the conclusion that they tend to decide the outcomes of war.

Indeed, the limits of motivational arguments, including those of Thompson and Howard, become clear when one thinks in comparative

---

[39] Jarvis uses the term "feel strongly" in "Bargaining and Bargaining Tactics," 281.

[40] See also the cogent discussion of subjective calculations of underdogs in Michael P. Fischerkeller, "David versus Goliath: Cultural Judgments in Asymmetric Wars," *Security Studies*, 7:4 (1998), 1–43.

terms. Let us first turn to counterfactual logic and then to evidence. Suppose that the balance-of-motivation formula is valid in the sense that it can be constructed a priori and without tautology. Had this been the case, we should expect to find that, on average, underdogs did better in the past than today. The point is that the average stakes in past insurgencies were ultimate – a likely extinction of the national or even physical existence of the rebelling community – whereas today they are much smaller in the genocide-sensitive world that legitimizes self-determination. In short, the interest and motivation of ancient underdogs should have been unrivaled, and this, according to the motivational theories, should have been reflected in relatively high rates of successful insurgencies. In fact, the logic of motivational arguments is such that the mere initiation of a challenge to empires in antiquity indicates the extreme motivation of underdogs.

Yet the historical record of success of underdogs in small wars seems abysmal in comparison with our age. And if so, then either past motivation was not particularly high, or material power and brutality (that is, realist arguments) overrode spirit and determination (that is, motivational arguments). In fact, the problems of motivational arguments do not end in their inability to explain outcomes across a temporal barrier (just like realist arguments, albeit in the opposite direction). Rather they also involve a failure to explain current variations in the outcomes of small wars. Is it reasonable to claim that the Algerians, Palestinians, Lebanese Shiites, and other communities that won independence – and drove foreign powers off their territory – displayed greater motivation than the Tibetans or the Kurds? After all, the Kurds and Tibetans faced (and still do) more misery and agony than did the Algerians and the Palestinians. And, if so, they should have developed greater interest and motivation than the latter, and consequently have enjoyed greater chances of success. Yet, the Kurds and Tibetans have failed, and the least one could say is that it was not for lack of courage or tenacity. In fact, the motivational logic fails to explain communal behavior, the timing of uprising, and conflict outcomes in yet another class of cases – those of occupation in which nationally cohesive communities preferred not to revolt at all. How is it that ethnically distinct East European, Baltic, and Central Asian nations waited passively until the collapse of the Soviet system before they discovered their interests, displayed motivation, demanded independence, and became ready to fight for it? Did their interest change, or was it the change in power relations that motivated them?

How Do Democracies Fail in Small Wars?

The puzzle of the outcomes of small wars can be presented in two ways. One can ask: "How do insurgents win small wars against democracies in spite of their military inferiority?" Or, one can ask: "How do democracies lose such wars in spite of their military superiority?" Motivational theories seem concerned with the first question. I was intrigued by the second.

My argument is that democracies fail in small wars because they find it extremely difficult to escalate the level of violence and brutality to that which can secure victory. They are restricted by their domestic structure, and in particular by the creed of some of their most articulate citizens and the opportunities their institutional makeup presents such citizens. Other states are not prone to lose small wars, and when they do fail in such wars, it is mostly for realist reasons. Furthermore, while democracies are inclined to fail in protracted small wars, they are not disposed to fail in other types of wars. In a nutshell, then, the profound answer to the puzzle involves the nature of the domestic structure of democracies and the ways by which it interacts with ground military conflict in insurgency situations.[41]

It is obvious that my argument relates to two major issues of international relations, besides insurgency wars: the issue of the relations between democracy and war, and that of the relations between domestic and foreign policy. As I address the discussion of democracy and war in the conclusion, I explain here only how my argument relates to three key statements about domestic structure and foreign policy – those of Robert Putnam, Andrew Moravcsik, and Thomas Risse-Kappen.

Robert Putnam proposed a two-level game framework for understanding the interaction between domestic and international politics in situations of international bargaining.[42] Putnam's work deals with efforts to achieve co-operation, but it is equally relevant for situations of confrontation, including war. Indeed, it is easy to explain my thesis in his terms. Democratic failure in small wars can be seen as reflecting a two-level game in which the "win set" – the international policy latitude that democratic leaders assume they enjoy – exceeds what a critical domestic constituency accepts. Because of the preferences of this constituency and its capacity to effectively exercise political power at home (through the media, the free marketplace of ideas, and political protest), the state's foreign policy is not "ratified" and the war-effort becomes unsustainable. To paraphrase Putnam, moves that were (perhaps) rational for the state at the international game board prove impolitic at the domestic game board.[43]

Andrew Moravcsik discusses the *liberal theory* (or paradigm) of international relations. His definition of one of the core tenets of the liberal theory – the assumption that politics is a bottom-up process that consists of a flow of influence from society to the state – is compatible with the end result of my

---

[41] See Andrew Mack, "Why Big Nations Lose Small Wars: The Politics of Asymmetric Conflict," *World Politics*, 27:2 (1975), 175–200.

[42] Robert D. Putnam, "Diplomacy and Domestic Politics: The Logic of Two-Level Games," *International Organization*, 42:3 (1988), 427–60.

[43] Ibid., 434.

arguments.[44] In this sense, I offer some confirmation for the paradigmatic liberal argument. However, the societal process I discuss does not fit any of the three specific variants of liberalism Moravcsik discusses (ideational, commercial, and republican), though it shares some attributes with each of them. Moreover, I assign the "state" a more significant role than his liberal theory seems to accept. Thus, in the final analysis, my overall argument probably fits best Moravcsik's definition of a synthetic model – a model that explains international choices (and outcomes in my case) as resulting from interaction between both realist and liberal variables.[45]

Thomas Risse-Kappen writes about the role of "domestic structure" and "coalition building processes" in the shaping of *international security policies* in *liberal democracies*. Not surprisingly, my argument has strong affinity to his. Indeed, I share with Risse-Kappen a number of tenets. In general, much like him I believe that the neglect of the "domestic environment" in the discussion of security is unreasonable. Similarly, I believe that "crucial events leading to [international outcomes] can only be examined in the context of domestic politics."[46] Finally, I agree that domestic politics influences state preferences.

However, I also diverge from Risse-Kappen and carry the argument further, primarily because my topic so permitted. Risse-Kappen studied the influence of West European and American societal forces on foreign policy-making in the context of arms control. Naturally, his work is about international cooperation, although in the context of competition. Moreover, the international outcomes of his case reflected both traditional state policy (hard-line bargaining position), societal pressures (for an accord), and a change in the adversary's position (the Soviet acceptance of the zero option by the new reformist, Gorbachev). My work leads in a different direction. I explain how societal preferences *undercut and defeat* state preferences, not ameliorate them, and I do so in the context of *bitter international conflict* that precludes cooperation. Indeed, I show how a process of societal "coercion" succeeds independently of favorable international developments such as cooperation by the adversary (the change from war to peace). These differences are of particular importance as they have permitted me to arrive at more radical conclusions. While Risse-Kappen found that "rarely does general public opinion directly affect policy decisions or the implementation of specific policies,"[47] I explain when and how the opinion of a *minority* among the public can defeat state policy in the toughest circumstances of national security: during war.

---

[44] Andrew Moravcsik, "Taking Preferences Seriously: A Liberal Theory of International Politics," *International Organization*, 51:4 (1997), 517.

[45] Ibid., 541–45, particularly p. 545, figure 1.

[46] Thomas Risse-Kappen, "Did Peace Through Strength End the Cold War? Lessons from INF," *International Security*, 16:1 (1991), 163–64.

[47] Thomas Risse-Kappen, "Public Opinion, Domestic Structure, and Foreign Policy in Liberal Democracies," *World Politics*, 43:4 (1991), 510.

State and Society

Because the discussion of domestic structure concerns the relations between two controversial concepts in political science, "*state*" and "*society*," I find it compelling to briefly clarify my position regarding both before I turn to the details of my argument.

My choice of the state as a variable is first and foremost a matter of conforming to an empirical reality.[48] Indeed, since its appearance on the world stage, the modern state has shown great institutional resiliency, and has increased its presence on both the international scene and in the life of its citizens.[49] Second, I chose the state as a variable because it is often the source of leaders' definition of their role and policy choices (or at least as they rationalize the latter by reference to the state).[50] In fact, this much can also be said of other state officials, and even ordinary citizens, who often "know" without any guidance what they should do on behalf of the state. Thus the use of the term "state" is justified because it dictates to officials and citizens a particular way to perceive events, a broad common action agenda, and a way to reason their actions.[51] My discussion of the state, then, differs from the approach one usually finds in the "state" literature in two ways. First,

[48] For (mixed) references to the state as a conceptual variable, see Gabriel A. Almond, *A Discipline Divided* (Newbury, CA: Sage Publications, 1990), 189–218; Yale H. Ferguson and Richard W. Mansbach, *The State, Conceptual Chaos, and the Future of International Relations Theory* (Boulder: University of Denver, 1989); Peter J. Katzenstein, in "International Relations Theory and the Analysis of Change," in Ernst-Otto Czempiel and James N. Rosenau (eds.), *Global Changes and Theoretical Challenges* (Lexington, MA: Lexington Books, 1989), 296–98; Stephen D. Krasner, "Approaches to the State: Alternative Conceptions and Historical Dynamics," *Comparative Politics*, 16:2 (1984), 223–46; Michael Mann, "The Autonomous Power of the State: Its Origins, Mechanisms and Results," in John A. Hall (ed.), *States in History* (Oxford: Oxford University Press, 1986), 109–36; Eric A. Nordlinger, "Taking the State Seriously," in Myron Weiner and Samuel P. Huntington (eds.), *Understanding Political Development* (Boston: Little, Brown, 1987), 353–90; Gianfranco Poggi, *The Development of the Modern State* (London: Hutchinson, 1978); and Theda Skocpol, "Bringing the State Back In: Strategies of Analysis in Current Research," in Peter Evans, Dietrich Rueschemeyer, and Theda Skocpol (eds.), *Bringing the State Back In* (Cambridge: Cambridge University Press, 1985), 3–37.
[49] See Ted Robert Gurr, "War, Revolution, and the Growth of the Coercive State," *Comparative Political Studies*, 21:1 (1988), 45–46; and Janice E. Thompson and Stephan D. Krasner, "Global Transactions and the Consolidation of Sovereignty," in Czempiel and Rosenau, *Global Changes and Theoretical Challenges*, 195–220, particularly pp. 206–14. State resiliency can be attributed to factors Samuel Huntington discusses in *Political Order in Changing Societies* (New Haven: Yale University Press, 1970), 1–93, particularly pp. 8–24.
[50] See Morgenthau, *Politics among Nations*, 5–7.
[51] In the "state" literature, the argument is often framed under the term "the autonomous power of the state," which refers to the presumed capacity of the state to transcend the interests of any particular faction. See Stephen Krasner, *Defending the National Interest: Raw Materials Investments and U. S. Foreign Policy* (Princeton: Princeton University Press, 1978), 5–34; and Eric A. Nordlinger, *On the Autonomy of the Democratic State* (Cambridge, MA: Harvard University Press, 1981).

whereas the state is often defined in a Weberian way as a centralized *set* of specialized institutions, I see it as a *single* super-institution. Second, I perceive the state as having a critical mental dimension.

The concept of "society" is at least as controversial as the concept of "state," in large measure because it refers to an entity that is far more amorphous. My use of the term "society" is primarily in contrast to "state." I regard society as a space where individuals think and act under no mental tyranny of the state. These individuals act on and promote preferences that are derived from liberal and non-statist values.[52] Not all citizens other than state officials, nor even their majority, use or take advantage of this space. Rather, it is exploited by a minority that is often led by members of the educated segment of the middle-class. These members share particular views regarding what is right and what is wrong, what are the obligations of the state, and what justifies support for and opposition to state-policy.

It is important to conclude this brief conceptual clarification with three caveats. First, the anti-war constituency cannot be equated with the entire, and rather amorphous, middle-class.[53] Second, state officials can join society, but they will then be crossing lines (indeed, having done so they are often considered as defectors or traitors). Third, as the boundaries between "state" and "society" have eroded in more recent times, and as individuals seem to cross them with relative ease, the demarcation between the two spaces is at times blurred, though not invisible.

Instrumental Dependence, Normative Difference, and Political Relevance

It is now possible to get into the details of my argument by defining its three conceptual building blocks: *instrumental dependence, normative difference,* and *political relevance. Instrumental dependence* refers to the state's degree of reliance on society to provide the resources, mostly manpower, needed to execute national security policies. *Normative difference* refers to the distance between the position of the state and that of the liberal forces (that give meaning to the term society) concerning the legitimacy of the demand for sacrifice and for brutal conduct. *Political relevance* refers to the inherent degree of influence societal forces have over policy-choices or their outcomes. Political relevance is attributed to groups, which among other things can make their preferences salient in the political discourse. Democracies, then, differ from other states in terms of the political power citizens routinely exercise and in terms of what agenda they may promote – that is, democracies

---

[52] See Adam B. Seligman, *The Idea of Civil Society* (NY: The Free Press, 1992), 1–15, particularly p. 5.

[53] See Frederick Pryor's essay on the middle class, "The New Class: Analysis of the Concept, the Hypothesis and the Idea as a Research Tool," *American Journal of Economics and Sociology*, 40:4 (1981), 367–79.

are unique in terms of the normative difference they may experience and the political relevance their citizens enjoy.

With these definitions in mind, the fundamental elements of my thesis can be presented. I submit that modern democracies lose protracted small wars because in situations of deep *instrumental dependence*, the *politically most relevant* citizens create a *normative difference* of insurmountable proportions. Essentially, what prevents modern democracies from winning small wars is disagreement between state and society over expedient and moral issues that concern human life and dignity. The details of this thesis and the consequences of my argument are discussed throughout this book. Nevertheless, I offer here a preview of the mechanism of democratic failure in small wars.

Let us proceed by considering the basic requirements of war in the abstract. It is almost obvious that the resort to a large-scale ground military action requires states to convert a significant amount of their societal resources into military power.[54] Ground wars then, almost by definition, lead to a substantial level of instrumental dependence. Yet it would be wrong to assume that warring states are limited only by the sum total of their societal resources or by their capacity to convert the latter into means of destruction. States are also bound by their political capacity to use and lose resources, particularly human beings. Therefore, states must secure their soldiers' and citizens' "acceptance" of the use of violence and the risk of being its victims. In other words, in order to fight, let alone win wars, states need their soldiers to be ready to harm others and be killed or maimed, and their citizens to accept the army's behavior and the risks their kin in arms face. Achieving a certain balance between these two requirements – the readiness to bear the cost of war and the readiness to exact a painful toll from others – is a precondition for succeeding in war.

This presumably trivial logic is illustrated in Figure 1.1. The vertical axis represents increasing levels of *tolerance to casualties* and the horizontal axis represents increasing levels of *tolerance to brutal engagement* of the enemy. The curve constitutes a theoretical continuum of combinations of tolerance for casualties and tolerance for violence that a state has to achieve in order to win a war. This curve of the balance-of-tolerance illustrates how the two types of tolerance are related. The less tolerance a society displays in one dimension, the more it must display in the other, if the state is to win the war.

The idea of a curve of the balance-of-tolerance can be understood even more clearly by considering the two theoretical extremes (or ideal types) that are represented in zones A and B. Zone A represents the place of a state whose society is ready to accept great sacrifices but vehemently opposes violence against others – that is, this society may be heavily bled by others, but being thoroughly pacific, it will refuse to shed their blood. Obviously, no matter how well endowed and equipped this "altruist" state is, it can hardly fight,

---

[54] Kugler and Domke, "Comparing the Strength of Nations," 39–71.

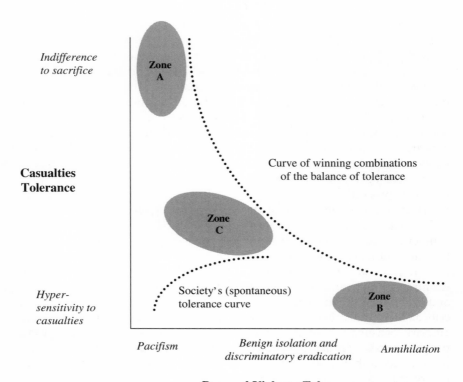

Casualties Tolerance

Indifference to sacrifice

Zone A

Curve of winning combinations
of the balance of tolerance

Zone C

*Hyper-
sensitivity to
casualties*

Society's (spontaneous)
tolerance curve

Zone B

Pacifism   *Benign isolation and
                     discriminatory eradication*   *Annihilation*

Personal Violence Tolerance

FIGURE I.I. The balance-of-tolerance

let alone win, conventional ground wars. Zone B represents a state whose
society is thoroughly unscrupulous but is also hypersensitive to casualties.
Such a "psychopathic" state is almost as unfit as the altruistic state to win
ground wars, though for an entirely different reason.

Figure I.I can also serve to illustrate the idea of normative difference. Zones A and B can be seen as each representing a case of such
one-dimensional difference. Zone A (of the "altruistic" state) represents a
morality-based difference, whereas zone B (of the "psychopathic" state) represents an *expediency-based difference*. In reality, the normative difference
democracies experience during protracted small wars is not entirely based on
either expediency or morality. Liberal societies are not thoroughly "spoiled"
nor utterly moral. Thus, the typical normative difference reflects gaps over
both which cost of war is expediently justified and what war objectives and
methods are morally acceptable. Zone C in Figure I.I illustrates this typical
normative difference. It consists of the gap between the winning balances-
of-tolerance that the state seeks to secure in order to win a war, and the cost
and violence society is ready to tolerate, *without* state intervention.

It is important to emphasize that the problem of achieving a balance-of-tolerance, and the success of politically relevant groups to create a debilitating normative difference, are relatively recent, and confined to democracies. Traditionally, the use of violence abroad did not involve difficulties at home. Subjects were often unwilling to sacrifice their money or life to underwrite their leaders' military adventures abroad, but they did not care about the fate of foreigners, be they insurgents or the civil population that supported the latter. Indeed, as long as attitudes toward inflicting violence externally did not change, the question of whether and how to conquer, subjugate, and pacify communities was dominated only by expediency – namely, the concern over the availability of resources. Ultimately, only the development of democratic political institutions and an educated liberal constituency in the West have changed this state of affairs.

It is also important to note that the normative difference in warring democracies is likely to be most pronounced in cases of small wars because they are not existential. And it is equally important to understand that it takes time for democracies to experience the full effect of the normative difference. It simply takes casualties to accumulate and brutality to increase and be "observed" by society before the anti-war constituency acquires a critical mass and acts with full force. Once these have been achieved, the anti-war constituency can take control over the agenda, shape the terms of the public debate, and shift the war's center of gravity from the battlefield to the marketplace of ideas, where the state's capacity to pursue its objectives is checked.

Finally, it is worth emphasizing the special role of instrumental dependence in this sequence of developments. The potential size and power of the anti-war coalition depends in large measure on the sort and number of people who are personally affected by the war – that is, the fate of the war depends on the nature and scope of military mobilization. Or, explained in conceptual terms, the relation between the autonomy of the state and instrumental dependence in democracies is negative in times of small wars. In the long run, a greater reliance on conscription and reservists reduces the capacity of the state to act in the battlefield with unrestrained force, to pursue far-reaching objectives, and to win the war.

The Destructive Dynamic of Lost Autonomy

Obviously the position of the state in small wars is not entirely given. Indeed, as I discuss in Chapter 4, states have tried to manipulate each of the three variables I discussed. Democratic states, however, are severely limited in their capacity to manipulate one of these variables – the political relevance of their citizens – because such an act undercuts their democratic identity. The two other variables, instrumental dependence and normative difference, can be, and often are, manipulated by democracies. For example, democracies can limit the size and nature of the fighting force and/or reduce the risks soldiers face in combat. In acting so, they either control their instrumental dependence or its impact. In either case, they can keep the expedient dimension of the normative difference from growing. However, in order to remain effective in spite of the reduction of the number and/or exposure to risks of soldiers, they must rely on higher and less discriminating levels of violence. In short, democracies can avoid the direct consequences of instrumental dependence without increasing the expedient dimension of the normative difference and compromising their counterinsurgency effectiveness, but they can do so only at the risk of increasing the potential size of the moral dimension of the normative difference.

I use the terms "at the risk" and "potential" rather than "necessarily" and "actual" because democratic states can theoretically control all aspects of the normative difference directly. By direct control, I refer to the presumable state capacity to shape the public agenda, raise the tolerance of society, and avoid a trade-off between higher levels of casualties and brutality. Reality, however, is different. Once casualties accumulate and/or tales of brutality reach society, the successful control of the normative difference in a *fair competition* in the marketplace of ideas becomes next to impossible. Thus, democratic states that aggressively pursue their objectives in small wars eventually face a choice only between succumbing to domestic pressure and trying to overcome it in deceitful and/or despotic ways. The latter, however, challenges the very foundations of democracy and thereby makes the war less sustainable. Inadvertently, then, efforts of democratic states to regain control at home end up expanding and exacerbating the normative difference as the state pits itself against society over an additional detrimental issue: the constitutional nature of the political order – that is, over the survival of democracy.

As if this dialectic process were not debilitating enough, democratic states have to confront an exogenous problem that magnifies the domestic challenge they face in small wars. The internal struggle in democracies does not escape insurgents.[55] Rather, it emboldens them, influences their feasibility calculations, and provides them with strategic targets outside the battlefield.[56] Indeed, insurgency leaders often follow the domestic developments within their enemies' societies, seeking to exploit the divisions they identify. They do so by trying to impose on their enemies a high enough casualty-rate in the expectation that the latter will trigger expedient opposition to the war. Occasionally, however, they also try to lure democratic opponents into

---

[55] See also Harry G. Summers, Jr., "A War Is a War Is a War Is a War," in Loren B. Thompson (ed.), *Low-Intensity Conflict* (Lexington, MA: Lexington Books, 1989), 37–41.

[56] Bui Tin, a former high commander and North Vietnam official, explained: "[The American anti-war movement] was essential to our strategy . . . the American rear was vulnerable. Every day our leadership would listen to world news over the radio . . . to follow the growth of the American anti war movement . . . it gave us confidence that we should hold on in the face of battlefield reverses . . . The conscience of America was part of its war-making capability, and we were turning that power in our favor. America lost because of its democracy. . ." Quoted in the *Wall Street Journal*, August 3, 1995.

behaving brutally in order to increase the moral opposition to the war. Both efforts are usually accompanied by well-tailored messages that are directed at the democratic society. All in all, then, insurgents try to add their spin to the course of events within democracies, seeking thereby to overcome their own battlefield inferiority.

Figure 1.2 illustrates the mechanism and key developments that constitute the process just described. It is a simplification because the developments

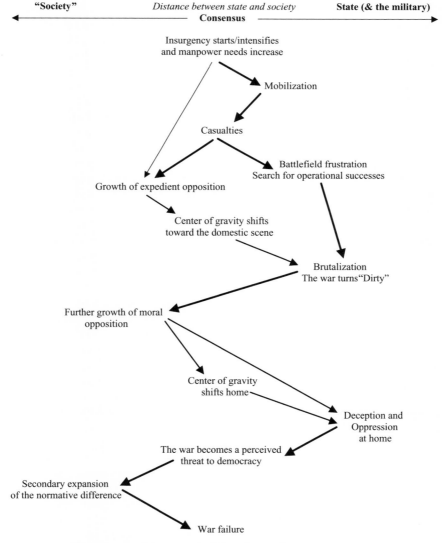

FIGURE 1.2. The process of democratic failure in small wars

may appear in less than a clear sequential manner. The whole process starts when the state takes military measures that require a substantial contingent in order to pacify an insurgent population. Stated in conceptual terms, the state engages or increases its instrumental dependence and thereby creates fertile soil on which a normative difference can grow. As time passes, society becomes better aware of the implications of the war, including its human cost. Because casualties, particularly in non-existential wars, threaten to undercut support for the war, the state is tempted to rely on more firepower and higher levels of brutality. In essence, state organs seek to prevent or minimize the expedient dimension of the normative difference. The ensuing brutality, however, invigorates moral opposition to the war. Depicted as immoral, the war objectives and casualties seem even less sensible. In the final analysis, then, events in the battlefield of small wars and the political requirements they entail create a front against the war that operates in the marketplace of ideas at home.

This front alone can convince democracies to relinquish the initiative and become defensive in the battlefield, if only in order to minimize the pressure at home. In such a case, the war initiative shifts to the insurgents, and retreat becomes only a matter of time. However, the state may decide to try to overcome the erosion of support for the war and remain aggressive on the battlefield. But then it must become more deceptive and/or coercive at home, and this in turn creates a secondary detrimental expansion of the normative gap. The war becomes synonymous with a threat to the democratic order, and the government consequently loses legitimacy. At the end, then, democracies fail in small wars because they cannot find a winning balance between the costs of war in terms of human lives and the political cost incurred by controlling the latter with force, between acceptable levels of casualties and acceptable levels of brutality. In summary, for democracies, the process that dooms the prospects of political victory in protracted small wars involves an almost impossible trade-off between expedient and moral dicta that arise from an intricate interplay between forces in the battlefield and at home.

The Role of Realist, Motivational, and Other Factors Reconsidered

Much as in other cases of interaction between international actors, the outcomes of small wars are not the result of a single cause originating in one level of analysis. Thus, having suggested the primary importance of the domestic structure, it is necessary to reconsider the possible influence of other factors, at different levels of analysis, on the outcomes of small wars that democracies and other states fight. Three such factors instantly spring to mind: the normative influence of the international community, and in particular of other democratic states, and two factors that I have already discussed: the balance-of-power and the underdog's motivation.

It is not unreasonable to argue that while democracies fail in small wars because they are unable to escalate the level of brutality at will, this and their eventual failure in such wars are the result of systemic rather than domestic factors. Specifically, it is sensible that democracies are constrained by their fear that a radical departure from accepted standards of behavior in war (standards that are embodied in the Nuremberg and Tokyo trials, the Geneva Convention, and so on)[57] will cost them dearly in their relations with other democracies. This, however, is a matter of time. Today, as I point out in the Conclusion, "sisterly vigilance" becomes increasingly important. But, until recently, it has not had all that much impact, although democracies did factor international standards into their calculations. Anti-war forces, rather than the authorities in warring democracies, usually expressed the fear of being depicted as pariahs. The democratic authorities tended to dismiss international condemnations of any origin as expedient and hypocritical. Indeed, there is little evidence to suggest that the governments of democracies were deeply committed to upholding high moral grounds elsewhere, or that their moral concern over the behavior of others was genuine before our own time. Rather, the condemnations of democracies of misbehaving sisters (as well as non-sibling states) often seemed to have been half-hearted or designed to score points in the domestic and/or international realms. And, in any event, sisterly vigilance was at least partially a result of domestic pressures within condemning democracies, and as such confirmed the importance of domestic structures (albeit in democratic third parties) no less than it indicated a normative international process.

As far as the role of relative power and the motivation of the underdog are concerned, I have already taken a stand: both are important but only in a qualified manner. Obviously, in their absence, small wars will not be protracted nor will incumbents be ready to even consider the demands of insurgents. In fact, the very shape of conflict does reflect the material capabilities, political organization, and motivation of the underdog, as indeed becomes clear from Doyle's study of empires and their decline.[58] It can also be easily comprehended once we take a second look at Figure 1.1. It is almost self-evident that realist and motivational factors are partially responsible for the place of the "winning curve" on the tolerance plane. The curve will shift inward or outward, according to elements such as the amount of resources available to the underdog, its social cohesion, and its motivation. Obviously, the more resources the underdog commands, the deeper the popular support

---

[57] See Geoffrey Best, *Humanity in Warfare* (London: Methuen, 1983); Richard S. Hartigan, *The Forgotten Victim: A History of the Civilian* (Chicago: Precedent Publishing, 1982); Michael Howard (ed.), *Restraints on War* (Oxford: Oxford University Press, 1979); Paul Gordon Lauren, *The Evolution of Human Rights* (Philadelphia: University of Philadelphia Press, 1998); Geoffrey Robertson, *Crimes against Humanity* (London: Allen Lane-The Penguin Press, 1999); and Donald A. Wells, *War Crimes and Laws of War* (NY: University Press of America, 1984).

[58] See Doyle, *Empires*, 131–35, 220–22.

it enjoys, and the stronger its resolve – the greater is the likelihood that the conflict will be more intense. The greater is the intensity, the greater are the chances that the battles will be brutal and more lives will be lost by both parties. Such conditions mean that if the strong power is to win the war, its society will have to display greater tolerance for both brutality and casualties. That is, the curve of the balance-of-tolerance in Figure 1.1 will shift up and to the right. In short, relative military capabilities, the underdog's level of motivation, and external support for its cause are not irrelevant.

All of these factors, however, do not undermine my criticism of realist and motivational explanations of the outcomes of small wars that involve *democracies*. In such wars, the focus on the balance-of-power and the bias toward the underdog's motivation remain unjustified. While both factors are relevant, their role is secondary. When one asks, "Which of the different explanations – the realist, the motivational, or that of the domestic structure – identifies the most fundamental reason for democratic failures in small wars?" the answer, I believe, is unambiguous. Realist and motivational factors constitute necessary causes, but not sufficient ones. In the final analysis, the nature of the *strong contender* – that is, its domestic structure – remains the most important determinant of the outcomes of small wars.

Research Design and Methodological Considerations

I have noted at the outset of the book that my objective was to expose and explain in detail a causal mechanism that has so far received little attention from scholars. That being the case, I have found the qualitative research approach to be the most appropriate.[59] Others might prefer to test propositions about the relations between domestic structure and the outcomes of small wars.[60] Their choice would then probably be to initiate a quantitative statistical analysis and/or a comparative study of a significant number of cases.[61] I have aimed elsewhere and chosen otherwise, but still

---

[59] For the logic of choosing qualitative research see Catherine Marshal and Gretchen B. Robertson, *Designing Qualitative Research* (London: Sage Publications, 1989), 46; Alexander L. George, "Case Studies and Theory Development: The Method of Structured Focused Comparison," in Paul Gordon Lauren (ed.), *Diplomacy: New Approaches in History, Theory, and Policy* (NY: The Free Press, 1979), 48–52; Keith F. Punch, *Introduction to Social Research* (London: Sage, 1998), 155–56; and Yale Ferguson and Richard Mansbach, "Between Celebration and Despair: Constructive Suggestions for Future International Theory," *International Studies Quarterly*, 35:4 (1991), 369.

[60] For a discussion of testing and case selection for that purpose, see Gary King, Robert O. Keohane, and Sidney Verba, *Designing Social Inquiry: Scientific Inference in Qualitative Research* (Princeton: Princeton University Press, 1994), particularly pp. 128–49.

[61] One can review the cases I briefly raise in Chapter 2 and compile a list for testing purposes. For example, one could focus on the French wars in Algeria in the 1830s and 1840s and 1954–1962, the American war in the Philippines in 1899–1901 and in Vietnam in 1964–1972, and so on. Still, testing would not capture the details of the relations between different causes

have tried to understand the phenomenon of democratic loss in small wars in its wider context. I have therefore relied on a mix of "positive" and "negative" techniques of comparative study. Neil Smelser's description of Max Weber's use of both techniques helps to explain the logic of my research design.

> Given that certain societies ... have developed the values of rational bourgeois capitalism, Weber asked what characteristics these societies had in common. In so doing he was using the *positive* comparative method – identifying similarities in independent variables associated with a common outcome. Then, turning to societies that had not developed this kind of economic organization ... he asked in what respects they differed from the West. In so doing he was using the *negative* comparative method – identifying independent variables associated with divergent outcomes.... The oriental societies that did not develop rational bourgeois capitalist ideals are logically parallel to control groups (because the crucial variable was not operative); the countries of the West are logically parallel to experimental groups (because the crucial variable was present).[62]

In essence, my research logic is similar, although the order of my comparisons is reversed. I asked – given that democracies repeatedly fail in protracted small wars in spite of their military superiority – what do they have in common? Yet, before answering this question, I studied what permitted other powerful states, including early or proto-democracies, to win small wars. Admittedly, my use of the negative comparative logic, in the first part of the book that deals with counterinsurgency strategies, is limited, because my focus was on the mechanism of democratic defeat in small wars rather than on testing cross-regime variations in the outcomes of these wars. Nevertheless, I note for the purpose of the negative comparison that the evidence suggests that powerful protagonists win small wars because, in the absence of foundations of a normative difference, they can escalate brutality indefinitely.[63] On the other hand, my use of the logic of positive comparison for the main empirical thrust of the book, in Parts II and III (Chapters 5 through 14), is rigorous. In these chapters, I discuss in depth two case studies

---

and mechanisms, nor the causal sequence of the process of failure of the strong party. See the general discussion of A. Michael Huberman and Matthew B. Miles in "Data Management and Analysis Methods," in Norman K. Denzin and Yvonna S. Lincoln (eds.), *Handbook of Qualitative Research* (Thousand Oaks, CA: Sage, 1994), 434–35.

[62] Neil J. Smelser, "The Methodology of Comparative Analysis," in Donald P. Warwick and Samuel Osherson, *Comparative Research Methods* (Englewood Cliffs, NJ: Prentice-Hall, 1973), 52–53.

[63] To further borrow from Smelser's arguments, my discussion of the emergence of the normative difference in proto-democracies and of its impact on military conduct in small wars (Chapter 3) explains how fluctuations in the key explanatory variable (the normative difference) influence a key intervening variable (battlefield behavior), if not the dependent variable itself (the outcomes of small wars).

that I had chosen instrumentally in order to expose the mechanism that leads democracies to lose small wars.[64] I analyze the two cases along a single set of variables, in a "structured and focused" way.[65] Rigor permitted me not only to find a broad common ground between the two cases, but also important variations that enabled me to further refine my arguments in the Conclusion. The methodical analysis also paid off in another way. It helped me dismiss some commonly held beliefs about democracies and small wars. In particular, I found that several arguments about the relations between casualties, economic cost, the electronic media, and democratic defeat were somewhat or grossly inaccurate.

The French and Israeli Case Studies

Two considerations dominated my choice of the case studies for the in-depth comparison: The cases had to maximize the potential for extrapolation (or generalization) and minimize the problem of confounding variables.[66] In order to have a maximal extrapolation potential, the cases needed to be representative of the class of cases for which the book developed an explanation. As such, they had to meet a few simple requirements. First, the strong protagonist had to be democratic and overwhelmingly superior in the battlefield *throughout the duration of the war*. These conditions matched the requirements of the first half of the puzzle that dealt with a nature of the strong protagonist and that of the military situation. Implicitly, however, the first half of the puzzle required a bit more. For example, that the war be least affected by forces that had the capacity to alter the asymmetry between the protagonists, whether by direct intervention, massive material support, or intensive diplomatic activity. Second, the cases had to meet the requirement of the second half of the puzzle – that is, the wars had to end in the failure of the democratic protagonist.

The need to minimize the effects of confounding variables in order to isolate the causal mechanism of democratic defeat added a few other requirements to the case-selection. For example, it was not sufficient to select cases in which the balance of power remained firmly in favor of the democratic protagonist. Rather, it was necessary to find cases in which the latter also enjoyed a relative international autonomy to fight as it pleased. Similarly, it was also necessary to find cases where the will and motivation of the underdog did not compensate for its material inferiority. In practice, the

---

[64] On the instrumental role of case studies see Robert E. Stake, "Case Studies," in Denzin and Lincoln, *Handbook of Qualitative Research*, 237. See also Punch, *Introduction to Social Research*, 152.

[65] George, "Case Studies and Theory Development," 43–68.

[66] On minimizing the effect of confounding variables, see Waltz, *Theory of International Politics*, 13. On "generalization" and "case-to-case transfer," see Punch, *Introduction to Social Research*, 155.

best proximate criterion to follow was to select cases in which the national cohesion of the insurgent was seriously fractured, or that its population suffered from other deep rifts.

The cases of France in Algeria (1954–1962) and Israel in Lebanon (1982–1986) best met these requirements. Both were fully compatible with the puzzle, and as I document in the first chapter of each case, they involved weak confounding variables. In both cases, systemic influence was marginal, the national cohesion of the underdog was at best limited, and yet the powerful democratic protagonist failed. In fact, the conditions were so much against the underdog that some might consider the cases as "hard" enough to support the plausibility of the thesis, rather than merely suitable for the development of theory.[67] In the final analysis, the advantage of the cases was that they led me to a relatively well-defined research path that focused the spotlight on the domestic structure of the democratic protagonist.

In fact, within this path, the cases I chose had particularly high instrumental value and extrapolation potential because in both the democratic state was "strong" and had decided to continue to invest resources and pursue its war goals in spite of mounting opposition at home. The significance of the particular strength of the French and Israeli states of the time is obvious. Conclusions regarding the capacity of society to control these states in the context of a small war are highly likely to be relevant for cases that involve democratic states that are weaker vis-à-vis their societies. In other words, the conclusions of this study are likely to be applicable to most other democracies.

The significance of state tenacity is less obvious, but it can be best understood by considering the opposite – cases in which the state pursued its war policy half-heartedly, was not ready to commit society, or quickly changed its course following initial signs of societal disenchantment. In such cases, it would have been harder to observe and follow any significant societal role because society would not have been "provoked" or given the chance to strongly oppose the war. In fact, realist and motivational arguments would seem to explain well the outcomes of such "under-invested" wars, although the fundamental cause of failure would still be the domestic structure that limited the military effort in the first place. In a nutshell, because the Israeli and French governments decided to commit their societies and material resources to a long and demanding struggle, in the face of mounting opposition, I was able to trace in detail the domestic process that doomed the war effort.

Finally, I would like to conclude the methodological discussion by noting that while the French and Israeli cases had much in common, they also differed in important ways that affected my research and presentation. The

---

[67] On the logic of using hard cases, see Waltz, *Theory of International Politics*, 125.

Algerian war ended over four decades ago, it lasted eight years, and it has since produced a solid body of scholarly works.[68] The Israeli war is now over a decade and a half old, it lasted only about three years, and the research about it is less thorough. Furthermore, the most important events in the Lebanon war, both in the battlefield and at home, took place in the first few months of the war. These differences influenced my research logistics, focus, and presentation style. In the French case, I relied largely on secondary sources, whereas in the Israeli case, I referred extensively to primary sources and newspapers in particular. Among Israeli newspapers, I relied most heavily on the *Ha'aretz* daily because it reflected and nourished the opinion of the mainstream anti-war segment of the Israeli society. Last, because of the shorter duration of the Israeli war and the much more condensed time frame of critical developments, I discuss social developments and trace the impact of the Israeli anti-war constituency in greater detail than in the French case.

Plan of the Book

Chapters 2 through 4 are devoted to laying the foundations for the study of the question of how liberal democracies lose small wars. These are based on deductive theorizing and historical induction. Chapter 2 explores a thematic question that is in essence a mirror image of my research question: How does military superiority often assure victory in small wars? The chapter revolves round a discussion of counterinsurgency strategies and propositions about the relations between the cost of war and the use of violence as a means to control it.

Chapter 3 seeks to account for the formation of political relevance and normative difference, two of the three variables that are at the heart of the failure of democracies to prevail in protracted small wars. The chapter discusses the emergence of the social and moral foundations of dissent and the gradual disappearance of the political, social, and cultural conditions that have permitted states to effectively use high doses of personal brutality in small wars.

Chapter 4 deals with the other half of the state-society equation – the state. Specifically, it reviews the prophylactic, reactive, and manipulative efforts of leaders and officials to contain society and thereby preserve the autonomy

---

[68] The already good state of knowledge concerning the Algerian War and its domestic aspects improved further particularly after the publication of Jean-Pierre Rioux (ed.), *La Guerre d'Algérie et les français* (Paris: Fayard, 1990); and Jean-Pierre Rioux and Jean-François Sirinelli (eds.), *La guerre d'Algérie et les intellectuels français* (Bruxelles: Éditions Complexe, 1991).

of the state. This chapter also puts together the two halves of the equation – the state and society – in an effort to address the question of timing and explain how democracies fail in small wars, particularly after 1945, rather than before.

Two caveats about bias and possible omissions are pertinent here. My analysis in Chapters 3 and 4 is biased toward developments that took place since the French Revolution – in France, England, and to a lesser extent Germany. The reason for this bias is twofold. First, these countries were the key powers of the time, and as leading imperialists, involved in multiple small wars. Second, the evolution of these countries, in terms of the social and political variables I discuss, is fairly representative of other democracies in Europe. My discussion is also biased in the sense that while it touches on the evolution of state and society, I could not provide a comprehensive account of all events and developments that may have had some relevance for my thesis. Instead, I chose to highlight only major developments that I thought were indispensable for the understanding of the creation of a social structure that proved inhospitable for certain military ventures. I am well aware that such an ambitious task carried great risks of omission. I had no other remedy than to progress carefully, and I hope that I have thereby managed to avoid blurring important idiosyncrasies and distorting the main course of events.

In Parts II and III of the book (Chapters 5 through 14), I discuss in details the French and Israeli case studies. In Chapters 5 through 9, I analyze the French war in Algeria, and in Chapters 10 through 14, I analyze the Israeli war in Lebanon. The two cases are discussed similarly. The first chapter of each case (Chapters 5 and 10) include an overview of the power relations between the protagonists, the development of the war, and its outcomes; a discussion of the political and bureaucratic attitudes to the war; and a review of the relations between the war and the economy. These chapters are designed to define both wars and to be informative. However, they also refute some domestic and international hypotheses that claim to explain the outcomes of these wars.

The following chapters of each case dissect the war according to my conceptual variables. Chapters 6 and 11 discuss the state, the war, and instrumental dependence. Chapters 7 and 12 discuss the development of normative difference. Chapters 8 and 13 discuss the reaction of the state and the consequent secondary expansion of the normative gap. Finally, Chapters 9 and 14 discuss some features of political relevance, and how the latter brought about the political end of the wars.

The concluding chapter (15) is devoted to the lessons of this study and the legacy of small wars. In particular I discuss three issues. First, I empirically reconsider the issue of multiple-level analysis by briefly discussing the Vietnam War. Second, I emphasize again the pivotal role of instrumental dependence, albeit this time by comparing a few characteristics of state policy and social response in the cases of France, Israel, and the United States (in Vietnam). Finally, I present the implications of this research for the study of the benevolence of democracies, and then I discuss how the legacy of failed small wars governs democratic decisions concerning military intervention.

2

Military Superiority and Victory in Small Wars

Historical Observations

The pattern of the outcomes of conflict between rivals of great military inequality remained unchanged from antiquity until well into the twentieth century. Control over superior means of destruction almost always promised victory, continuous domination, or successful pacification. Weak protagonists – and insurgent populations – did not always accept this state of affairs, nor did they always assess correctly the balance of power or the might and determination of their powerful conquerors or rivals. Nevertheless, when the military superiority of oppressors was unquestionable, so were the results.

In this chapter, three issues are discussed. First, I note why, under conditions of acute military inferiority, weak protagonists chose an insurgency strategy in order to fight domination. Second, I explore how military superiority was traditionally employed in pacification, and I define strategic prototypes of counter-insurgency. Third, I expose the key variable that guaranteed that military superiority would be translated into effective domination or pacification.

Fighting Small Wars: Insurgents and Oppressors

Much of what is known about military aspects of armed struggle against foreign domination comes from the study of guerrilla warfare.[1] Communities and nations choose to fight a guerrilla war against oppressors because it proves to be "frugal" and because it makes their own forces less vulnerable. Guerrilla warfare turns out to be the only form of violent resistance that

[1] See Robert B. Asprey, *War in the Shadows: The Guerrilla in History* (NY: William Morrow, 1994); Gerard Chaliand (ed.), *Guerrilla Strategies* (Berkeley: University of California Press, 1982); and John Ellis, *A Short History of Guerrilla Warfare* (NY: St. Martin's Press, 1976), and *From the Barrel of a Gun*; and Walter Laqueur, *Guerrilla* (Boston: Little, Brown and Co., 1976).

has any chance of surviving repeated encounters with a militarily superior oppressor. Its advantages can perhaps be best understood by considering the burden associated with conventional warfare of pitched battles. In conventional warfare, armies seek to marshal their forces for decisive battles. They therefore rely on a great deal of logistic support, fixed bases, and a few wide supply lines. These require a great deal of centralization, investment of material and human resources in infrastructure, and in its defensive maintenance. Ultimately, these offer good targets, particularly for the militarily superior side. Guerrilla warfare, by relying on small independent formations, and on supply and shelter from an existing, widely decentralized infrastructure – the general population – can avoid much of the burden, as well as a single knockout blow.[2] To use Mao's famous words, guerrilla warriors are fishes, while oppressed communities are the latter's sea. In these communities, guerrilla warriors find a vast and dispersed support and shelter system, and thus a base for great mobility and reduced vulnerability.[3]

The primary goal and best hope of insurgent movements has always been that they will manage to dissuade their powerful rivals from continuing to fight by imposing on the latter a high enough cost for a long enough period. Until roughly the second part of the twentieth century, however, and in spite of its obvious advantages, guerrilla warfare rarely proved to be a way to solve the political problems of oppressed communities. Indeed, both students and practitioners of insurgency warfare tend to agree that the success of guerrilla warfare depended primarily on the nature of the oppressor and the context of war, rather than on the particular advantages it provided to the oppressed.[4]

Oppressors hardly ever intended to let insurgency wars drag on or bleed them so much as to make their losses unacceptable. Rather, they devised a number of ways that individually, or in combination, could circumvent the insurgents' guerrilla strategy and defeat armed insurrection. The crudest strategy was to target the popular base of insurgency and *eliminate it indiscriminately*, thereby destroying the ability of populations to produce and support insurgency. Alternatively, oppressors targeted the link between guerrilla forces and the popular base, trying to render insurgents ineffective by *isolating* them from their external and internal supply sources. Finally, oppressors targeted guerrilla leaders and fighting formations in an effort to *surgically eradicate* the military potential of oppressed communities. These traditional strategies were targeted at different aspects of the complex that

---

[2] See also Krepinevich., *The Army and Vietnam*, 7–10.

[3] For excerpts from Mao's work, see Mao Tse-Tung, *On Guerrilla Warfare* (NY: Praeger, 1961); *Selected Military Writings* (Peking: Foreign Languages Press, 1963); and *Strategic Problems of China's Revolutionary War* (Peking: Foreign Languages Press, 1954).

[4] See J. Boyer Bell, *The Myth of the Guerrilla* (NY: Knopf, 1971), 52, 54; Arthur Campbell, *Guerrillas* (London: Arthur Barker, 1967), 3; and Laqueur, *Guerrilla*, 382–84. See also Mao's analysis of the war against Japan in Mao Tse-Tung, *On Guerrilla Warfare*, 50, 113, and *Selected Military Writings*, 154–85, 277–88.

makes insurrection work. Each required a particular conduct, incurred various costs, and derived different benefits. However, they were not exclusive. Rather, these strategies were often used in a complementary manner, or in succession. Still, for analytical purposes, it is worthwhile to deal with each strategy separately.

Targeting the Popular Base: National Annihilation

The strategy of national annihilation is rarely used these days.[5] In past times, however, it was part of the common political repertoire of conquerors. Empires, for example, expanded and retained their control over subjugated peoples by relying on the deterrent and actual effect of force. They often faced stiff opposition in conquest and insurrections thereafter. Occasionally, empires solved these problems by eliminating the population or national identity of their weak rivals. Such radical steps were taken by empires, not only in order to ensure that insurgents would not pose a similar problem again, but also in order to convince other subjugated peoples to calculate their behavior in a predictably docile manner. In short, the superior military power of empires presented weaker foes with a painfully limited choice between survival under subjugation or annihilation. The extreme outcomes of encounters between mighty military powers and their proud yet imprudent victims can be illustrated with great lucidity in historical cases, including those of the Melian refusal to accept Athenian hegemony (assuming that Thucydides' account is either real or representative), the Jewish Bar-Kokhba revolt against the Roman empire, and Cromwell's war against the native Irish.[6] In these and many other instances, military superiority was used indiscriminately and without inhibition, and as a result, these confrontations were decided in a conclusive, and occasionally, irreversible manner.

The little community of the island of Melos was asked by Athens to switch sides and join its empire against Sparta. The Melian leadership was aware of its military inferiority, yet, for reasons that do not concern us here, decided not to commit itself to anything beyond neutrality. The cost of rejecting the Athenian quest for domination was the extermination of the entire adult male population – the actual and potential leaders and warriors – and the deportation of the remaining people as slaves.

The Jewish Bar-Kokhba revolt ended almost as tragically. The Romans had already fought a major Jewish revolt fifty-eight years earlier (in 66 A.D.) and they had faced continuous low-intensity insurrection thereafter.

---

[5] For a discussion of annihilation strategy in all but name, see Summers, "A War Is a War Is a War Is a War," 38–39.

[6] See Thucydides (trans. Rex Warner), *The Peloponnesian War* (London: Penguin Books, 1954), 358–66; Edward T. Salmon, *A History of the Roman World* (London: Methuen, 1957), 194–97; Bernard W. Henderson, *The Life and Principate of the Emperor Hadrian* (Rome: "L'Erma" di Bretschneider, 1968), 215–21; Yigael Yadin, *Bar Kokhba* (Weidenfeld and Nicolson, 1971), 17–23; and Edgar O'Ballance, *Terror in Ireland* (Novato, CA: Presidio Press, 1981), 7–8.

However, the Bar-Kokhba uprising presented them with a more recalcitrant and costly challenge, and thus they took more extreme measures. They liquidated Bar-Kokhba's rebellious bands together with their popular base of support, and in this way made sure that the Jews would be unable to revolt again. Indeed, when the Romans had finished with the revolt, the Jewish nation was in ruins. A relatively conservative estimate suggests that the Jewish population was decimated by about 50 percent.

A millennium and a half later, the Irish revolt against England resulted in a similar outcome. Cromwell annihilated the garrisons of rebellious cities ruthlessly, in the process often killing the innocent civilians and the Catholic clergy. Many of the native Irish who survived were sent into exile in concentration camps, and Ireland's population was thoroughly blended with English colonists. Within ten years of the repression, the Irish population was decimated by an estimated one-third (the loss amounted to some half a million people).

Extreme brutality and little discrimination in dealing with empire-building and maintenance are not characteristics of the distant past only. Until well into the twentieth century, European states used to practice measures of extermination in various parts of the lands they ruled. The German wars in East and South-West Africa (1904–1907) left history some textbook examples of strategic annihilation.[7] In August 1904, for example, General Lothar von Trotha pushed the rebellious Herero people into the Omaheke desert in South-West Africa (today's Namibia) and sealed off the west and southwest ends of this arid territory for about a year, in order to destroy the Hereros. The official 1906 history of the German General Staff noted that "the arid Omaheke was to complete what the German army had begun: the annihilation of the Herero people."[8] Moreover, after von Trotha had isolated the Herero, he issued the following proclamation on October 2, 1904:

[T]he Herero are no longer considered German subjects. They have murdered... and now refuse to fight on, out of cowardice ... [they] will have to leave the country. Otherwise I shall force them to do so by means of guns. Within the German boundaries, every Herero, whether found armed or unarmed, with or without cattle, will be shot.[9]

---

[7] See Horst Drechsler, "South West Africa 1885–1907," in Helmuth Stoecker (ed.) [trans. Bernd Zöllner], *German Imperialism in Africa* (London: C. Hurst, 1986), 53–58; Helmut Bley [trans. Hugh Ridley], *South-West Africa under German Rule 1894–1914* (London: Heinemann, 1971), 149–69; John Iliffe, "The Effect of the Maji Maji Rebellion of 1905–1906 on German Occupation Policy in East Africa," in Prosser Gifford and Wm. Roger Louis, *Britain and Germany in Africa* (New Haven: Yale University Press, 1967), 560–61; Helmuth Stoecker, "German East Africa 1885–1906," in Stoecker, op. cit., 111–13; and Thomas Pakenham, *The Scramble for Africa, 1876–1912* (London: Weidenfeld & Nicolson, 1991), 602–28.

[8] Drechsler, "South West Africa," 58.

[9] Pakenham, *The Scramble for Africa*, 611.

Indeed, the results of this and other German campaigns were devastating. Large portions of the original populations of rebellious African tribes – between 50 and 80 percent – perished. Most, it seems almost unnecessary to emphasize, were innocent civilians who had never taken up arms. Many of those who survived the systematic hunting-down operations and the premeditated food and water deprivation were sent to labor camps or into exile under harsh conditions.

As Iraq's treatment of the Kurds and its southern Shiite population indicates, annihilation is for some states still an acceptable method of oppression, if not a final-solution strategy. The foundations of the Kurds' communal identity, cohesion, and national survival (and their capacity to oppose Baghdad) are and were in the ethnically homogeneous pastoral base and the proximity to external sources of supply. After the collapse of the Kurdish revolt of 1974–1975, Iraq launched a calculated campaign in order to destroy these foundations. Large Kurdish groups were forced to move into specially constructed and easily accessible villages near cities or major roads. Other Kurds were resettled, often in groups of up to five families, in Arab villages in southern Iraq. Hundreds of Kurdish villages were either destroyed or repopulated with Arabs.[10] In 1991, following the Shiite insurgency in southern Iraq, one of Saddam Hussein's leading henchmen, Ali Hassan Magid, appeared in an Iraqi army film explaining to his lieutenants that the way to handle the rebellious Shiite villages was to annihilate them altogether.[11]

Finally, the Chinese approach to Tibet and the Indonesians' dealings with the native residents of East Timor seem also to fall within the general pattern of national destruction. Communist China brutally subdued Tibet in 1950, and since then has spared no effort to eliminate all symbols and feelings of Tibetan nationalism and identity.[12] Similarly, Indonesia invaded East Timor (following the departure of Portugal), annexed the territory (July 1976), and declared it to be the country's twenty-seventh province. Since then, and until it granted East Timor independence in 2002, Indonesia has engaged in a continuous struggle against the native people, avoiding no brutal method of oppression.[13]

[10] See Marion Farouk-Sluglett and Peter Sluglett, *Iraq since 1958* (NY: KPI, 1987), 167–70, 187–88; Peter Sluglett "The Kurds," in CARDRI (Committee Against Repression and for Democratic Rights in Iraq), *Saddam's Iraq* (London: Zed Books, 1990), 197–99.

[11] Shown on CBS's evening news and the *MacNeil-Lehrer News Hour*, January 31, 1992; and on CBS's *60 Minutes* on February 23, 1992. Accounts of the 1988 Iraqi "Al Ansal" campaign were broadcast in the *MacNeil-Lehrer News Hour*, February 28, 1992; *Frontline* program "Saddam's Killing Fields," March 31, 1992; and ABC's *Nightline* program, May 11, 1992.

[12] See Chris Mullin and Phuntsong Wangyal, *The Tibetans* (London: Minority Rights Group, Report No. 49, 1983), 16–17, 21. China crushed rebellions in 1956 and 1959 with great brutality. It orchestrated mass executions, deportations of tens of thousands, and the admission of Tibetan children to [re]education centers.

[13] See Alan J. Day (ed.), *Border and Territorial Disputes* (Detroit: Gale Research Co., 1982), 296–302; Peter Carey, *East Timor: Third World Colonialism and the Struggle for National*

Targeting the Social Bonds: Mild and Extreme Strategies
Indiscriminate annihilation, scattering, or exile can be replaced by a less radical strategy of isolation. Oppressors can effectively respond to insurrection
by targeting the political base of the guerrillas, which constitutes the vital
link between the warriors and the population. The commander of the French
forces in Algeria during the late 1950s (and the leader of the Army coup of
April 1961), General Challe, explained:

The theory, the famous theory of water and fish of Mao Tse-tung, which has achieved
much, is still very simple and very true: If you withdraw the water, that is to say, the
support of the population, fish can no longer live. It's simple, I know, but in war only
the simple things can be achieved. . . .[14]

In principle, isolation can be achieved both by benevolent conversion and
by intimidation. Furthermore, benevolent and coercive methods are not necessarily incompatible. They can be complementary, as Galliéni recommended
in 1900 and as Magsaysay proved in his struggle against the Hukbalahaps
in the Philippines (in the early 1950s).[15] Or they can be applied differen-
tially to various segments of the same population. A mixed application of
methods becomes particularly useful in dealing with ethnically, politically, or
otherwise heterogeneous populations. Indeed, imperial powers often quite
shrewdly calculated the dosage and application of coercion according to
the internal divisions in enslaved provinces. The British imperial policy of
"divide and rule," and particularly the emphasis on winning "hearts and
minds," demonstrate that benevolence can indeed be integrated into isolation strategy without giving up coercion.[16]
  Still, benevolent conversion is rarely the dominant method of isolation.
Moreover, benevolent isolation can easily regress into intimidation and terror, and the escalation of brutality does not necessarily end there. In other
words, coercive isolation can very well lead to annihilation. Indeed, the line
between coercive isolation and annihilation, analytically clear as it may be,

Identity (London: RISCT, Conflict Studies 293/294, 1996); and Amnesty International, *East
  Timor Violations of Human Rights* (London: Amnesty International Publications, 1985). The
  Indonesian army was apparently responsible for the death of one-quarter to one-third of the
  native population of East Timor in the late 1970s alone.
[14] Peter Paret, *French Revolutionary Warfare from Indochina to Algeria* (London: Pall Mall,
  1964), 42
[15] Ellis, *From the Barrel of a Gun*, 147–48; and Blaufarb, *The Counterinsurgency Era*, 27–37.
[16] Robert G. Thompson, *Defeating Communist Insurgency: The Lessons of Malaya and Vietnam*
  (NY: Praeger, 1966), 50–62 and particularly 55–57; and Edgar O'Ballance, *Malaya: The
  Communist Insurgent War* (Hamden, CT: Archon Books, 1966), 168. Still, it is important to
  note that the British authorities often relied on coercion. See Asprey, *War in the Shadows*,
  281.

is in the real world not all that distinguishable.[17] The French struggle against
the Algerian rebel Abd al-Qadir in the mid-nineteenth century illustrates the
degenerative nature of isolation.[18] With the prolongation of war and the
accumulation of frustration, the French increasingly resorted to policies of
terror and devastation. One of their preferred and most savage techniques
was the *razzia* – an indiscriminating raid involving the killing of people, destroying and plundering property, and burning the crops of tribes that joined
the insurrection. Soon enough, however, the *razzia* did not seem satisfactory,
and some of the officers developed an even more brutal mode of thinking.
By 1843, one of them recommended: "Kill all the men over the age of fifteen,
and put all the women and children aboard ships bound for the Marquesas
Islands or elsewhere. In a word, annihilate everyone who does not crawl at
our feet like dogs."[19]

The British employed a milder form of isolation policy during the Boer
War. They relied on a static network of blockhouses and on a scorched-
earth policy, which resulted in the destruction of property, the killing of
cattle, and the burning of crops. They also executed rebels and incarcerated
Boer families in concentration camps.[20] Although by the counterinsurgency
standards of the time their policies were quite restrained, they still imposed
an appalling cost on the Afrikaners. Thus, while the Boer warriors lost an
estimated 2,500 people out of some 60,000–65,000 who were involved in
the guerrilla stages of war, an additional 20,000 people, mostly children,
perished in the concentration camps.[21]

Of course, the idea of isolating indigenous populations from the insurgents by concentrating the former in controlled areas, was not new, nor
applied only by the British Empire. The Spanish used concentration camps
in Cuba, and the Americans, who denounced them for doing so, had earlier concentrated the Indians in reservations. The Americans also treated the
Philippines in much the same manner later on. Then the Mexicans did so

[17] Mugabe is quoted as saying: "Where men and women provide food for the dissidents, when
we get there we eradicate them. We don't differentiate when we fight, because we can't tell
who is a dissident and who is not." See Summers, "A War Is a War Is a War Is a War," 38.
[18] See Anthony T. Sullivan, *Thomas Robert Bugeaud* (Hamden, CT: Archon Books, 1983), 122–
26; and Raphael Danziger, *Abd al-Qadir and the Algerians* (NY: Holmes & Meier, 1977),
223–37.
[19] Sullivan, *Thomas Robert Bugeaud*, 125. Bugeaud explained: "To conquer [the Berbers] one
must attack their livelihood ... destroy the villages, cut down the fruit trees, burn or dig up
the harvest, empty the granaries, scour the ravines, rocks and grottos to seize their women,
children, old men, cattle and possessions..." Quoted in Ellis, *From the Barrel of a Gun*, 139.
[20] See S. B. Spies, *Methods of Barbarism?* (Cape Town: Human & Rousseau, 1977), particularly
pp. 183–201; and Byron Farewell, *The Great Boer War* (London: Allen Lane, 1977), 348–65,
392–420.
[21] See Farewell, *The Great Boer War*, 392; and Eversley Belfield, *The Boer War* (Hamden, CT:
Archon Books, 1975), 10, 165–68.

domestically, and the Japanese followed suit in Manchuria.[22] In fact, the same ideas were also behind the U.S.-South Vietnamese "strategic hamlet program" that was inaugurated in February 1962 and assumed larger proportions during 1966–1970. In this case, the goal was to drive villagers, often by means of bombing, into the hamlets' perimeters, so as to deny support to the Vietcong insurgents and create free-fire zones where greater firepower could be applied indiscriminately.[23]

Iraqi policy toward the Kurds also comes to mind as containing a component of isolation through terror. For example, during clashes between the government and the Kurds in 1963, the military governor of Northern Iraq declared:

We warn all inhabitants of villages in the provinces of Kirkuk, Sulaimaniya and Arbil against sheltering any criminal or insurgent and against helping them in any way whatsoever. We shall bomb and destroy any village if firing comes from anywhere near it against the army, the police, the National Guards or the loyal tribes.[24]

It is important to conclude the discussion of isolation by emphasizing that the shift from less to more coercive methods of isolation is not only the product of the frustration created by the dynamics of insurgency war, or the need to deter the indigenous population from cooperation with the insurgents.[25] Rather, brutalization is also the result of the fact that the war is fought over a commodity that no antagonist fully controls: perceptions of the future (which explains the French and American emphasis of psychological warfare).[26] In that sense, both parties fight over not only the current, but also over future, relations with the population. They try to convince the people that they alone will be in power once the struggle is over. That is precisely one of the major reasons why such struggles tend to involve high levels of brutality against civilians. After all, the legitimacy of rulers is intimately related to the perceived degree of institutional monopoly over coercive power. Thus insurgents try to prove that they can break the rulers' monopoly of coercive power, and rulers try to prove just the opposite. Both are ready to remorselessly punish any cooperation with their antagonist in

---

[22] See Brian Aldridge, "'Drive Them till They Drop and then Civilize Them': The United States Army and Indigenous Populations, 1866–1902," paper presented to the conference on *Low Intensity Conflict: The New Face of Battle?*, University of New Brunswick, Fredericton, Canada, September 27–28, 1991, particularly p. 29; and Edward E. Rice, *Wars of the Third Kind: Conflict in Underdeveloped Countries* (Berkeley: University of California Press, 1988), 95–98.

[23] See Guenter Lewy, *America in Vietnam* (NY: Oxford University Press, 1978), 25, 226.

[24] U. Zaher, "Political Developments in Iraq 1963–1980," in CARDRI, *Saddam's Iraq*, 63.

[25] After a massacre of an American army infantry company on the Philippine island of Samar, General Smith instructed: "I want no prisoners, I want you to burn and kill; the more you burn and kill, the better it will please me." Quoted in Asprey, *War in the Shadows*, 131 (perhaps surprisingly, the orders were not carried out).

[26] See Thompson, "Low-Intensity Conflict: An Overview," 4; and Paret, *French Revolutionary Warfare*.

order to prevent perpetual erosion in their public position. This tendency to resort to the extremes of brutality was immortalized in the absurd words of an American officer in Vietnam. The latter explained, after the bombing of Binh Tri (during the Tet offensive), that "it became necessary to destroy the town in order to save it."[27]

Targeting the Military and Political Cadres: Decapitation and Eradication

Modern military forces often prefer to deal with military opponents rather than with the civil population. The military objective of hostilities (irrespective of the type of war) is usually to engage and destroy the enemy's fighting formations or render them ineffective by eliminating their military and political command. Counterinsurgency forces regularly consider isolation not as an end in itself, but rather as a means of forcing on insurgents a military showdown. In such cases, insurgency-warriors and insurgency-leaders are the prime targets for eradication. Such eradication is carried out in several ways. Benign methods include, for example, apprehension, incarceration, and deportation. Less benign methods include the use of bounty hunting, murder, and executions – sometimes following judicial procedures, but more often without any consideration for laws.

During the Italian campaign in Abyssinia in 1935–1936, Mussolini's orders were straightforward: Shoot all rebels.[28] In 1947, the French in Indochina launched the failed operation "Lea" that was designed to eradicate the Vietminh's fighting force and leadership.[29] Even in the British Empire, perhaps the most benevolent of all modern colonial and oppressive systems, the execution of insurgent leaders and warriors was considered a legitimate means to fight and deter insurrections.[30] The American *search and destroy* missions (such as in operations Attleboro, Cedar Falls, and Junction City), *body-counting* policy, and project Phoenix during the Vietnam war are but a few of the latest examples of policies that were designed to eliminate the backbone of insurrection.[31] The Israeli operation of special hunting squads against the Palestinian insurgents during the early 1970s (Rimon) the first Intifada (Shimshon and Duvdevan) and the incursions into Palestinian cities and villages in April 2002 and after are other modern examples of efforts designed to eliminate the fighting backbone of insurrection.

[27] Air Force Major Chester L. Brown quoted in Wells, *War Crimes and Laws of War*, 104.
[28] See V.G. Kiernan, *From Conquest to Collapse: European Empires from 1815 to 1960* (NY: Pantheon Books, 1982), 202–03.
[29] See Bernard B. Fall, *Streets Without Joy* (London: Pall Mall, 1964), 27–28. See also, on French coercion in Syria, A. L. Tibawi, *A Modern History of Syria* (NY: St. Martin's, 1969), 340.
[30] See Farewell, *The Great Boer War*, 330–34; Laqueur, *Guerrilla*, 179; and O'Ballance, *Terror in Ireland*, 28–29.
[31] See Lewy, *America in Vietnam*, 50–56, 78–82, 279–85; Rice, *Wars of the Third Kind*, 93–95; Blaufarb, *The Counterinsurgency Era*, 246–48, 250, 274–76; Krepinevich, *The Army and Vietnam*, 190–91.

Eradication has occasionally failed. But, as demonstrated by the killing of
Mahmadou Lamine by the French in Western Africa (1887), the capture of
Aguinaldo in the Philippines by MacArthur's forces (March 1901), and the
war in the *Vendée* (1793–1794 phase), decapitation can work well, at least
in some cases and for some time.[32] In fact, the idea of fighting insurgency
by eradication was apparently so attractive, that totalitarian states carried it
to monstrous extremes, as a preventive rather than reactive doctrine, which
was designed to assure quick submission following a conquest. The Soviets
proved this in their massacre in the spring of 1940 of Polish POWs and other
subjects in Katyn, and the Nazis proved this in their plans and conduct in
Poland and other conquered Eastern territories.

Lavrenti Beria, the ruthless NKVD chief, wrote to Stalin regarding the
25,700 Polish prisoners who were held by his organization after the liqui-
dation of Poland, that they were "all ... bitter enemies of the Soviet power,
filled with enmity to the Soviet system ... Each ... plainly awaits liberation,
thereby gaining the opportunity to actively join the battle against the Soviet
authorities."[33] Accordingly, and since the NKVD considered "all of them
[as] hardened enemies of the Soviet power with little expectation of their reform,"[34] Beria found it "essential" to "apply towards them the punishment
of the highest order – shooting."[35] Hans Frank, the Nazi Governor General
of Poland, summed up Hitler's objectives for the "Extraordinary Pacification
Action" in Poland in much the same terms. "The men capable of leadership
in Poland," Frank told his officers, "must be liquidated. Those following
them ... must be eliminated in their turn."[36]

Violence and Counterinsurgency: Brutality as a Means of Cost Management

Violence is not only the primary means of getting the desired results of war.
Rather, it is also a way of managing its costs. In other words, states resort
to greater and less selective methods of brutality in pacification wars not
only because these prove to be effective, but also because they prove to be

---

[32] See Ellis, *From the Barrel of a Gun*, 141; Asprey, *War in the Shadows*, 18, 130–31; and Laqueur, *Guerrilla*, 24–25.

[33] Quoted in Wojciech Materski (ed.), *Katyn: Documents of Genocide* (Warsaw: Institute of Political Studies, Polish Academy of Sciences, 1993), 19. The prisoners included former police officers, clerks, landlords, policemen, intelligence agents, military police, immigrant settlers, prison guards, manufacturers, former Polish officers, and others who were either trained state officials or potential leaders.

[34] Ibid., 23.

[35] Ibid. The recommendation was accepted literally in the Politburo, was returned to Beria, and was promptly executed. See ibid., 11.

[36] William L. Shirer, *The Rise and Fall of the Third Reich* (NY: Simon & Schuster, 1960), 662. See also ibid., 660–65.

efficient. Higher levels of violence can cut down on the investment and loss of manpower and material, both through the destruction involved and the fear generated. This instrumental logic was succinctly encapsulated in the order of General Keitel to the Nazi occupation forces in Eastern Europe:

In view of the vast size of the occupied areas in the East, the forces available for establishing security in these areas will be sufficient only if all resistance is punished not by legal prosecution of the guilty, but by the spreading of such terror by the Armed Forces as is alone appropriate to eradicate every inclination to resist among the population ... Commanders must find the means of keeping order by applying suitable draconian measures.[37]

Of course, the Nazi concept of violence and the consequent atrocities German soldiers perpetrated in occupied territories were not entirely innovative.[38] Conquest and pacification, including those involving European powers up to the mid-twentieth century, were often based on high doses of indiscriminate violence. Brutality was perceived as a pragmatic, and often as the only, way of solving the problem of the shortage of resources. Indeed, as Michael Howard reminds us, early European conquests outside the Continent were often obtained in spite of great numerical inferiority, precisely because of a superior ability to employ violence.[39] The Spanish conquests in America, Howard writes, were owed to the "single-minded ruthlessness ... desperation, and ... fanaticism" of the Spanish soldiers.[40] Technological inventions such as artillery, later the machine gun, and eventually air power, only improved the ability to manage cost through violence.[41] Modern European powers simply continued an old imperial tradition. They conquered, and then prevented the deterioration of their rule through the

[37] Leon Friedman (ed.), *The Law of War: A Documentary History*, Vol. II (NY: Random House, 1972), 948. See also Field Marshal List's directions to the forces in Yugoslavia, in Paul N. Hehn, *The German Struggle Against Yugoslav Guerrillas in World War II* (NY: East European Quarterly, distributed by Columbia University Press, 1979), 33. For accounts of German brutal execution of instructions see ibid., 29, 55, 56, 58, 64–65, 69, 90, 97, 138; Omer Bartov, *The Eastern Front, 1941–45, German Troops and the Barbarization of Warfare* (NY: St. Martin's Press, 1986).

[38] See, on the Japanese expedient calculus, China in Yung-fa Chen, *Making Revolution* (Berkeley: University of California Press, 1986), 96–97.

[39] See Michael Howard, "The Military Factor." See also Ian F. W. Beckett and John Pimlott, *Armed Forces and Modern Counter-Insurgency* (NY: St. Martin's Press, 1985); and Christopher Duffy, *The Military Experience in the Age of Reason* (NY: Routledge & Kegan Paul, 1987), 280–81. For the realities of colonial wars, see Kiernan, *From Conquest*, 160–66, particularly pp. 160–61.

[40] Howard, "The Military Factor," 35.

[41] Hugh Trenchard, the Chief of the RAF Staff, explained to his Middle East commander that "the air force is a preventative against risings more than a means of putting them down. Concentration is the first essential. Continuous demonstration is the second essential. And when punishment is intended, the punishment must be severe, continuous and even prolonged." Quoted in Asprey, *War in the Shadows*, 279.

instigation of short and particularly violent actions. In the 1898 Omdurman battle in the Sudan, for example, the British forces led by Kitchener subjugated the upper Nile river region, losing 48 soldiers while killing some 11,000 Dervish.

The use of brute force in order to control the costs of imperial/colonial wars continued throughout the first half of the twentieth century even as some European powers became more liberal and democratic.[42] According to observers, the French strategy during the Druse revolt in Syria in the 1920s seems to have been to crush the rebellion "by the maximum use of every mechanical contrivance [but] with the minimum use of French soldiers."[43] On May 7, 1926, the French turned a whole quarter in Damascus into rubble in a twelve-hour period. The death toll was estimated at between 600 and 1,000. During the 1916 Easter rebellion in Ireland, a British four-day military repression in Dublin resulted in 1,351 Irish dead. In Africa, the British felt free to pacify Somaliland through air bombardment. The Italian army added the use of poison gas to these practices in its war in Abyssinia (1935–1936). In both the British and Italian campaigns, violence, as Michael Howard observes, "achieved its purpose in terrorizing resistance into rapid submission and so diminishing the requirements for a prolonged land campaign."[44] In May 1945, in Setif, Algeria, the French are estimated to have killed at least 6,000 people. During the 1947–1948 revolt in Madagascar, 60,000 people are estimated to have been killed.

The relationship between the level of violence and the material and human cost of conquest can be clearly illustrated by a brief comparison of the strategy, cost, and outcome of the 1904–1907 German pacification of rebellious African tribes and the British pacification of the insurgent Boers in 1899–1902. Admittedly, the variance between the cost and outcome of the two cases cannot be attributed solely to the difference in the strategic choice (which includes the methods of pacification and the consequent levels of violence and degrees of discrimination). Still, the variance is so remarkable that it would be unjustified to deny the role played by the strategy and methods of violence. While the Germans chose to indiscriminately annihilate, the British chose to isolate and eradicate selectively. The Germans used altogether some 18,000 troops in East and West Africa. The British deployed some 449,000 troops (though only about 50,000 were used for offensive operations). The cost for the Germans was 22 million pounds sterling. The British spent

[42] Paragraph data are from Donald Cameron Watt, "Restraints on War in the Air Before 1945," in Howard, *Restraints on War*, 64; John Ellis, *The Social History of the Machinegun* (Baltimore: Johns Hopkins University Press, 1975), 79–109; Tibawi, *A Modern History of Syria*, 340–48; Philip S. Khoury, *Syria and the French Mandate: The Politics of Arab Nationalism 1920–1945* (Princeton: Princeton University Press, 1987), 97–244; and John Pimlott, "The French Army: From Indochina to Chad, 1946–1984," in Beckett and Pimlott, *Armed Forces*, 47.
[43] Quoted in Khoury, *Syria and the French Mandate*, 192.
[44] Howard "The Military Factor," 41.

220 million pounds. The Germans were responsible for the death of perhaps as many as 400,000 people, the British for some 25,000. In the battles of the Boer War, the British lost some 7,900 soldiers. Yet the total of British dead amounted to 22,000. Moreover, the ratio of battle fatalities was almost one to two in favor of the Boers. Overall, then, the British lost more soldiers in the Boer War than the Germans used in their campaigns in Africa.[45]

With these and additional examples in mind, one can form some generalizations about the use of violence. From the perspective of unscrupulous oppressors, the removal of the popular base of insurrection or the destruction of the national identity of subjugated peoples are simple and cost-effective measures. Indiscriminate annihilation requires relatively little investment and military skills, and produces long-lasting results.

Much as with annihilation, no particular genius is necessary for the exercise of isolation, particularly when it is based on coercion. Isolation does require greater investment and patience than annihilation. But its requirements can be minimized through the escalation of the level of brutality.

Admittedly, the use of less-discriminate methods of violence in the pursuit of isolation is not risk-free. The attitude of the target population can presumably be hardened, the pool of material and human resources available to insurgent forces may grow, and the readiness of oppressed people to fight and endure sacrifice can also increase with additional suffering.[46] The implications of such potential developments, however, should not be exaggerated. Beyond a certain threshold of coercion, the emboldening effect of brutality may very well be offset by the fear it creates. Oppressed communities may become too fearful to let their feelings of humiliation, insult, and vengeance guide their behavior. Moreover, oppressors may be indifferent to the counterproductive effects of coercive strategy, assessing that it is still easier and cheaper to base their rule on crude terror rather than compromise, seduction, or careful application of brutality. Japanese conduct during the 1930s and early 1940s in China and East Asia[47] – the largest population base on earth – illustrates that great violence and brutality do not necessarily create a problem of unmanageable proportions, nor do they necessarily turn out to be self-defeating.[48]

---

[45] Data are from Farewell, *The Great Boer War*, 351; Belfield, *The Boer War*, 10, 165–68; and Pakenham, *The Scramble*, 614–15, 622. Numbers are approximations.

[46] This may have been indicated by the larger 1919–1921 upheaval in Ireland, after the British brutally repressed the 1916 Easter rebellion. See Ellis, *From the Barrel of a Gun*, 159. Indeed, this point did not escape the attention of oppressors as, for example was suggested by the German Foreign Office Plenipotentiary in South-East Europe, in 1943. See Best, *Humanity in Warfare*, 243.

[47] See Chen, *Making Revolution*, particularly pp. 331–38, 78–84.

[48] Ibid., 513–14. For details on the extremity of Japanese terror in Asia, see ibid., 109; and Friedman, *The Law of War*, 1070–76, 1083–88.

Successful eradication of insurgent forces, be it one against the leaders, the warriors, or both, requires a greater investment and more talent than other strategies of counterinsurgency. It depends on such factors as timely and accurate intelligence, highly competent mobile-forces, and a widespread deployment (that can provide logistics for the gathering of intelligence, the strike operations, and the defense of local communities). Investment is not limited to the creation of infrastructure, but rather includes continuous maintenance as well. The more "surgical" the eradication effort, the greater the patience, skill, and investment required.

Yet even "surgical" eradication does not eliminate the need to rely on brutality. The hasty acquisition of intelligence, often from sources unwilling to supply it, necessarily involves a great deal of personal violence. The eradication of guerrilla forces, whose culpability can hardly ever be proven in a proper and cost-effective legal manner, is also inherently brutal. In short, while selective and careful counterinsurgency is more costly than other pacification strategies, it does not eliminate the need to rely on extreme violence. The application of violence could very well be more selective, but almost unavoidably the methods – torture and summary executions, for example – are not so selective or legitimate (and nothing has been said about the brutality involved in "preventive" eradication, as was revealed, for example, in Hitler's and Stalin's treatment of the Polish elite during World War II).

Conclusion

In the face of military superiority, conflict between conquerors and oppressed communities tends to regress into guerrilla and counterinsurgency struggle. Guerrilla strategy offers the underdog a cheap, efficient, and often the only way to remain militarily active in spite of logistical, numerical, and material inferiority. It provides the insurgent with a chance for a prolonged struggle by relying on the support of indigenous population. Conquerors and oppressors who refuse to compromise with the political demands of their weak rivals can nevertheless deal with insurgencies in one of several ways. They can annihilate the popular base of insurgency, isolate the population from the insurgents, or selectively eradicate the insurgents and their leaders. Each of these strategies requires a readiness to resort to violence against a civilian population, and violence indeed proves to be effective and efficient. It reduces the amount of human and material resources invested and lost in conquest and pacification. All in all, then, our discussion in this chapter reveals a vicious principle: If the oppressors are uninterested in reconciling their interests with those of the oppressed, then the incentive to escalate the level of violence is compelling. The chances are that a less selective use of violence will cut the costs and reduce the time of planning and executing each of the strategies of pacification. From an expedient point of view, then, the movement on the strategic scale from selective eradication to indiscriminate annihilation is tempting. In that sense, counterinsurgency is inherently de-generative. Benevolent isolation can easily give way to coercive isolation, and the latter contains the seeds of annihilation. Indeed, the most disturbing conclusion from our current moral vantage point is that brutality pays. The logistical parsimony of guerrilla warfare can be met with the parsimony of uninhibited violence, at least as long as altruistic moral restraints are absent.

3

The Structural Origins of Defiance

The Middle-Class, the Marketplace of Ideas, and the Normative Gap

As I noted in Chapter 2, all other things being equal, the readiness of strong powers to escalate the level of brutality is the key to winning small wars. This readiness, however, is a necessary, not a sufficient, condition. States, as I further noted in the Introduction, need also to be able to mobilize and convert societal resources into military might, and then use the latter with little, if any, restraints. Thus, leaders must secure the readiness of the military forces to meet the "requirements" of the battlefield, and ensure the people's "acceptance" of the military strategy and the costs it involves. The capacity of strong powers to win small wars, then, is almost by definition a function of their domestic structure. Or, formulated otherwise, society cannot a priori be overlooked. If the political order in a country leaves room for society to intervene in politics, then the capacity to win small wars becomes a function of the state of normative alignment (or conversely, the magnitude of the normative gap).

In this chapter, I review the foundations of the "space" that was opened for social forces to influence policy in liberal democracies. I combine again inductive and deductive logic, and I draw on historical observations in order to explain the origins of social defiance of the conduct of state in small wars. Finally, I refer almost exclusively to social and ideological developments that are related to the divide between the state and society: the political rise of the educated middle-class, the function of the free marketplace of ideas, and the emergence of a normative gap over issues that concern the state, society, and violence abroad.

Power Out of Feebleness: The Rise of the Middle-Class and the State

"The same industrial and commercial revolutions that touched off intense domestic political conflicts in the 1800s and 1900s," writes Benjamin Ginsberg, "also resulted in the creation and diffusion of wealth, organizational skills, communications techniques, and a host of other politically relevant resources that, in effect, increased the potential for opposition to state power and diminished the state's coercive capabilities."[1] Similarly, the German scholar Meinecke notes that "the great event of the eighteenth century . . . was the fact that under cover of the ruling absolutism, the middle-class gained in strength both intellectually and socially, and began to exploit the riches of Rational and Natural Law for their own class interest which was also now gradually acquiring a political tinge."[2]

Indeed, these observations succinctly capture the most important developments that were at the root of the power of society to check the state in liberal democracies. Nevertheless, they need to be complemented by one additional, presumably paradoxical, observation – the rise of the power of the middle-class was not simply the result of favorable change in its relative power, but rather it was also the consequence of its own weakness. However, in order to understand why such was the case, we must first recall the key challenges that rulers in Europe confronted in the late eighteenth century.[3] Social life was becoming unstable as an embittered "underclass" was crowding the cities; collective economic, social, and political demands were on the rise; and the capacity to meet the latter was strictly limited by the growing demands of international competition and the inadequate structure of the state. Thus, as the French Revolution suggested, the life of dynastic and aristocratic rule was at the mercy of social challenges that could not be adequately met without major adjustments of the order at home. Or, formulated more elegantly, the coercion-extraction cycle of dynastic absolutist monarchism was reaching its limit toward the end of the eighteenth century.[4]

Thus we have the origin of many of the political changes that restructured the domestic political scene in nineteenth-century Europe, including the limited reliance on the middle-class for the daunting task of governing after the French Revolution and the Napoleonic wars. This is not meant to argue that the tacit "alliance" between the monarchy and the middle-class was free of risks and costs for the king's court and the aristocracy. Rather, the point is that of all possible groups, the middle-class best met the rulers' description of a strategic partner because it possessed skills and resources that could

[1] Benjamin Ginsberg, *The Captive Public* (NY: Basic Books, 1986), 25.
[2] Friedrich Meinecke, *Machiavellism: The Doctrine of Raison d'Etat and its Place in Modern History* (London: Routledge & Kegan Paul, 1954), 346.
[3] See also the analysis in Bruce D. Porter, *War and the Rise of the State* (NY: The Free Press, 1994), 121–45.
[4] In Samuel Finer's terms, it was the time of transformation from extraction-coercion to extraction-persuasion. See "State- and Nation-Building in Europe: The Role of the Military," in Charles Tilly (ed.), *The Formation of National States in Western Europe* (Princeton: Princeton University Press, 1975), 96–97, 155–56. For example, Spain lost Latin America in the 1820s partially because its soldiers refused to fight for the king overseas. See Edward R. Tannenbaum, *European Civilization since the Middle Ages* (NY: John Wiley & Sons, 1971), 401.

contribute significantly to the capacity to govern, because it shared some basic interests with the ruling elite, and because, best of all, it was too weak to pose a real challenge to dynastic and aristocratic rule. Thus, while the middle-class certainly wanted some political change, its members had little taste for violent action and clearly they preferred bargaining and compromise to confrontation. In fact, it was only when the middle-class was totally ignored or its rights abruptly suppressed – as happened under the reactionary monarch Charles X in 1830 and toward 1848 – that the middle-class became dangerous. Even then, the middle class alone was no match for the rulers. Indeed, as has occasionally happened since then, when rulers felt that the middle-class challenge was actually or even potentially exceeding their own tolerance threshold, they cut it down to size or dealt it painful blows.[5]

In any event, whereas one part of the story consists of the weakness of the middle-class, the other part consists of the interests that rulers and the middle-class shared. For one thing, while the members of the middle-class distrusted and loathed the dynastic aristocracy and monarchy, the former were troubled as much as the latter by the populist legacies of the French Revolution. "Popular sovereignty" and "egalitarian order," for example, were almost as menacing for the petit bourgeoisie as they were for the nobles and the court (indeed, bourgeois liberalism stopped short of being progressive, particularly as the urban working-class gained self-consciousness).[6] Moreover, even issues that could have presumably set the two apart encouraged their cooperation under the prevailing circumstances. Thus the middle-class was ready to help extend the power and role of the nation-state domestically, although that could undercut its own power, and the rulers were ready to accept the principle of meritocracy and promote the rule of law, even though these undercut the power of genealogy and the privileges of the ruling class.

At first blush, the rulers–middle-class partnership may seem a simple case of trade-off. In reality, however, this tacit partnership was formed on a more complex basis. The internal expansion of the state, as long as it included meritocracy, opened to the middle-class an expanding job market, provided it with an opportunity to promote its social and economic interests, and increased its relative power. On the other hand, the value of meritocracy for state-building and preservation, and its value as a way of buying social peace and the cooperation of the middle-class, outweighed the threat it posed to upper-class privileges and the strain it put on the court–nobility coalition, in

---

[5] There is a solid line connecting (a) the *Carlsbad Decrees* that Metternich orchestrated with the German monarchs in 1819 against nationalism and liberalism in the students' associations, and (b) schemes to eliminate the educated progenitors of liberal dissent in Nazi Germany, Soviet Russia, Communist China, Pol Pot's Cambodia, and even in small and little noticed places such as Equatorial Guinea (during the 1960s–1970s under the rule of Macias).

[6] John Stuart Mill accurately captured capitalist contempt toward the lower classes in *Principles of Political Economy*. See Reinhard Benedix, *Nation Building and Citizenship* (NY: John Wiley & Sons, 1964), 40–41.

large measure because meritocracy was introduced when the public sector was expanding.[7]

Both education and law provide good examples. As we shall see in the next chapter, elementary state-controlled education was indispensable in the process of transforming subjects into docile and loyal citizens, and secondary-level schooling was necessary in order to produce lower- and middle-level bureaucrats and managers for the public and private sectors. For obvious reasons, education also appealed to the members of the middle-class, particularly as secondary schooling was biased in its favor and became the yardstick of competence for the meritocratic system. Similarly, the expansion of the rule of law served well both the rulers and the middle-class (although it also threatened to limit the power of the former). Thus, as the law transcended the individual ruler, it depersonalized the state, obfuscated inequalities that were inherent in the class-based social order, and eliminated some of the antagonism that rule by decrees created.[8] At the same time, as the adjudication and litigation system became more complex and required learned expertise, it provided the middle-class with yet another expanding job market. Moreover, control of the legal system insured a hospitable environment for economic and other values that were dear to the middle-class.[9]

The Political Relevance of the Middle-Class

As I have just noted, the cooperation of rulers and the middle-class benefited both sides. In the process of state expansion, monarchic rule gained a lease on life, and the middle-class was empowered, in part due to its inherent weakness. Yet it is not clear at all where the gains of the middle-class were reflected most forcefully, or in which institutions. It is often argued that the extension of suffrage and representation are the ultimate expressions of the rising relevance and power of different social groups, and primarily the middle-class. After all, these political rights were more often refused than granted. Moreover, when they were granted – be it as a down payment for the readiness to bear the cost of state policy, as compensation for sacrifice, or

---

[7] See, for example, the reasons for the introduction of meritocratic principles into Napoleon's army and its vanquished Prussian enemy in Geoffrey Best, *War and Society in Revolutionary Europe, 1770–1870* (NY: St. Martin's Press, 1982), 70; and William H. McNeill, *The Pursuit of Power* (Chicago: University of Chicago Press, 1982), 216. See also Willhelm Friedreich's use of a balance between the middle-class and the nobility in the bureaucracy, to prevent a challenge to the crown, in Colin Mooers, *The Making of Bourgeois Europe: Absolutism, Revolution, and the Rise of Capitalism in England, France and Germany* (NY: Verso, 1991), 118. See also ibid., 117–20.

[8] See also Gianfranco Poggi, *The Development of the Modern State* (London: Hutchinson, 1978), 74.

[9] The French 1804 Civil Code ("Code Napoleon") is a good example of the legal protection of middle-class and property rights.

as a last measure to avert forced change – it was presumably due to various pressures and weakness of the central authority.[10]

Yet this common view seems to exaggerate the initial significance of franchise and representation. In reality, both were terribly flawed for a long time after their introduction, and their influence on the conduct of state affairs was quite negligible. In essence, rulers maximized the benefits they gained from the people's feeling of representation and minimized the risk involved – that is, they prevented the possibility that societal forces that were excluded from the real political process would be in a position to intervene in it.[11] In fact, rulers sometimes extended the franchise, or appealed to "the people" at large, on their own initiative, because they did not consider such initiatives to be too risky.[12] In part at least, they did so because it was a way to bypass the liberal middle-class much as did the Jacobins during the French Revolution when they appealed directly to the masses as a ploy against the Girondins.

France and Prussia are both good examples of such extension of the franchise. In France, the franchise was extended (again) in 1840 under middle-class pressure to include 200,000 citizens who paid at least 200 francs per year in taxes. Eight years later (following the 1848 Revolution), universal male suffrage was granted without too much debate.[13] In December 1851, the young Second Republic became a dictatorship under Louis Napoleon (Napoleon III), and suffrage was revoked. Soon, however, suffrage was restored in the new constitution of January 1852. Of course, Napoleon III never intended to submit himself to real parliamentary control, let alone popular control. Rather, like his uncle before him, he sought legitimacy by orchestrating the semblance of popular control, including the submission of

---

[10] See Benjamin Ginsberg, *The Captive Public*, 13–23, and *The Consequences of Consent: Elections, Citizen Control, and Popular Acquiescence* (NY: Random House, 1982), 9–21. Limited reforms in Prussia following the defeat by Napoleon, and the democratic reforms of Napoleon III, following fears of a rising Prussia, are also good examples. See Finer, "The Role of the Military," 153; and Alfred Vagts, *A History of Militarism* (NY: W.W. Norton, 1937), 221–22. See also Morris Janowitz, "Military Institutions and Citizenship in Western Societies," *Armed Forces and Society*, 2:2 (1976), 189–93.

[11] In general, the approach of rulers to voting was reminiscent of Cicero's description of the Roman system. "Our law," he wrote, "grants the appearance of freedom, retains the authority of the aristocrats and eliminates the causes of strife..." Quoted in Peter T. Manicas, *War and Democracy* (Cambridge: Basil Blackwell, 1989), 66. Indeed, representative institutions were not open for real mass-participation until well into the twentieth century. For the chronology of suffrage extension and change of voting procedures in European countries, see Stein Rokkan, *Citizens Elections Parties* (NY: David McKay, 1970), 33, figure 3. For examples of how rulers manipulated the voting process see Benedix, *Nation Building and Citizenship*, 95–99.

[12] Ginsberg, *The Captive Public*, 13–16.

[13] In 1817, only 90,000 men had the right to vote in France, and only 1,652 were entitled to run for election. See Roger Price, *A Social History of Nineteenth-Century France* (London: Hutchinson, 1987), 358, 360.

TABLE 3.1 *Extension of the franchise in the U.K.*

Year	Registered electors (%)[a]	Next electoral reforms
1831	5.0	1832 Reform Act
		1867 Reform Act
1868	16.4	1884 Reform Act
1886	28.5	1918 Representation of the People Act
		(universal male suffrage and suffrage for females over 30)
1921	74.0	1928 Equal Franchise Act
		(women's voting age reduced to 21)
1931	96.9	1949 Abolition of last plural voting rights of members
		of the university and business community, mostly males
		(502,000 until 1945 and 266,000 until 1948)

[a] Of the population over 20 years old.

Source: Bédarida, *A Social History of England*, 142, table 3; and David Butler, "Electors and Elected," in Halsley, *British Social Trends*, 297–98.

his absolutist constitution to a plebiscite. In Prussia, suffrage and representation were equally mocked. In 1866, Bismarck, who was entangled in a power struggle with middle-class liberals in the Landtag, declared universal male suffrage (though for the Reichstag only). Yet he and the Kaiser governed Prussia, and then Germany, much as Napoleon III governed France. They ran a regime that was authoritarian in all but name, while at the same time their people enjoyed universal male suffrage.[14]

Nineteenth-century England, on the other hand, was far slower than either Germany or France in extending suffrage (see Table 3.1). Nevertheless, the English system offered social forces more opportunities to have a political significance. By mid-century, the right to associate and freedom of press were secured, and the elites were genuinely in agreement over individual rights and parliamentary government (which could not be dissolved at the executive's will). Indeed, politics in England was conducted in a genuinely competitive, if also limited, manner, with rival parties alternating in control of the state through a routine of elections.

In summary, initial political reforms, including the introduction of suffrage and representation, helped to domesticate the masses, rather than alter fundamentally the structure of political power.[15] Rulers often granted political rights under pressure, but they soon realized that certain "liberal" reforms could actually improve their hold of power, increase civic compliance,

---

[14] See Vagts, *Militarism*, 217 (the policy of inclusion was extended, for example, in England through the Poor Law, and in Germany through Bismarck's social legislation, to also include entitlements).

[15] See the discussion in Ginsberg, *The Consequences of Consent*, particularly pp. 29–31; and in *The Captive Public*, particularly pp. 208–14.

secure a larger and more docile tax base, and condition people to accept mobilization for military purposes less grudgingly. Moreover, these political reforms, which were implemented on a mass base, offered the bonus of checking rather than promoting the power of the liberal middle-class.

The Marketplace of Ideas

Although rulers succeeded in manipulating political reforms, the middle-class still managed to gain political clout via other ways than franchise and representation.[16] Thus the middle-class gained some power because its members entrenched themselves in the accelerating capitalist economy, swelling administration, and the expanding legal and education systems. Yet it gained even more power in the emerging marketplace of ideas.[17] Indeed, it was the free marketplace of ideas that proved to be the most significant of all the conveyer belts of political influence, and it is in this market that eventually the fate of small wars was decided. Now, much like representation, rulers (and states) expediently embraced the marketplace of ideas and "public opinion" because both fostered the purpose of nation-building and consequently state-building.[18] Being a "national" institution (as opposed to regional or local), the marketplace of ideas had the advantage of nourishing subjects of different allegiances with messages that strengthened the collective of the nation-state. In the national marketplace of ideas, provincial identities and loyalties were blurred, mentalities were changed, and differences among, rather than within, national communities were emphasized.

Still, the benefits that states reaped from a national marketplace of ideas should not obfuscate its unique importance for the standing of the middle-class, nor should it leave us with mistaken ideas about the subjective feelings of rulers. The fact that the marketplace of ideas operated outside the strict control of rulers provided members of the middle-class with a unique opportunity to assert their views at a time when no other means of influence, besides organized violence (which the middle-class understandably detested), was available to its members. Members of the middle-class, being better educated and more articulate than those of other groups, had the advantage in this market, where they could convert their economic and ideological power into political clout. Indeed, precisely for this reason, rulers (and states) were very reluctant to accept the idea that the marketplace of ideas should be

---

[16] The changing of the guard between the nobility and the bourgeoisie in England and France was essentially slow. For details, see François Bédarida (trans. A. S. Forster), *A Social History of England, 1871–1975* (London: Methuen, 1979), 125–32; and Price, *A Social History of France*, 114–16, 136, 362–63.

[17] For an analysis of the creation and role of the marketplace of ideas, see Ginsberg, *The Captive Public*, 36–40, 86–107.

[18] Poggi and Ginsberg make such arguments but do not deny the market's value to the middle-class. See Poggi, *The Development of the Modern State*, 83; and Ginsberg, *The Captive Public*, 87.

free. Their indignation only grew further, since economic and demographic forces crowded ever larger numbers of people into alienating urban conditions, which helped the dissemination of criticism of the government.[19] Thus, if rulers did not crackdown hard enough on the free marketplace of ideas, it was not for lack of will or insight, but rather out of weakness and lack of choice. In particular, it was problematic to act with force in the middle of a move from coercion to persuasion, which became even harder and more dangerous once a marketplace of ideas was established.

Nineteenth-century France is a good example of the relations between the power of rulers and the state and the vitality of the free marketplace of ideas. In post-Napoleonic France – where the political order was far from liberal, and individual rights and freedoms were not well protected – press freedom oscillated in direct relation to the despotic powers of the regime and its sense of confidence. Overall, the state tried to tame the press and regulate the production and circulation of newspapers. Licenses to publish newspapers were often expensive, the newspapers themselves were taxed, censorship was essentially political, and the libel laws and sanctions were designed to make attacks on the state very costly. In essence, state policy selected who could afford to own a newspaper and who could buy one because the costs of production of capitalist ventures are, as a rule, passed on to consumers. Nevertheless, the authorities still were disgruntled, and complained that the people had access to the press in coffee-houses (especially in the aftermath of the 1848 Revolution).[20]

During the Second French Empire, Napoleon III eventually provided the bourgeoisie and the working-class with some pressure-release valves that included the right to organize strikes (1864) and the tolerance of public meetings and the relaxation of the censorship laws (1868).[21] His rule ushered in the beginning of a great increase in the number and the circulation of newspapers in France.[22] The cost of production was shifting to advertisers, the press became cheaper, more people became literate, and the demand for newspapers grew. Yet the political impact of the proliferation of newspapers was mitigated and delayed. Overall, the press became less critical, as editors adjusted the content of newspapers to their new consumers, who were less

---

[19] Certainly the monarchy in England did not like the discussion and caricaturing of George IV's family quarrels, mistresses, and lavish life style, particularly as such press reports antagonized the middle-class, which became increasingly Victorian. The mockery and critique of the Orlean or Bourbon regimes in France was probably not liked any better.

[20] Price, *A Social History of France*, 353.

[21] Mooers, *The Making of Bourgeois Europe*, 88–89.

[22] The rapid growth in circulation of newspapers in France between 1880 and 1910 (although the quality press accounted for the smaller part of total circulation), the declining cost of newspapers in general, and the growing number of book titles in England from 1900 onward are good measures of the growing marketplace of ideas. See Price, *A Social History of France*, 353–55; and Bédarida, *A Social History of England*, 240.

educated and whose taste for information was different from that of previous readers. Much of the new readership regarded the press as a form of
entertainment, was less interested in political issues, and did not consider the
press as an agent of social control over the government and the state. Thus
much of the press became populist, emphasizing sensational news reporting
and banal stories rather than political analysis and criticism.[23]

But the free marketplace of ideas and the independent press still remained
essentially irritants for the state. Perhaps the best indication that such was
the case – that the marketplace of ideas was the favored arena of the liberal middle-class and that it grew out of state weakness – becomes clear
when one contrasts printed and broadcast communications. Unlike newspapers, broadcast communications were introduced when the state was much
more powerful and better organized. Joseph Goebbels, who monopolized
the German marketplace of ideas on behalf of the Nazi regime, told his ac-
complices that the written press was an "exponent of the liberal spirit, the
product and instrument of the French Revolution," whereas broadcasting
was "essentially authoritarian" and therefore a suitable "spiritual weapon of
the totalitarian state."[24] In fact, states far less despotic than Nazi Germany
also prevented broadcasting from becoming a medium in the free marketplace of ideas (for as long as they could, at least). The nineteenth century
"surrender" of the written press to the bourgeoisie was not repeated. Broadcasting was tightly controlled through licensing or forthright monopolization. Even in leading Western democracies, such as France and Britain, broadcasting was monopolized (and, worth noting, from the perspective of this
book, in Israel as well).[25]

Foundations of a Normative Gap

As long as there were no firm conceptions of both *state* and *society*, there
could obviously be no meaningful difference of opinion between the two
over foreign policy and security matters. Much of the friction during the
period before "subjects" became "citizens" was over narrow issues of taxation of human and material resources, food shortages, and food distribution. As far as foreign policy was concerned, the state of affairs was one
of indifference. As state expansion and middle-class formation accelerated,
however, important changes took place: The middle-class acquired a distinct
view about the destiny of the state, and intellectuals became more engaged
in developing utilitarian and moral standards for the conduct of foreign
policy.

[23] See Price, *A Social History of France*, 353–55.
[24] Quoted in Z.A.B. Zeman, *Nazi Propaganda* (London: Oxford University Press, 1964), 48.
[25] On the creation and state monopolization of broadcasting in England, see A.J.P. Taylor,
English History 1914–1945 (Middlesex, UK: Pelican, 1970), 297–99.

Whereas rulers, and those who manned the swelling state apparatus, increasingly regarded the people as a resource designed to serve the national interest, the intellectual harbingers of enlightenment, and then the members of the middle-class, perceived the state as responsible for the promotion of a rational agenda of progress for the benefit of society and the individual.[26] "Men began to look at the state purely from beneath," writes Meinecke, "and [the state] began to be treated, even more decisively than in earlier times . . . as a purposive institution aiming at the happiness of the individual."[27]

In a nutshell, the fundamental difference between those committed to the "nation-state" and those committed to "society," as both constituencies were being formed, were rooted in a dispute among classes over who should control the state, and between ideological camps over the purpose of the latter. The dispute over political power was expressed in the struggle to gain access to politics – that is, over franchise and representation. The dispute over the purpose of the state was expressed in the agenda of the different sectors, and most notably in the clash between the "etatist" call for altruism and the "bourgeois" ethos of materialistic individualism.[28] In fact, having developed simultaneously, these antagonistic positions fed on each other, and the tension between both was only exacerbated over time as a result of two structural developments: (1) the inadvertent introversion of the state, and (2) the escalation of warfare. First, in order to assure social peace, bind yesterday's subjects to the "new" national state, and extract more from the former, the state assumed additional functions, extended benefits to the people, and made the life of the citizens-to-be less arbitrary. However, such developments undermined the etatist agenda because they supported the view that the state existed for the benefit of the individual and in order to promote progress.[29] Second, the escalating nature of warfare blurred the demarcation lines between the army and society, and between the front and the rear, and consequently made war more horrendous and the demands of the state from society, greater.

Indeed, in the wake of the escalating nature of warfare and the parallel consolidation of etatist ideology and middle-class materialism, fears that the bourgeois ethos would weaken the state's capacity to meet external

---

[26] See David Kaiser, *Politics and War* (Cambridge, MA: Harvard University Press, 1990), 205.
[27] Meinecke, *Machiavellism*, 346.
[28] The argument that bourgeoisie is pacific was made by Joseph Schumpeter in *Imperialism and Social Classes* (NY: Augustus Kelley, 1951), particularly pp. 90–99. See also Daniel Pick, *War Machine: The Rationalization of Slaughter in the Modern Age* (New Haven: Yale University Press, 1993), 162.
[29] The changing state-allocation of resources from defense to civil purposes is a good expression of this development. It is the process Charles Tilly described as the "civilization of government." See *Coercion, Capital, and European States AD 990–1990* (Cambridge: Basil Blackwell, 1990), 122–24, and particularly table 4.4. See also Michael Mann, *States, War and Capitalism* (Oxford: Basil Blackwell, 1988), xi.

challenges flourished. During the debate over the future of the French education system (in the late nineteenth and early twentieth centuries), one conservative critic lamented that "to form a 'citizen' or to form a 'soldier'... is to
teach them the art of subordinating something of themselves and their 'natural rights' to the interests and rights of the community. Without that, no
'army', no 'fatherland', no 'society'... "[30] In fact, the anxieties over the impact of the materialistic ethos led conservatives to the extreme belief that war
was the ultimate means of maintaining an essential level of national cohesion
in the face of the degenerative effects of peace. "Without war," maintained
Moltke, "the world would stagnate and lose itself in materialism."[31]

At the same time that etatist and middle-class factions clashed over issues
of obligation, the morality of war and intervention became a source of contention, and part of the public discourse. Moreover, morality and political expediency occasionally intertwined. States were thus at times urged to
intervene abroad in the name of morality, and moral argument acquired a
political value. Certainly, the British military intervention in the Russian-
Ottoman conflict that led to the Crimean War (1854–1856) was encouraged
by jingoistic journalism that drew attention to the backwardness of the Russian order. Next, it was Disraeli's indifference to the massacre of Bulgarian
Christians by the Turks in 1876 that brought Gladstone, riding on the wings
of a moral agenda, back into political life. Similarly, on the other side of
the Atlantic, press and public resentment against the Spanish oppression in
Cuba, and in particular against the use of concentration camps, helped push
McKinley to intervene against Spain.[32]

These instances aside, the "public's" taste for intervention did not hold
for long, but rather receded in the face of more expedient calculations. As
the Crimean War dragged on, it became unpopular in England, and the
army command was criticized by the press for incompetence. The Zulu War
(1877–1879), two decades later, was unpopular, demonstrating the lack of
public enthusiasm for distant wars. Yet political expediency still converged
with morality, at least when sacrifice was not called for. Indeed, Gladstone's
moral agenda, including the call for the respect of the rights of small states
(such as Belgium and Italy), humanity, and international justice, helped him
bring about the Liberal landslide in the general election of 1880.[33]

Even more important, a vocal minority that formed a noticeable challenge during the eighteenth century to the etatist agenda was gaining ground

---

[30] Quoted in Fritz Ringer, *Fields of Knowledge: French Academic Culture in Comparative Perspective, 1890–1920* (Cambridge: Cambridge University Press, 1992), 148.

[31] Quoted in Best, *Humanity in Warfare*, 145.

[32] See Allan Keller, *The Spanish-American War* (NY: Hawthorn Books, 1969), 9–26, particularly
p. 13.

[33] See, on the United States, Larry H. Addington, *The Patterns of War since the Eighteenth Century*
(Bloomington: Indiana University Press, 1984), 114, and on England, Walter L. Arnstein,
Britain Yesterday and Today (Lexington, MA: D.C. Heath, 1976), 93–98, 140–43.

in the public sphere.[34] This minority included Liberals, Socialists, and Anarchists, who criticized war and the militaristic state on both rational and moral grounds. Richard Cobden, Herbert Spencer, Pierre-Joseph Proudhon, Norman Angell, and Ivan Bloch, to name a few, peddled various anti-war ideas – that war interfered with commerce (which they considered an agent of progress), war drained the finances of states, and war was plainly evil and reprehensible.[35] Moreover, whereas thinkers tended to discuss the evils of war in the abstract – that is, without reference to the state – toward the end of the nineteenth century more of them became convinced that the source of evil may just have been the excessive power of the state.

One incident in particular – the Dreyfus affair (1894–1906) – captures the significance of the developments I have discussed in this chapter. Obviously this affair did not directly concern war, small or otherwise. Yet it did call into question the presumption that the "state" was inherently right when matters of national security were concerned, and thus also the idea that the state should preserve its absolute powers in the latter issue area. Moreover, the affair proved that the liberal, educated class could successfully challenge powerful state institutions in the free marketplace of ideas.[36] In fact, in the wake of the Dreyfus affair, a "new" concept of civic virtue – at odds with that advocated by etatists – was articulated. Historian Gustave Lanson, one of the leading proponents of educational reform in France, drew the outlines of this "new model" citizenry, when he wrote the following "seditious" words:

Education in a democracy that wants to guide itself must form men capable of guiding themselves. Hence [we need] free minds, with a passionate love of the truth . . . Further, [we need] free consciences, free inner as well as outer servitudes, incapable of finding the good outside the truth, and able to act in the name of justice, of love, and of solidarity.[37]

---

[34] On the development of the "peace movement" prior to World War I, see John Mueller, *Retreat from Doomsday* (NY: Basic Books, 1989), 17–36; Best, *Humanity in Warfare*, 131–34; and Michael Howard, *War and the Liberal Conscience* (Oxford: Oxford University Press, 1978), 31–72.

[35] See Pick, *War Machine*, 19–27, 42–47, 77–79; and John F. C. Fuller, *The Conduct of War: 1789–1961* (Westport, CT: Greenwood Press, 1981), 128–30.

[36] While academics were split over the affair, it was clear that younger academics, particularly in history and philosophy, and from the Sorbonne and Ecole Normale, tended to be Dreyfusards. Those from the professional schools tended to be etatists. Famous Dreyfusards included Durkheim, Jaurès, Anatole France, Zola, and Clemenceau. The struggle in the marketplace of ideas is best remembered from the famous " *J'Accuse*" letter of Zola. However, there were other public actions that set the terms of reference for future intellectual involvement, particularly during the Algerian war. See Ringer, *Fields of Knowledge*, 219–25, 283.

[37] Quoted in Ringer, *Fields of Knowledge*, 227. In 1904, Emile Durkheim draw even more clearly the demarcation line between the intellectuals and the state, in "The Intellectual Elite and Democracy." See ibid., 310.

A single case, important as it may have been, obviously, should not lead us to exaggerate the power of the free marketplace of ideas, nor bolster the consequences of the growing challenge that part of the educated middle-class posed to the state. Nevertheless, it is clear that part of the empowered middle-class and the emerging normative difference between that part and the state, became – as the actions and rhetoric of rulers suggest – politically relevant. Thus, European rulers consistently exempted the better educated and more affluent from the obligation to risk themselves or their children in wars, until the power of the state was consolidated and international competition left little room for exemptions. Moreover, politicians and military elites sanitized their messages when they discussed war and military power in public. In particular, large armies were marketed as insurance policies in the name of peace, and wars were depicted as defensive and designed to bring lasting peace. Of course, all that did not mean that militarism was vanishing toward the end of the nineteenth century. Rather it meant that the peace movement, as Geoffrey Best aptly noted, was simply "far too strong not to make its mark in national politics and international affairs."[38] In fact, leaders in Europe may have become over-sensitive to the presumed peaceful proclivity of "the people." In 1908, the German Chancellor, Bernhard von Bülow, explained that "above all we ought never to forget that nowadays no war can be declared unless a whole people is convinced that such a war is necessary and just. A war, lightly provoked, even if it were fought successfully, would have a bad effect on the country."[39] Indeed, in line with this logic, wars were planned, and promised, to be short.[40] For example, World War I was originally so designed, partly because military and political elites assessed that if it were to be prolonged, they would face a serious popular rift – that is, a normative gap of too large proportion.

Colonial Wars: The Domestic Dimension of Brutalization

It was not long before the changing relations between the "state" and "society" – the combination of a budding normative gap and a free marketplace of ideas – expressed themselves in the context of colonial wars. Anthony Sullivan, a historian who studied the French pacification of Algeria in the 1830s and 1840s, argues that by 1843, the conduct of the French army there became so violent as to "make the hair on the head of an honest [French] bourgeois stand straight up."[41] Indeed, when in the summer of

---

[38] Best, *Humanity in Warfare*, 139–40.
[39] Quoted in Kaiser, *Politics and War*, 271. "Bad effect," for Bülow, meant that "every great war is followed by a period of liberalism, since a people demands compensation for the sacrifices and effort war has entailed."
[40] See Vagts, *Militarism*, 379–81, 391–96, particularly p. 394.
[41] Sullivan, *Thomas Robert Bugeaud*, 127.

1845 a French force asphyxiated some 500 men, women, and children who had refused to surrender and had taken refuge in a cave, revolted bourgeois opinion demanded action. A military report on the incident was released to the Chamber of Peers, and as a result, General Bugeaud, the commander and governor of Algeria and his forces were condemned. Furthermore, a delegation of the Chamber of Deputies, headed by Alexis de Tocqueville, paid a visit to Bugeaud, only to return to France with a broader denunciation of the colonial regime and a recommendation to export *the Continental standard of conduct* to Algeria.[42]

Anglo-Saxon states also encountered cultural objections to their brutal conduct in small wars. While campaigning against the native Indians of the Trans-Mississippi West (1866–1881), both Sheridan and Sherman noted the power of the humanist camp that demanded that they act in a civilized manner toward the Indians.[43] In fact, toward the turn of the nineteenth century, it became clear that the criticism of brutal behavior in wars of conquest and pacification was no longer sporadic. Douglas Porch observes that "a second drawback of harsh measures [employed in small-wars] was that their application 'shock[ed] humanitarians'... plenty of journalists and war protesters were prepared to recount the brutality of colonial warfare, which made small wars a periodic source of criticism and even scandal in Europe and America during the decades prior to World War I."[44]

The reaction in England to the conduct of the army in the Boer War exemplified this trend. The scorched earth policy of the army was perceived in London as unacceptably barbaric, and consequently evoked sharp criticism. The army, well aware of the potential damage of tales of brutality, tried to prevent the flow of information from the battlefield to the rear. Its efforts, however, were in vain, and stories about atrocities reached the British public through the books and diaries of officers and the letters that soldiers sent home. Admittedly, the criticism (which came mainly from the minority of well-educated upper-middle-class citizens)[45] never turned British public opinion against the war. But it is noteworthy that a cohesive opposition could be organized outside the established political arena and make a colonial war, and particularly the troops' conduct, a public issue. In fact, the British authorities responded to this sign of cultural incompatibility and

[42] See ibid., 126–41. Not surprisingly, the authorities and state agents became increasingly impatient with their critics. "Very few people in France," lamented Bugeaud, were "capable of understanding the necessity of total war" (ibid., 129). In 1848, this hostility culminated in the belief that only the army could save France from itself, a "lesson" that would haunt France again over a century later, in very similar circumstances, in Algeria.

[43] See Aldridge, "Drive Them till They Drop," 5–6.

[44] Quoted in the Introduction to Callwell's *Small Wars*, xiv.

[45] See Arthur Davey, *The British Pro-Boers 1877–1902* (Cape Town: Tafelberg, 1978), particularly pp. 52–60, 121, 122, 161, 162–66; Farewell, *The Great Boer War*, 314–17, 353–54; and Kiernan, *From Conquest*, 167–73.

independence by setting limits to the army's combat conduct. Thus the
retaliatory farm-burning policy of the first British commander, Lord
Frederick S. Roberts, was somewhat restrained by London, while the ideas
of General Horatio Herbert Kitchener (his successor) – including the advice
to confiscate property, force mass Boer emigration, and deport certain Boer
women – were rejected.[46]

Moral objection to brutality, one should note, did not form only in the
period's beacons of humanism. Even in Germany and Japan, where rampant
nationalism was combined with a strong commitment to "the necessity of
the state,"[47] brutal pacification methods created a certain measure of political and moral liability. During the German campaigns in Africa in the 1890s,
the Socialists complained in the Reichstag that the soldiers did not take prisoners, and when von Trotha's October 1904 *extermination order* against the
Herero became public (see the discussion in Chapter 2), even officials in the
government were disturbed. In fact, Chancellor von Bülow went so far as
to ask Wilhelm II to cancel the order because he believed that it constituted
a crime against humanity and a threat to Germany's "standing among the
civilized nations of the world."[48] Moreover, even von Schlieffen, who supported von Trotha's brutal policy, came to the conclusion that it "could not
be carried through successfully in the face of present opinion."[49] Even in
Japan, which was geographically and culturally rather remote from Europe,
and notorious for its unscrupulous conduct in the territories it conquered,
brutality was perceived as carrying a potential political liability. Thus, during
the 1930s, the Japanese authorities, fearing both domestic and international
criticism, did their best to stifle the stories that soldiers told about the army's
conduct in China.[50]

In fact, the impact of the changing balance between state and "society," and of the new cultural disposition toward brutality, was so serious
that the inherent immunity that states extended to their agents began to
erode. The American army indicted several officers, including a brigadier
general, for brutal conduct during the 1899–1902 pacification war in the
Philippines.[51] The French High Commissioner in Syria, Maurice Sarrail,
was recalled and replaced following national and international protest over
the brutal 1926 repression of the insurrection in Damascus. In Britain, the
Hunter Commission forced the resignation of General Reginald Dyer after
the bloody April 1919 repression of riots in Amritsar (which resulted in 379
Indian fatalities). Some years later, when the RAF reported the results of an

[46] See Spies, *Methods of Barbarism?* 111, 115, 185–86, 300. In fact, Roberts himself rejected the extreme ideas of General Hunter.
[47] On the German understanding of "necessity," see Best, *Humanity in War*, 145–46, 172–74.
[48] Pakenham, *The Scramble*, 612.
[49] Bley, *South-West Africa*, 165.
[50] Friedman, *The Law of War*, Vol. II, 1066.
[51] See Wells, *War Crimes*, 68–69.

early experimentation with "air control" against some "exceptionally unruly tribe," Winston Churchill (then Secretary of State for the Colonies) sharply reproved Air Marshal Hugh Trenchard: "I am extremely shocked. . . . To fire willfully on women and children is a disgraceful act, and I am surprised you do not order the officers responsible for it to be tried by court-martial. . . . By doing such things we put ourselves on the lowest level."[52]

Conclusion

Nineteenth-century Europe witnessed the continuous ascent of the same social layer that already challenged the *anciens régimes* in the late eighteenth century. The intellectuals, artists, and literary figures developed a distinct social identity of a "public," disseminated knowledge and ideas, and energized and radicalized the bourgeoisie through the press, clubs, societies, and salons. As Gianfranco Poggi notes, when this social layer was not inhibited by censorship or repression, its members discussed political issues, including the conduct of foreign affairs, in a critical manner.[53] For reasons that I will discuss in greater detail in the next chapter, foreign policy was not altered in any radical manner. Most certainly, the size and power of the social forces that opposed small wars were too limited, in part because major institutions that open opportunities for critical participation – namely, genuine comprehensive suffrage, the multi-party system, and representative political bodies – were flawed. Suffice to note that 50 percent of the population was denied suffrage because of gender, and that representation was rendered ineffective or was dramatically biased toward the more affluent.

Nevertheless, one particular institution was open to the more articulate elements in society, giving them indirect yet effective access to the political process: the free marketplace of ideas. It was in this market that the best educated exercised their power and promoted their demands at a time when other avenues of influence were largely closed to them. In fact, in the coming chapters we will observe how important was the role of the free marketplace of ideas in the case of small wars, long after access to politics was granted to all adult citizens.

One additional important development occurred in nineteenth century Europe. During most of history, the despotic power of rulers, the cultural indifference to brutality, and the minimal dependence of the proto- and infant-states on society guaranteed the kind of compatibility that made military superiority the single most important factor in conquest and pacification. In the West however, cultural and political changes culminated in the formation of a normative difference between the state and educated segments of society. The latter increasingly questioned the power and purpose of the state, on

---

[52] Quoted in Asprey, *War in the Shadows*, 280.
[53] Poggi, *The Development of the Modern State*, 81–82.

both utilitarian and moral grounds, and their criticism expanded to include issues of foreign policy, and in particular the ways states conquered, subjugated, and pacified foreign societies.[54] Indeed, etatist impulses to achieve imperial objectives in the most expedient and efficient way became antagonistic to the cultural disposition of a small but important social elite. That, in turn, imposed a cost on the offensive and suppressive use of violence overseas. Tales of brutality kindled public debate, parliaments started to interfere with military conduct in war, and even the immunity of state-agents was occasionally suspended. In short, conquest and pacification wars began to be shaped not only by military considerations but also by the assessment of how compatible political structures and the cultural disposition of "society" were with particular strategic choices.

Most importantly, there were the first clear signs of a downward movement on the scale of brutality of counterinsurgency strategies. In that respect, the resort to isolation in concentration camps, to a selective scorched earth policy, and even to the execution of rebels following legal, if dubious, procedures, should not go unnoticed. Compared with the indiscriminate and unrestrained conduct of previous conquerors, such practices, horrible as they may have been, still indicated real progress. This change is thus of great significance, and more so only because it occurred at the height of jingoism and when the middle-class was hardly asked to pay anything for imperial and colonial wars. Indeed, it is perhaps the best indication of a critical qualitative development in the emerging liberal democracies of Europe that in time foreclosed the opportunity of Western powers to win protracted small wars against militarily inferior insurgents.

---

[54] On the development of norms and laws of conduct in war, see Richard S. Hartigan, *Lieber's Code and the Law of War* (Chicago: Precedent, 1983), and *The Forgotten Victim*; Best, *Humanity in Warfare*; Howard (ed.), *Restraints on War*; and Wells, *War Crimes and Laws of War*. The use of uninhibited violence was rejected on legal and moral grounds in the early sixteenth century (without success) by iconoclastic intellectuals such as Vitoria and Las Casas, who criticized the Spanish repression of Indians. See Quentin Skinner, *The Foundations of Modern Political Thought* (NY: Cambridge University Press, 1978), Vol. 2, 169–17; and Mario Góngora (trans. Richard Southern), *Studies in the Colonial History of Spanish America* (NY: Cambridge University Press, 1975), 56–59. There is evidence of incompatibility between "state necessity" and the "conscience" of democratic societies from the time of ancient Greece, as suggested by the Athenian change of heart with respect to the fate of the population of rebellious Mytilene. See Thucydides, *The Peloponnesian War*, pp. 180–91; and Manicas, *War and Democracy*, 38–39.

4

The Structural Origins of Tenacity

National Alignment and Compartmentalization

In Chapter 3, I noted the key developments – the rise of the educated middle-class, the emergence of the free marketplace of ideas, and the birth of a normative difference between state and society – that are responsible for the eventual inability of liberal democracies to win small wars. Still, we know that Western powers continued to fight and win small wars past the turn of the nineteenth century, even though the underlying conditions supporting the use of unbridled violence were continuously eroded. My discussion this far cannot fully explain this period of the twilight of Western capacity to win protracted small wars, for one major reason: In Chapter 3, I discussed only half the story, that which concerned "society." If one wants to gain a full picture of the forces that shaped the fate of "democratic" small wars, then one also needs to consider the other half of the story: that which concerns the measures that rulers and states took in reaction to social changes and challenges.

Thus, this chapter is devoted to the "state" perspective. Specifically, four issues will be addressed. First, I will briefly discuss, in the abstract, possible institutional reaction to internal challenges. Second, I will review certain developments, most notably in the realm of formal education, that permitted the state to wage war in general and small wars in particular, in spite of the emergence of the normative gap. Third, I will discuss particular ways that rulers used in order to minimize the normative gap and contain its consequences. Finally, I will pull together the findings from my discussion in Chapters 2 through 4 in order to address, with additional arguments, the question of timing. That is, I will venture an explanation as to why Western powers lost their capacity to overcome the effects of the normative gap and started to fail in small wars only at some point in the middle of the twentieth century.

Institutional Incongruity and Strategic Preferences

I have noted in Chapter 2 that if states wish to fight effectively and win small wars, then they must secure the readiness of their military forces to meet the "requirements" of the battlefield, and their citizens' consent, apathy, or support for their policies and the cost those incur. I have further argued that the capacity of Western powers to fulfill these "requirements" was increasingly being eroded and challenged by the progenitors of "society." At the same time, however, I have noticed that Western states continued to apply the high levels of brutality that were necessary in order to secure victories in imperial wars of conquest and domination. The explanation of this disjunction between domestic developments and international outcomes can be constructed from an abstract insight into the functioning of multiple process or multiple function systems.

Multiple process and function systems are almost bound to face occasional incongruity among their objectives, actions, and constituting parts. Such incongruity, however, can be resolved in one of several ways. A system's first choice would presumably be to realign and condition its parts to fulfill its tasks in full, and without reducing its overall effectiveness and coherent functioning. If alignment is impossible, a system can still achieve its functions in full by isolating processes, or their consequences, from each other. When neither the option of alignment nor that of isolation exists, a system would start to compromise or trade-off objectives, all according to established preferences. Only then should a system be expected to make the decision to unconditionally abandon some or all of its objectives. In summary, realignment is the first systemic reaction to a state of disharmony or internal conflict, compartmentalization the second, some sort of compromise the third, and abandonment the last.

Social institutions can be perceived as systems. As such, they should not be expected to alter or abandon any particular objectives or practices just because these happen to irritate some of their constituent members. Instead, those in control of institutions would logically first try to align their constituents with their objectives and the methods used to achieve them. That is, the authorities would try to mobilize support and marginalize opposition to their policies. If the desired level of alignment cannot be achieved, then social institutions are likely to try to manage the costs that policies entail (including the costs associated with cost-management policies themselves) through compartmentalization. Only then should one expect institutions to consider a compromise, a trade-off, or, if all else fails, an abandonment of objectives.

The state is but a particular type of social institution, and as such it will first try to condition its citizens and agents to support its objectives, demands, strategies, and methods. When the incompatibility between state policies and the preferences of significant segments of society is unavoidable, state leaders should be expected to initiate actions of compartmentalization. Only then are they likely to consider a compromise, or if that too proves insufficient, abandon the contentious pattern of behavior or objective. Specifically, if the military conduct of a small war becomes a source of political dispute, leaders are likely to try and rebuild as much domestic support for their war policies as possible. If that proves unobtainable, they will not necessarily alter the conduct or combat methods of their military forces (themselves forms of cost management) just because certain social forces are able to impose a political cost on this behavior. Rather, they will try to minimize the political cost that such behavior entails by manipulating and compartmentalizing the factors that are involved in the genesis of the objection to that behavior.

The Modern Foundation of the State's Autonomy: Popular Alignment

The fact that Western powers continued to fight and win small wars for quite a significant period of time after they had first met objections to their ways of small-war making was first and foremost the result of an elaborate effort of alignment. While elements of the educated middle-class labored hard to establish a separate and independent realm of "society," thereby laying the foundations of the normative difference, various other forces worked forcefully to align the people with the needs and objectives of the state. Some of these forces promoted this alignment inadvertently, while others did so in a calculated manner. Either way, they all worked toward the single purpose of nation-building.[1]

On the one hand, nation-building was achieved as a result of a bottom-up process largely because nationalism turned out to be emotionally seductive in a period that witnessed the destruction of traditional feelings of identity and belonging as a consequence of major social, technological, and ideological changes. On the other hand, nation-building was also promoted through a top-down process since it was perceived as both inevitable and rewarding. Inevitable (and risky) in the sense that once the power of nationalism was released by the French Revolution, the issue became whether it would be harnessed, or consume whomever stood in its way (as it did Louis XVI). Rewarding because it provided rulers with a means to contain inter-class tensions and address more cheaply the new manpower demands of international competition in the age of mass conscription.[2] In short, nation-building owed

---

[1] On other aspects of nation-building, nationalism, and militarism, particularly cultural and symbolic, see George L. Mosse, *The Nationalization of the Masses* (New York: Howard Fertig, 1975), and *Fallen Soldiers* (NY: Oxford University Press, 1990); Anne Summers, "Militarism in Britain before the Great War," *History Workshop*, No. 2 (1976), 104–23; and John MacKenzie (ed.), *Popular Imperialism and the Military, 1850–1950* (Manchester: Manchester University Press, 1992), and *Propaganda and Empire* (NY: Manchester University Press, 1984).

[2] Prussia is a compelling example. After the defeat at Jena, a small group of reformist officers concluded that the French success was owed to the spirit of sacrifice that nationalism had

much to design. It was achieved through a concerted effort that included the provision of protection, arbitration, and public order through legislation, the courts, and the police, investment in communication means, and the extension of entitlements and other forms of patronage to selected groups and to society at large.

Yet the most important aspect of securing the allegiance of people to the "new" nation-state, at least from our point of view, was the reconstruction and spread of education. From an institutional perspective, the first two important things to note about state-controlled education in nineteenth century Europe is that it was quite slow to develop, but that eventually it succeeded in becoming comprehensive and effective. By and large, European states began to support elementary education, force local communities to establish schools, and develop interest in the curricula only in the first half of the nineteenth century.[3] In fact, education did not become a primary concern of most states until sometime later. Thus, with the exception of Prussia (which made elementary education compulsory in 1763 and declared schools and universities state institutions in 1794), elementary education became effective and comprehensive only in the last quarter of the nineteenth century.[4] In Germany, compulsory elementary education was established following unification, in 1871. In England, state "expropriation" of the education system gained great impetus in 1870 with the Forester Education Act. School attendance became mandatory in 1880, and all fees for elementary schools were abolished in 1891. In France, the great leap forward took place when Jules Ferry introduced compulsory, non-clerically controlled, free elementary education in the laws of 1881–1882.[5] Prior to the Ferry reform, education

---

spawned. They convinced Friedrich Wilhelm III, against strong Junker opposition, that in order to save Prussia from extinction and/or revolution, an alliance with society must be forged. It was necessary "to give the people a fatherland," reasoned Gneisenau, "if they are to defend the fatherland effectively." Quoted in Michael Howard, *War in European History* (NY: Oxford University Press, 1976), 87. When the monarch declared war on Napoleon, he indeed called on all Prussians – subjects who were now "citizens" (*Staatburger*) – to fight. The badges of the *Landwehr* militia promptly read "With God for King and Fatherland." See Best, *War and Society*, 166.

[3] Napoleon tried to establish firm control over education, but his centralized system, and particularly the *grandes écoles* (and the *lycées*), were directed mostly toward state-building (rather than nation-building) from top to bottom (rather than bottom-up). See Ringer, *Fields of Knowledge*, 40–41.

[4] Denmark introduced a three-day week for elementary education for seven years, in 1814. Sweden enacted compulsory education in 1842, but only in 1878 did it set the duration at six years (Prussia had already specified seven to eight years in 1763). Austria introduced compulsory education successfully only in 1869 (it tried first in 1774). Data are from Peter Flora et al., *State, Economy and Society in Western Europe, 1815–1975* (Chicago: St. James Press, 1983), Vol. I, 555, 567, 584, 613; and Benedix, *Nation Building and Citizenship*, 88–89.

[5] State participation in covering the total cost of "school expenditures" in France decline from 22.7 percent in 1834 to 14.2 percent in 1866. But then it rose sharply to 67.2 percent in 1896. This growth reflected the 1889 decision to pay public school teachers more, in part as a way

TABLE 4.1 *Education levels in France, Germany, and England*

A. Percentage of students in the population[a]

	Elementary			Secondary			University		
	FRA	GER	ENG[b]	FRA	GER	ENG[b]	FRA	GER	ENG[b]
1900	14.4	15.9	14.6	0.26	–	–	0.08	0.08	–
1920	10.8	14.5	13.7	0.6	1.7	0.9	0.13	0.19	0.16
1935	12.8	11.8	11.7	1.1	1.4	1.1	0.18	0.11	0.16
1950	11.3	12.4	9.2	1.9	1.6	4	0.33	0.23	0.23
1960	12.5	9.4	9	3.3	2.2	6.15	0.46	0.38	0.27
1970	10.1	10.5	10.4	5.4	3.7	6.5	1.28	0.68	0.52

Sources: Mitchel, *European Historical Statistics*, 4, 8, 397, 400, and Flora, *State Economy and Society*, vol. 2, 53, 57, 58, 79. Slightly different numbers (which reflect the same trends) appear in Fritz Ringer, *Fields of Knowledge*, 48, table 1.1; and Halsey, "Higher Education" in Halsey, *British Social Trends*, 270, table 7.2.

B. Percentage of students per age group[a]

	Elementary (5–14 years)			Secondary (10–19 years)			University (20–24 years)		
	FRA	GER	ENG[b]	FRA	GER	ENG[b]	FRA	GER	ENG[b]
1900/01	86	73	74	–	2.8	1.4	0.9	0.7	–
1910/11	86	72	79	–	5	2.7	1.3	1	–
World War I									
1930/31	84	75	81	3.5	7.5	6	2.3	1.5	–
1935	–	–	84	5	–	7	2.7	–	1.2
World War II									
1950	91	80	68	9.1	8	34	4.2	2.2	2.3
1960	70	69	60	15	11	45	7	3.4	3.1
1970	58	68	65	18	17	49	15	7	5

[a] All numbers are rounded and based on rounded figures.
[b] Including Wales.
Source: Flora, *State Economy and Society*, vol. 1, 578, 580, 582, 587, 589, 624, 626.

in France was under-funded and backward, particularly in the countryside, where some 65 percent of the population lived.[6] Thus, as Table 4.1 indicates,

of making them accountable to the state. See Price, *A Social History of Nineteenth-Century France*, 317.

[6] On the state of education and schools in nineteenth century France see Eugene Weber, *Peasants into Frenchmen* (Stanford: Stanford University Press, 1976), 303–38.

only by the end of the nineteenth century did the European states finally have, under their control, a widespread elementary education system.

The second important aspect to note about education in Europe is that the majority of children were exposed to a curriculum that was heavily nationalized when they were young and therefore particularly vulnerable to indoctrination. Indeed, the objective of primary education was to produce loyal and obedient citizens.[7] Prussia, for example, was at the forefront of the efforts to instill obedience in its subjects (and continued this very effort long past the elementary level of education). Its institutions of education emphasized the great days of the Roman Empire, the wickedness of France, and Hegelian themes such as the divinity of the state and the merit of obedience.[8] Indeed, Prussian success in building and preserving a garrison state was in part due to this "mental" preparation of its younger subjects in schools and universities. Other states used the education system for the purpose of state- and nation-building almost to the same degree. Even in England, the curriculum increasingly became nationalist, militarist, and etatist, although the government did not need to defend the social order, nor develop a large and reliable field army, to the same extent that Continental states did.[9]

The third important fact to note about the system of education in Europe is that beyond the elementary level, and until well into the twentieth century, it was neither inclusive nor progressive. That is, the education system did not absorb a high proportion of the relevant age group (see Table 4.1), nor a significant percentage of the lower classes. Secondary education was often expensive, and it was intended for the privileged few. Indeed, it both reflected and preserved social boundaries. The English system was particularly notorious for such bias, and the French system was almost equally discriminatory. The key difference was that French higher education was biased toward the new elites of the affluent bourgeoisie and the bureaucracy, while in England it was biased, through the system of "public schools" (Eton, Harrow, and so on) toward the more traditional aristocracy.[10] Secondary education, then, was intended to prepare the more affluent to lead the state and to assume in due time the administrative and managerial positions their parents

---

[7] As Weber explains, the French elementary school was defined as "an instrument of unity," an "answer to dangerous centrifugal tendencies," and the "keystone of national defense." See ibid., 333.

[8] A.J.P. Taylor, *The Course of German History* (London: Methuen, 1985), 60.

[9] See MacKenzie, *Propaganda and Empire*, 5–6, 10–11.

[10] On the English education system, see Bédarida, *A Social History of England*, 153–58, 235–41. See also A.H. Halsey, "Schools," in A.H. Halsey (ed.), *British Social Trends since 1900* (London: Macmillan, 1972, 1988), 227–67. On education in France, see Price, *A Social History of Nineteenth-Century France*, 307–56, particularly pp. 139, 307, 339–48; and Ringer, *Fields of Knowledge*, particularly pp. 55–62.

occupied. Fritz Ringer, describes the French education system in the following way:

In primary schooling, the emphasis was mainly on reaching the whole population, on social integration and on the encouragement of patriotism. In secondary education, access to elite position was the major issue. Educational reformists and conservatives alike explained again and again that France was no longer a caste society, and that it could only be led by an "aristocracy of merit." Educational conservatives ... were satisfied that the traditional structure of secondary education adequately selected this elite, [and] their meritocratic rhetoric served primarily to confirm the existing social hierarchy.[11]

In summary, toward the end of the nineteenth century the European masses were thoroughly immunized against anti-etatist ideas, and this was expressed not only in the extent of elementary education, but also in other ways, including a more favorable attitude toward national armies.[12] The nation-state was well entrenched in the minds of people (in spite of the mistrust of leaders of "public opinion"), and the common citizen was inclined to believe that he should support his country, "right or wrong." In fact, European states, England included, used the education system not only to promote nationalism but also to propagate imperialist ideologies.[13]

Institutional Responses to Threats against the Normative Alignment

The conditioning of the masses to support state policy most certainly helped the powers of nineteenth century Europe to expand their empires and use their military forces overseas with little restraint. Imperial power politics was also supported by a significant portion of the middle class, whose sons were offered economic security, an avenue for upward mobility, and adventure as officers and bureaucrats overseas. Nevertheless, as I noted in Chapter 3, the legitimacy of imperial oppression and of brutal military conduct in small wars drew an increasing amount of criticism, and thus became a source of contention. In short, the structural advantages that the state enjoyed as a result of its investment in elementary education, and the other factors that supported state autonomy in national security and foreign policy matters, proved insufficient to remove the issue of brutal behavior overseas away from the "public" discourse.

---

[11] Ringer, *Fields of Knowledge*, 213.

[12] In France, the military was "accepted" by rural society only after the defeat of 1871, and it had become a fully accepted national institute only by 1890. Positive attitude changes toward the British army occurred during the second half of the nineteenth century. See Weber, *Peasants into Frenchmen*, 298, and MacKenzie, *Propaganda and Empire*, 5.

[13] See MacKenzie, *Propaganda and Empire*.

This reality meant that if states were to preserve and further extend their empires, they had to exploit all avenues and develop direct ways to counter the possible effects of the growing normative difference. Indeed, as Western powers remained firmly committed to their imperial agenda, the leaders and their administrative agents spared no effort to reduce the incentive to criticize small wars and the possible consequences of such criticism both within the army and in society.

As a first measure, states segregated the objects of their violence from the human race. In fact, this segregation was all the more "successful" because it was grounded in a framework of presumed progress. For several centuries, a few European scholars labored to "civilize" war by developing a set of legal and normative tenets. Eventually, they articulated "standards of civilization" that specified when and how armies could kill, and who could be killed, and thus presumably restrained the use of violence. However, the boundaries of "civilization" also left out most of the people who were the subject of European violence overseas. Thus, the standard of civilization served as a means of compartmentalization that gave Europeans a short ethical blanket that denied others sanctuaries that Western culture had developed for those involved in war, and hence sanctified the use of extreme brutality against insurgent populations and warriors.[14] Thus, when European empires met and pacified Indians, East Asians, Blacks, Arabs, indigenous Irish, or other native populations, these people were defined as *barbarians, savages, wild men,* or their equals, and sure enough they soon became the victims of extreme brutality. The 1899 Hague conference, for example, banned the use of dum-dum bullets (which expand on impact) in battle, but it did so only as far as *civilized* people were concerned, not *wild* men.

British imperial behavior is a good example. Overall, the British are considered to have run the most lenient of all European colonial systems. Nevertheless, they did not hesitate to use the most brutal methods when they applied the principle of normative segregation to their enemies in small wars. The 1920 edition of the *Field Service Regulations* contained a chapter titled *Warfare against an Uncivilized Enemy,* which discussed, among other things, the general principles of "savage warfare."[15] And the text that continued to guide the British army in small wars in between the two world wars, Callwell's book *Small Wars,*[16] was predicated on the assumption that the "races" of the imperial world deserved brutal treatment.[17] "Uncivilized

---

[14] See Gerrit W. Gong, *The Standard of Civilization in International Society* (Oxford: Clarendon, 1984); and Pick, *War Machine,* 153.

[15] T. R. Moreman, " 'Small Wars' and 'Imperial Policing': The British Army and the Theory and Practice of Colonial Warfare in the British Empire, 1919–1939," *The Journal of Strategic Studies,* 19:4 (1996), 109.

[16] Ibid., 109–10.

[17] See also Douglas Porch in the introduction to Callwell's *Small Wars,* xv; and MacKenzie, *Popular Imperialism and the Military,* 6–7.

races attribute leniency to timidity" the book stated, and "a system adapted to La Vendée" (which was hardly a model for leniency in counterinsurgency), the book continued, "is out of place among fanatics and savages, who must be thoroughly brought to book and cowed or they will rise again."[18] In that respect, the Nazis' vicious dealing with insurgents, after the latter were segregated from the human race, fell well within a pattern of behavior that had been developed previously in Germany and elsewhere in Europe. But in the Nazi case, the compartmentalization was not applied to some remote people, but rather to insurgent populations right in the European backyard.[19] Of course, the Nazi racist ideology was not invented for small-war purposes. However, the German leadership understood that its foul ideology provided it with a superb means to shape its agents' conduct abroad and control reactions to the latter at home.

This state of affairs, in which the brutality of Western powers outside their own perimeter was considered legitimate, was supported by yet another means that survived until after World War II – the international conventions that articulated the laws of conduct in war. The latter were based on the 1863 seminal text on military conduct in war (*Jus in bello*), U.S. Army General Orders No. 100 (known as the Lieber Code, after their author). In fact, these orders were rather progressive as far as war conduct and the rights of non-combatant enemy civilians were concerned. However, they were also extremely harsh toward irregulars who lead double lives and "commit hostilities" or "rise in arms against the occupying or conquering army."[20] These were to be regarded as pirates, and their deeds were punishable by death. The Hague and Geneva Conventions that followed the Lieber code granted insurgents no better rights. "The Hague Rules" writes Geoffrey Best, "left little room for the guerrilla fighter, and displayed no sympathy towards populations which might rise in arms against an occupying power."[21] Indeed, as the very legitimacy of insurgency was denied, counterinsurgency forces continued to be free from moral restraints.

The position of Western states vis-à-vis domestic opposition to overseas brutality was influenced by more than the mental preparation of the masses and the cunning manipulation of the definition of legitimate violence. The

---

[18] Callwell, *Small Wars*, 148.

[19] The Germans lumped together guerrilla insurgents, communists, and Jews, and as the Jews (and Slavs) were defined as *untermenschen*, the Germans thereby conditioned their soldiers to treat insurgents on the Eastern Front in ways they may have had trouble inflicting on their "own kind." Hitler explained in his October 2, 1941 "order of the day" that the "enemy" in the occupied Eastern territories did not "consist of soldiers but to a large degree only of beasts." Quoted in Zeman, *Nazi Propaganda*, 161.

[20] See Hartigan, *Lieber's Code and the Law of War*, 60–61; and Francis Lieber's own discussion "Guerrilla Parties Considered with Reference to the Laws and Usages of War" in ibid., 31–44.

[21] Best, "Restraints on War by Land Before 1945," 35.

incentive and opportunity of social groups to oppose small wars were also limited because the state manipulated the level of instrumental dependence and segregated military contingents that fought small wars from society. Again, the key principle underlying such manipulation was compartmentalization. Only in this case it was the deeds and the losses of the military forces that were compartmentalized from society rather than the objects of violence from the human race. The compartmentalization itself was achieved in several ways. First, some of the cost of small wars, and human losses in particular, was imposed on societies other than that of the oppressor. Thus, for example, 26,000 out of 76,000 French soldiers occupying Morocco in 1913 were African, while in Syria, the French used Senegalese and other soldiers from local origin – namely, from among the Circassian, Armenian, and Kurdish ethnic minorities. Similarly, in 1935–1936, the Italian colonial army in Abyssinia included about 20 percent indigenous Ethiopian soldiers, who also happened to suffer most of the battle casualties.[22]

Second, the nature of standard military forces was such that it helped states to compartmentalize small wars and isolate them from society. Most notably, states relied on professional and long-term service conscripts. For example, the British imperial peacetime tour of duty was up to sixteen years for soldiers in infantry units. For soldiers in cavalry regiments, it was slightly shorter, fourteen years, but cross-posting and volunteering made it even longer for at least some officers and soldiers. Between the two world wars, service in India was six to ten years.[23] Obviously, professional and long-term conscripts were less attuned to the mood of society than reservists or regular troops would have been. They were fit to fulfill almost any brutal "requirement" of the battlefield, they developed loyalty to the institutional interest of the army and the state, and they occasionally helped distance from society those members who were considered troublesome to begin with.[24] Of course, colonial expeditions occasionally included regular conscripts. However, the selective nature of conscription – often leaving out the middle-class – made fresh conscript armies somewhat detached from international matters. Indeed, conscripts were mobilized – as indicated by the peasant base of some nineteenth-century European armies and the Portuguese colonial army many years later – mostly from social strata that tended not to question foreign policy or combat conduct, and whose political relevance was limited. In a way, the conceptual foundation underlying such army format was somewhat similar to that behind regressive taxation: Groups pay in inverse relation to their ability to resist and impose a political cost on the taxing authority. Indeed, such conscript armies had the advantage of tolerating staggering

---

[22] Kiernan, *From Conquest*, 138–41, 202–03.
[23] Moreman, " 'Small Wars' and 'Imperial Policing,'" 106–08, 112.
[24] See Douglas Porch on the French Foreign Legion in *The French Foreign Legion* (NY: Harper/Collins, 1991), xiii, 5, 6.

ratios of casualties (in battle or from disease) without evoking the political firestorm that would have been all but certain in contemporary democracies in similar circumstances. Let us consider, for example, the following cases: between 1819 and 1828, the British military mortality rate overseas was 5.7 percent a year, about four times higher than the 1.5 percent at home. During the 1859–1860 war in Morocco, the Spanish lost 66 percent of their 7,000-strong army. In 1862, Florence Nightingale discovered that the death rate in the British army in India was 6.9 percent (that is, a number of soldiers equal to one company in each regiment died every twenty months). In 1895, about a third of the 15,000-strong French expeditionary force in Madagascar perished (mostly from diseases). Between January 1919 and April 1920, the French lost 150 officers and 3,432 soldiers in Syria. In the first stage of the Herero War (in 1904), battle and disease reduced the German Eastern detachment from 534 to 151 soldiers (a loss of 72 percent).[25]

A Question of Timing

Having explained what factors permitted Western states to continue to fight and win small wars in spite of the rise of the middle-class, the development of the free marketplace of ideas, and the division of views over what justified sacrifice and brutality, I can now turn to the last piece of our puzzle and address the question of timing. Obviously this is an ambitious task that calls for humility. In most cases, it is difficult enough to associate long-term changes to particular outcomes, let alone pinpoint the precise historical moment in which these changes start to yield new patterns of outcomes. Nevertheless, I believe that on the basis of the information that I have assembled thus far, and a few additional facts, I can address the timing question adequately. The key, however, will be our capacity to combine quantitative, qualitative, and conjectural thinking.

As I have argued in the Introduction, and intend to demonstrate in the coming empirical chapters, powerful states fail in small wars when the domestic opposition to a small war succeeds in shifting the center of gravity from the battlefield to the marketplace of ideas, and in controlling the agenda there. However, in order for the latter process to occur, several preconditions must be fulfilled, of which three quantitative ones are critical. First, the core opposition to the war has to be of some minimal size and voice. Second, the second-tier audience, which is receptive to the criticism and ready to take action, must also be of some minimal size. Three, the war policy must threaten, or be perceived as threatening, the core interests of a significant portion of the population. The synergetic impact of the interaction of these

---

[25] Data are from ibid., 286–87; Kiernan, *From Conquest*, 130–31; Tibawi, *A Modern History of Syria*, 341; and Drechsler, "South West Africa," 55–56.

conditions is what makes a small war difficult for a liberal democracy, more than the added impact of each factor separately. Put otherwise, small wars started to fail because of social pressure, only after instrumental dependence, political relevance, and normative difference reached certain minimal levels.

The level of all three variables, however, was not significant enough until well into the twentieth century. Thus, while instrumental dependence had peaked already at the turn of the century, when European powers had finished building mass-conscript armies, the prospects of success in small wars remained high because the level of the two other variables was kept low. On the one hand, society was by and large kept out of small wars because Western powers managed to fight many of their imperial wars with native, foreign, and professional troops that were socially unimportant at home. And on the other hand, the level of objection to the cause, the cost, and the conduct of small wars, and the size of the politically relevant community that could champion an anti-war agenda, were rather minimal. In other words, the state encountered only a small minority that was ready and able to challenge the morality or rationality of foreign policy objectives and military conduct in war, particularly since small wars involved little threat to society at large.

A good way to demonstrate the quantitative aspect of the equation and evaluate the "structural advantage" of the state is to reconsider the students' per-population and age-group ratios that appear in Table 4.1. The number of elementary school graduates indicates the level of nationalization. The number of university students, perhaps in combination with some portion of the secondary-school graduates, correlates to the size of the popular base of "attentive" citizens who could opt for dissent. In fact, the number of university students also indicates the potential size of the critical mass (the core opposition) that generates and articulates anti-war and anti-state ideas, as it correlates to the size of the academic staff and as both groups correlate to the overall size of the intellectual community. Table 4.1 indicates, then, that by the early 1900s, the majority of the population was thoroughly nationalized, whereas the groups that were most likely to generate opposition to small wars and be attentive and receptive to criticism, were rather small before 1914. This ratio of national to opposition power is all the more striking once one considers the levels of education ratios in the period between the world wars, and even more so after World War II.

To be sure, the quantitative aspect of our timing equation is but one among several. The expansion of the normative gap was also checked before World War II by factors that cannot be inferred from the data in Table 4.1. First, a substantial part of the finite educated elite of late nineteenth- and early twentieth-century Europe was not disposed against war. Rather, many intellectuals developed admiration for the state, became chauvinistic, and regarded war as a necessary and noble activity that helped redeem society from peacetime decadence.[26] In fact, academics all over Europe endorsed perverted theories of social Darwinism, which glorified war and legitimized the resort to extreme brutality, particularly when "inferior uncivilized races" were involved.[27] Moreover, most of those who were employed in, or associated with education, were "captives" of the state, mentally or otherwise. As universities and schools were financially dependent on the state, teachers and professors were not in the best of positions to openly criticize state policy. At the same time, many of the graduates of higher education were also prevented from joining the second tier of anti-war and anti-state protest because some of them were recruited for state service, while most of them were protected by virtue of their class and vocations from the hazards of small wars.

Quantitative and class- or education-based arguments should not obfuscate or preclude the consideration of conjectural factors. Qualitative changes occasionally take place in the wake of catalytic events. Two such events come to mind as most critical for the development of the Western capacity to conduct small wars successfully: World Wars I and II. These two wars left in Western societies a lingering legacy of new standards that emerged from the changed images of war and of legitimate conduct in war. World War I "robbed" war of glamor. It demonstrated the monstrosity of organized violence in the industrial and national age, and destroyed some of the romantic perception of war as an agent of personal purification and convalescence (though many veterans treasured their war experience).[28] During this war, and largely because of Wilson's Fourteen Points, international conventions concerning the right of self-determination, and by inference legitimate insurgency, began to develop.[29] World War II further strengthened these trends. It shook a large number of people out of indifference to inhuman military conduct and it cured Europeans (and in particular members of the educated elite) of social Darwinism, at least for a while. It also helped bring about new formal conventions that redefined the legitimacy of insurgency and the rights

[26] This intellectual propensity was of course attacked in Julian Benda's 1928 book. See Benda (trans. Richard Aldington), *The Treason of the Intellectuals* (NY: Norton, 1969).

[27] See Pick, *War Machine*, 16, 75–106, 149; Mueller, *Retreat from Doomsday*, 37–52; and Best, *Humanity in Warfare*, 135–38.

[28] John Mueller presents the bloodshed in World War I as a turning point against war, in *Retreat from Doomsday*. Skeptics would argue that his thesis is short on reference to sociological and cultural studies of that war's generation. They would also point out that the war marked the beginning of the rise of popular fascism, and all its nationalistic and militaristic messages, in all major Continental powers except Communist Russia. Finally, they would point out that support for socialism after the war, arguably the most peaceful and anti-nationalist ideology, declined sharply. See Robert Wohl, *The Generation of 1914* (Cambridge: Harvard University Press, 1979), particularly pp. 230–33.

[29] See Jan W. Schulte Nordholt (trans. Herbert H. Rowen), *Woodrow Wilson: A Life for World Peace* (Berkeley: University of California Press, 1991), 251–80.

of occupied populations.[30] Above all, it helped establish not only juridical definition of war crimes, but also widespread popular terms of reference concerning unacceptable conduct during war.[31]

The short period between the world wars was relatively uneventful as far as democratic involvement in small wars was concerned. European democratic powers stopped expanding, and their success in previous colonial and imperial wars gave oppressed communities little reason to believe that they could win small wars outside the battlefield. Western powers seemed likely to continue to find the "golden mean" between societal commitment, violence, and cost, and thus the chances that small wars could be protracted seemed minute. Thus, oppressed communities were confronted with seemingly unchanging conditions that were reminiscent of those conditions that they faced (and still face) when they fought enemies other than liberal democracies. That in itself was enough to reduce the enthusiasm of underdogs for violent encounters with democratic oppressors. Indeed, between the two world wars there were hardly any significant clashes between Western powers and rebelling communities, excluding the ongoing struggle in Ireland, the Riff rebellion against the French in Morocco, and the Druse rebellion against the French in Syria. The British, however lost much of their appetite for imperial wars, and relaxed their control over Ireland. The French managed to win, not without difficulties, in Morocco, and to break the relatively small scale rebellion in Syria. Authoritarian and totalitarian regimes, however, fought protracted small wars successfully in Manchuria, China, and Ethiopia, disregarding the half hearted efforts of democracies to stop them.

Conclusion

Once the normative difference between state and society over issues of violence and war started to develop, the winning odds of Western powers engaged in small wars became increasingly dependent on their capacity to effectively compartmentalize different aspects of the war effort and its costs. Western powers did not shy away from this challenge. They tried continuously and quite successfully for a long while to compartmentalize every possible aspect of small wars. They dehumanized insurgents and delegitimized insurgency and thereby made insurgent populations victims of a double stigmatization. The latter had no protection against what are now considered war crimes, because the only form of war they could wage was

---

[30] See *Geneva Conventions of August 12, 1949, for the Protection of War Victims* (Washington: U.S. Government Printing Office, 1950), 163–216.

[31] See the effect of Nazi horrors on American soldiers in Robert H. Abzug, *Inside the Vicious Heart: Americans and the Liberation of Nazi Concentration Camps* (NY: Oxford University Press, 1985), particularly pp. 169–73. Richard Rosecrance considers World War II to be a turning point, marking the beginning of the undermining of the "the military political system." See *The Rise of the Trading State* (NY: Basic Books), 69.

defined as unlawful. Moreover, they were denied the sanctuaries that Western culture created for civilians and warriors because they were descendants of the "wrong" ancestors and located in "unfortunate" geographical areas. At the same time, Western powers also tried to isolated overseas wars, their armed forces, and their military conduct from their own societies. Indeed, the absolute number of soldiers involved in conquering, pacifying, and maintaining imperial and colonial territories was relatively small,[32] and they could act without restraint and take extremely high rates of casualties without creating a major political crisis at home.

In fact, the same methods of compartmentalization continued to serve states even after their efficiency had been seriously reduced. Thus, armed forces continued to segregate insurgents and thereby free themselves from the moral restraints that prevent the use of excessive brutality. Thus, for example, the derogatory term "gook," which American forces used in Vietnam in order to describe Vietcong warriors (note that the term "Vietcong" was itself derogatory) apparently encouraged an attitude that condoned indiscriminate killing.[33] Similarly, states continued to rely, as the history of French and Portuguese colonialism suggests, on foreign troops and professional soldiers. France relied on a mix of indigenous, colonial, and professional soldiers as late as the 1940s and 1950s in Indochina. Moreover, while its army's command there was largely French, the officer corps still fell within the general colonial pattern because officers often spent much of their time overseas and thus were rather detached from society. The Portuguese used a mix of conscripts and indigenous troops during their 1960–1970 colonial wars in Africa. Indeed, their death toll is assessed to have reflected a casualty ratio of 1:1 metropolitan to native forces.[34]

In any event, after World War II, the high levels of instrumental dependence of Western powers became somewhat less open to manipulation than before, and the levels of political relevance and normative difference differed significantly from what they were only a few decades earlier. The need to rely increasingly on national troops, the inability to monopolize the supply of information to society, the failure to isolate the battlefield for long periods of time, and the continuing spill-over of liberal values from civil society to the military and the administration reduced the ability of the state to "hide" wars from society. All in all, then, as a result of changes in the capacity to manipulate instrumental dependence, in the size and nature of

---

[32] The most significant British colonial overseas detachment – in India – consisted of some 70,000 soldiers in the 1920s and some 45,000 in the late 1930s. See Moreman, " 'Small Wars' and 'Imperial Policing,' " 112.

[33] The term *gook*, which originated in the Korean War, was used in Vietnam along other terms such as *dinks*, *slopes*, and *slants*. See Lewy, *America in Vietnam*, 310.

[34] Ian Beckett in Beckett and Pimlott, *Armed Forces*, 150–51. Basil Davison argues that the Portuguese forces concealed the real number of casualties from society; see *The People's Cause* (London: Longman, 1981), 179.

the politically relevant forces in society, and in the magnitude and nature of the normative gap, the fate of such wars ceased to be exclusively dominated by military considerations.

Still, in spite of the new reality, etatist impulses to achieve state objectives in the most expedient and efficient way – by resorting to the extremes of violence – continued to guide state policy for a while. All that, however, while the conduct of small wars became increasingly incompatible with the political and cultural attributes of leading sectors of society. The consequences of these contradictory developments and the ensuing clash between state and society over the conduct of small wars are the subject of the following chapters. I will thus start the analysis of the military, cultural, and political aspects of the experience of France in Algeria (1954–1962) and Israel in Lebanon (1982–1985).

PART II

5

The French War in Algeria

A Strategic, Political, and Economic Overview

French exploitation, occasional repression, and an international climate favoring self-determination drove nationalist Algerians to start (again) an uprising in late 1954. France's choice of reaction to this violent wave of Algerian nationalism was almost preordained, The two major policy alternatives to war – rapid and egalitarian integration, or disengagement – were politically and psychologically unsustainable. On the one hand, hardly anybody in France would have agreed to a massive transfer of resources and an honest power-sharing program that equitable integration would have required. Ethnic, religious, cultural, and racial divisions between French and Algerians, and a French preoccupation with the standard of living in France, combined to preclude any such attempt.[1] On the other hand, no government could seriously contemplate rapid disengagement.[2] The most powerful politicians in France were deeply committed to French Algeria. The Algerian lobby and the army were sure to evoke the bitter memories of World War II and Indochina, raise a challenge no government could meet, and veto what they perceived as "abandonment." Besides, with few exceptions, repression proved to have been effective throughout the entire colonial history of France, including in Algeria. Thus, the only political option open to the French government in late 1954 was to keep Algeria by means of force.

The Strategic Dimensions of the War in Algeria

The Algerian war can be divided into two military phases. From late 1954 until 1957–1958, the Algerian Front for National Liberation (FLN) increased

---

[1] The French also had the phenomenal German growth rates for comparision.

[2] Already in 1918, the liberal governor of Algeria, Jules Cambon, had concluded that Algeria was too different to permit full assimilation, but too important to be permitted self-government. See Ian Lustick, *State Building Failure in British Ireland and French Algeria* (Berkeley: Institute of International Studies, University of Berkeley, 1985), 65.

its presence and activity throughout Algeria, dictated the pace of the war, and established itself as the leading force of the Algerian national struggle. Thereafter, however, the military situation was dramatically reversed. The FLN failed not only to produce the decisive battle envisaged by revolutionary Maoist theory, but it also found itself increasingly on the defensive, rapidly losing ground on both the urban and rural fronts.[3]

Algiers, the capital, was particularly important for the FLN because there it could best discredit French rule and France's capacity to guarantee a peaceful order. Therefore, hoping to make Algiers the site of the decisive battle of the war, the FLN leadership considered the capital a separate operational zone, subordinated it directly to its executive committee, and instructed its commander there, Yacef Saadi, to launch a relentless campaign of urban terrorism.[4] The symbolic importance of the capital, the gravity of the nationalist threat there, and the potential consequences of a loss of control over events made Algiers equally important for the French. In January 1957, after it had become clear that Algiers was out of control, the French Minister-in-Residence in Algeria, Robert Lacoste, gave General Jacques Massu and his 10th Paratrooper Brigade full powers, and ordered him to restore order in the capital.

In the first month, Massu was able to swiftly break a general strike called by the FLN. After nine months of an intense and brutal campaign, Massu's forces virtually destroyed the FLN terror network in Algiers. Most of the FLN activists – those who provided logistical support, the warriors, the bombers, as well as the leadership, including Yacef Saadi – were either captured or killed. The military results of the battle of Algiers were clear cut: The FLN lost decisively and violent FLN actions in Algiers ceased almost completely.[5]

In the countryside, the turning point came in the summer of 1958, but it was less visible than in Algiers. In 1957, the French command became more methodical in its approach to fighting the FLN in rural areas. Better military performance and initial French successes increased the dependence of the FLN forces, which operated inside Algeria, on external sources of

[3] On the war, see Philippe Tripier, *Autopsie de la guerre d'Algérie* (Paris, Éditions France-Empire, 1972); Alf Andrew Heggoy, *Insurgency and Counterinsurgency in Algeria* (Bloomington: Indiana University Press, 1972); Alistair Horne, *A Savage War of Peace: Algeria 1954–1962* (NY: The Viking Press, 1978); John Talbott, *The War Without a Name: France in Algeria, 1954–1962* (NY: Alfred A. Knopf, 1980); and Pierre Montagnon, *La Guerre d'Algérie: Genèse et engrenage d'une tragédie* (Paris: Pygmalion/Gérard Watelet, 1984).

[4] See Tripier, *Autopsie*, 131. Abane Ramdane, the charismatic FLN leader of the early war years, is quoted as saying: "Our Dien-Bien-Phu – we shall conquer it in Michelet Street [in Algiers]." See Jean Planchais, *Une histoire politique de l'armée*, Vol. 2, 1940–1962 (Paris: Seuil, 1967), 305.

[5] See Horne, *A Savage War*, 189–219; Heggoy, *Insurgency*, 230–44; Tripier, *Autopsie*, 127–44, 631.

supply. The French, realizing this vulnerability of the FLN, took the necessary measures to isolate Algeria. The sea routes to Algeria and its airspace were dominated by the French air force and navy. Nourished by good intelligence, the French were able to score big successes, such as the prevention of arms from reaching the FLN on board some ten vessels, including the *Athos* (October 1956) and the *Slovenija* (January 1958).[6] They also succeeded in intercepting an airplane carrying Ben-Bella and other FLN leaders (October 1956). On land, however, the French faced a more formidable problem. The borders of Algeria with Morocco and Tunisia were remarkably permeable. By 1958, however, this problem was largely solved. The French army finished constructing effective barriers along both borders, composed of fences, electric wires, and mine fields. Furthermore, the inhabitants in regions along the borders were "regrouped" – gathered, transferred, and resettled away from the borders. As a result of both actions, the momentum shifted in favor of the French forces. The formations of the National Liberation Army (ALN), which were previously able to cross the borders with relative ease, avoid immediate combat, resupply the FLN forces, maintain communication, and harass their colonizers, increasingly fell prey to the better-equipped French forces. The FLN was deprived of much of the weapons, fresh manpower, and other essential logistics it needed in order to replace its losses and continue to initiate actions.

In 1959, General Maurice Challe, who succeeded General Raoul Salan as the commander of the French army in Algeria, stepped up operations against the FLN. French forces had already demonstrated in 1958, in several localities, that they were able to repress FLN activity. Building on earlier experience, Challe concentrated his forces and moved systematically to mop up, one at a time, different sectors in Algeria. Effective intelligence, highly trained units with strong *esprit de corps*, and the mobility provided by a fleet of French- and American-made helicopters enabled him to use his forces relentlessly.

The French strategy had a devastating impact on the Algerian insurgents. Whole FLN and ALN formations were hunted down in their previous sanctuaries – the night and the mountains. The units of the nationalist movement became younger, less experienced, badly supplied, and often short-lived. Their morale, critical for a prolonged campaign, was deeply shattered. Local FLN commanders inside Algeria began to realize the magnitude of the damage inflicted by the French force, even before 1959. Algerian communities rejected the FLN authority more vigorously than before, its forces' efforts to extract money failed more frequently, and the organization's overall revenues

---

[6] Retired Rear Admiral Bernard Estival contends that the overall navy catch was 1,350 tons of FLN-bound military equipment, which was equal to the total quantity of weapons the FLN forces of the interior possessed in 1958. See Estival, "The French Navy and the Algeria War," *Journal of Strategic Studies* 25:2 (2002), 80–84.

diminished accordingly.[7] The exiled nationalist leadership was confronted by the hard-pressed leaders of the forces inside Algeria who bluntly questioned its motives and objectives as well as the prospects of success in the struggle. In the summer of 1960, the problems facing the Algerian national leadership reached a critical dimension.[8] In June, the commander of one of the six FLN operational zones (*Wilaya*), *Algéroise* – which was located in central coastal Algeria – initiated probing negotiations with the French. At the same time, Ferhat Abbas, the head of the Provisional Government of the Algerian Republic (GPRA), gloomily reported to his government in exile:

It becomes increasingly impossible to penetrate the barriers in order to nurture the revolution in the interior ... unless directed, supplied with fresh troops, effective weaponry, and money in great amounts, the underground forces will not be able to live for a long time let alone achieve victory.... The organic infrastructure has been dismantled in the urban centers, and it is increasingly nonexistent in the countryside.[9]

As has already been noted, the success of the war in the provinces was never as comprehensive as in the capital. The FLN managed to retain some degree of fighting capability in the countryside throughout the war. However, considering the nature and vastness of the countryside, and the resulting operational complexities, the French success was still impressive. Whereas prior to May 1958 the FLN was able to recruit more warriors than it lost, and amass a growing number of weapons, FLN activity and power after April declined steadily (the trend was reversed only after de Gaulle revealed that France would not force a solution on Algeria). Indicators such as the ratio of gained to lost weapons over time, the number of FLN and ALN warriors, and the number of attempted actions strongly support such a conclusion.[10]

The FLN performance on the third front – inside France – seems to have followed the general pattern of its actions in Algeria. Its most "productive" years – the years in which its activity in France produced most casualties – were 1957 and 1958. Thereafter, its activity there declined. Furthermore, much of its combat activity in France – excluding some spectacular plots such as the failed September 1958 attempt to assassinate Jacques Soustelle, the Minister of Information and former Governor of Algeria – was directed against the rival National Algerian Movement (MNA, formerly MTLD)

---

[7] Tripier, *Autopsie*, 179–80.

[8] See ibid., 434–47, particularly p. 436; and Talbott, *The War without a Name*, 191–96.

[9] Quoted in Montagnon, *La guerre d'Algérie*, 323, see also pp. 183, 285–97; and Horne, *A Savage War*, 330–40.

[10] See Tripier, *Autopsie*, 167–69, 665. While in 1955 and 1956, the ratio of gained to lost FLN/ALN weapons was roughly 1:1, in 1957 it was 1:4!

rather than French institutions or individuals. Indeed, most casualties of FLN activity in France were Algerian.[11]

Other indicators than the mood of FLN commanders and FLN activity also underscore the French success. First, French losses were relatively small: Altogether, between 25,000 and 30,000 soldiers and officers perished during eight years of war.[12] These fatalities were particularly small in comparison with French casualties in previous wars, and with those of the Algerian national forces.[13] Second, the French succeeded in recruiting many Algerians to fight on their side. In fact, the number of these allies dwarfed the assessed FLN order of battle in Algeria. Whereas in May 1957, 42,000 Muslims fought on the French side, a year later the number had more than doubled to 88,000. In 1961, it reached an all-time high of some 200,000. In comparison, in May 1958, at the peak of its activity, the number of FLN regulars (*moujahidine*) stood at some 20,000. These were backed up by another 20,000–30,000 auxiliary and irregular troops (*moussebiline*) armed mainly with inferior weapons.[14] Finally, the FLN degree of influence over Algerian society at large indicated a relative weakness. In eight years of war, and in spite of great terror, the FLN failed where other nationalist movements often succeeded – it never managed to organize a general strike or a massive popular uprising. In fact, the FLN did not even succeed in insuring adequate compliance with its instructions. For example, the FLN called for a boycott of the September 1958 referendum, which legitimized the constitution of the

[11] During the war, some 4,000 Algerians were killed in over 10,000 violent actions in France. Most died in clashes between the FLN and MNA movements. In continental France, the French police lost only 53 officers. See Charles-Robert Ageron, "Les français devant la guerre civile algérienne," in Rioux, *La Guerre d'Algérie et les Français*, 55.

[12] See Montagnon, *La Guerre d'Algérie*, 404; Patrick Éveno, and Jean Planchais, *La guerre d'Algérie: Dossier et témoignages* (Paris: Le Monde/La Découverte, 1989), 321; Guy Pervillé, "Bilan de la guerre d'Algérie," in *Études sur la France de 1939 à nos jours* (Paris: Seuil, 1985), 297–301. The official 1968 total count of fatalities is 24,614 army and air force fatalities, and overall 1,000 prisoners and MIAs. 35,615 soldiers were wounded in action and 29,370 in accidents. See Henri Alleg et al., *La guerre d'Algérie* (Paris: Temps actuels, 1981), Vol. 3, 570.

[13] About 30 percent of the total fatalities (9,000) were French officers and non-Commissioned Officers (NCOs). These, however, included *pied-noir* soldiers, Legionnaires, and French Muslims. A conservative assessment of FLN fatalities stands at 140,000. Other assessments are between 300,000 and 500,000. In Indochina, the French had about 92,000 fatalities, but most soldiers were foreigners. Still, over 20,000 of the fatalities, mostly officers and NCOs, were French. See Montagnon, *La guerre d'Algérie*, 404; Éveno and Planchais, *Dossier et témoignages*, 321; and Pervillé, "Bilan de la guerre d'Algérie," 297–301.

[14] Data from Tripier, *Autopsie*, 182, 206, 428; and Heggoy, *Insurgency*, 179. John Ambler puts the highest number of FLN warriors inside Algeria at 30,000–40,000. See *The French Army in Politics 1945–1962* (Columbus: Ohio State University Press, 1966), 156. In Tunisia and Morocco, the ALN had additional contingents of 2,000 regulars and some 5,500–7,000 recruits in training. The number of warriors in Tunisia and Morocco almost doubled in 1960; however, because the borders were tightly guarded, these forces had little military significance.

Fifth Republic and de Gaulle's widespread powers, and failed. In January 1961, the FLN called again for a boycott, and failed again. This time it was a referendum on "self-determination." On both occasions, large numbers of Muslims voluntarily went to vote in open defiance of FLN orders.[15]

Yet, in spite of the French military success, the political course of the war differed sharply from the realities of the battlefield. The FLN achieved its main political objectives in full: It was recognized as the sole representative of the Algerians, Algeria was granted independence, and its national unity and territorial integrity were secured.[16] France on the other hand, was forced to withdraw from Algeria all vestiges of its institutions as well as its entire civil presence there.

The French Political System, the State, the Army, and the War

Political analysts are occasionally tempted to explain policies that in hindsight seem to reflect an irrational persistence as the results of an irresolution that they attribute to deadlocked political systems. Indeed, the splintered and unstable political system of the Fourth Republic – *le système* – was depicted as a classical harbinger of sterile policies, including the long and futile war in Algeria. However, such an explanation, which implicitly assumes that a stable political system would have performed better in response to similar stimuli in the 1950s, is a historical exercise of questionable merit. Unstable political systems do not necessarily perpetuate irresolute, ambivalent, or failed policies, much as stable political systems – as the American war in Vietnam suggests – do not necessarily avoid them. Indeed, the unstable political system of the Fourth Republic managed to produce the government of Mendès France, which was able to take very decisive steps. This government ended the Indochina war and initiated the process, which was completed during the life-time of the Fourth Republic, of getting France out of its Tunisian and Moroccan protectorates.

It would also be imprudent to suggest that the effort to keep Algeria French was utterly, or even largely, irrational. There were quite a few sound reasons, at least from a Realpolitik point of view, to preserve the empire, and particularly Algeria, under French control. In general, the empire provided human and strategic resources and global bases that gave France an opportunity to be involved in world affairs. Within the empire, Algeria was the "crown jewel." In fact, it proved its strategic value during World War II as

---

[15] For example, in the 1961 *self-determination* referendum, 2,800,000 people voted in Algeria. The majority of voters were necessarily Muslim because there were only about 1,000,000 *pieds-noirs*. Furthermore, as 1,920,000 adults voted "yes," the decisive majority of the positive vote was Muslim as well.

[16] FLN objectives were stated in the Soummam conference. See Tripier, *Autopsie*, 571–601, particularly pp. 583–84; and Éveno and Planchais, *Dossier et témoignages*, 114.

a depository of oil and gas reserves and as a nuclear test site during the later years of the Algerian war.[17] Besides, it had been part of France for some 130 years, with some one million French settlers, most of whom were descendants of French families that had lived in Algeria for several generations.

Of course, the association of the empire, and particularly Algeria, with the international status of France, and the belief that both were vital, were not entirely rational. Reasonable perceptions of vitality were amplified by a wounded national psyche. A decade and a half of military humiliations had left the French elite with an obsession to reverse France's misfortunes – an obsession that was acute enough to convince French politicians that unless they restored the empire, France would inevitably degenerate.[18] In that respect, this self-entrapment was compounded further because Algeria did not simply constitute a problem; it also provided a seductive opportunity. As it was associated with the glorious past of the empire, it also became a means and an opportunity to resurrect this past.

This imperialist thinking, other scholars have already observed, transcended political cleavages in France.[19] For much of the duration of the war, and with few exceptions, the most powerful politicians in France were firmly committed to the idea of preserving French Algeria, if need be by every means necessary. The divisions France had experienced over the war were largely within, rather than between, the different parties. Indeed, the list of those who supported the policy of preserving Algeria included ranking figures from almost every part of the French political milieu.[20]

[17] Oil was "exported" from Algeria as of January 1958. No wonder that Mollet depicted the Sahara as "French Texas." See Gérard Bossuat, "Guy Mollet: la puissance française autrement," *Relations Internationales*, 57 (1989), 28. See also Jacques Soustelle in "The Wealth of the Sahara," *Foreign Affairs*, 37:4 (1959), 626–36. For the range of pro-French Algeria arguments, see Charles-Robert Ageron, "L'Algérie, dernière chance de la puissance française," in *Relations Internationales*, 57 (1989), 113–39.

[18] For example, in 1947, de Gaulle rationalized the French commitment to the empire in the following way: "For us in the world as it is, and as it moves along, to lose the French Union would be a fall that could cost us up to the point [of losing] our independence. To keep it and make it live, is to remain great and consequently, to remain free." Quoted in Raoul Girardet, *L'idée coloniale en France* (Paris: La Table Ronde, 1972), 200.

[19] See Tony Smith, *The French Stake in Algeria, 1945–1962* (Ithaca: Cornell University Press, 1978); and Odile Rudelle," Gaullisme et crise d'identité républicaine," in Rioux, *La Guerre d'Algérie et les Français*, 180–201, particularly p. 187. For an analysis of the initial positions of French parties, see Serge Berstein, "La peau de chagrin de 'l'Algérie française'," in ibid., 202–08.

[20] Consider the following partial list: Pierre André (*Independents*); Bourgès-Maunoury and Félix Gaillard (*Radicals*); André Morice (*Moderates*); Michel Debré (*Gaullist, UNR*); Jacques Soustelle and Georges Bidault (*Gaullist, MRP*); Alfred Coste-Floret (*Christian Democrats*); Max Lejeune, Robert Lacoste, and Guy Mollet (*SFIO*); and Edgar Faure (the hodgepodge *RGR*); Field-Marshal Join, General Weygand, President Coty, Soustelle, Bidault, Georges Pompidou, Edmond Michelet, and Cardinal Saliège (archbishop of Toulouse). See also Girardet, *L'idée coloniale*, 235–48, 264; and Berstein, "La Peau de Chagrin," 212–16.

This national consensus over the need to fight and win in Algeria, and particularly the support of the Left and Center of this policy, are particularly intriguing from the point of view of this book. They corroborate the contention that, at least in times of national security crises, the institution of the state overrides the parochial and ideological proclivities of the political parties in power. Indeed, the rhetoric of, and more so the actions taken by Left-led governments in France, leave little doubt as to whether Algeria evoked etatist impulses. As soon as the war started, Pierre Mendès-France, who led France out of Indochina, firmly declared to the National Assembly:

There is no compromise when the defense of the internal peace of the nation, its unity, and the wholeness of the French Republic are concerned ... The departments of Algeria constitute a part of the French Republic. They have been irrevocably French for a long time ... between it [Algeria] and the Metropolis, there is no conceivable secession ... never will France, nor any government, nor any French parliament, whatever their specific tendencies, ever give up this fundamental principle ...[21]

Other partners in Mendès-France's government proved just as consumed by etatist ideology. The reaction of François Mitterrand, the interior minister, to the violence in Algeria was typical: No to "negotiations with the enemies of the fatherland ... the only negotiation [is] war," *"L 'Algérie, c'est la France."*[22] Things did not change after the victory of the Socialist-Mendèsist bloc, the Republican Front, in the January 1956 elections, even though the Front ran on a peace-in-Algeria platform (which some believe, vague as it may have been, was responsible for its victory).[23] If anything can be said about the post-election Socialist position it is that it hardened. Once in power, and after being bombarded by tomatoes during a February 1956 visit to Algiers, Guy Mollet – the same Mollet who had previously argued in public that the war was "stupid and leading nowhere" and who contemplated "immediate [Algerian] independence" – suddenly suggested that "France without Algeria will be nothing."[24] Yet, above all, the Left proved that it was thoroughly etatized by deeds. It was Mollet and his Socialist and centrist partners – Maurice Bourgès-Maunoury, the Minister of National Defense, and Max Lejeune, the State Secretary for the Armed Forces – who gave the army full powers in Algeria, saturated it with French conscripts, and turned the

[21] Quoted in Montagnon, *La guerre*, 127–28. This was Mendès-France's genuine opinion as indicated by an April 1956 letter, quoted in Berstein, "La peau de chagrin," 205.
[22] Montagnon, *La guerre*, 128. See also Mitterrand's other statements in the National Assembly in Hervé Hamon and Patrick Rotman, *Les porteurs de valises: la résistance française à la guerre d'Algérie* (Paris: Albin Michel, 1979), 25.
[23] See, for example, Jean Lacouture (trans. George Holoch), *Pierre Mendès-France* (NY: Holmes and Meier, 1984), 355–56. The platform never elaborated how, or what type of, peace would be achieved, but Mollet specifically condemned the repression policy as a major obstacle to peace. See Smith, *The French Stake*, 132–33.
[24] Quotes are from Lacouture, *Mendès*, 356, 358, and Michel Winock, "Pacifisme et attentisme," in Rioux, *La guerre d'Algérie et les Français*, 16.

war into a national crusade. Indeed, as journalist Jean Daniel noted in one of his 1960 articles: "The French left [was] not essentially anti-colonialist, undoubtedly it [was] egalitarian, that is, opposed to economic exploitation: But ideologically it [was] imperialist, which [led] it to be for integration in Algeria."[25]

Even the French Communist Party (PCF), which opposed colonial rule in principle and was against the methods French forces used in Algeria, supported the war effort for over a year and a half. It is true that much of the protest against the mobilization waves in the fall of 1955 and spring of 1956 was Communist. But the protest in France and anti-war actions initiated in Algeria were led by local-level young militant cadres. The central party establishment did not condone these actions. Rather, it voted in favor of the March 1956 "special powers" Mollet requested for the administration in Algeria, and in June abstained (rather than voted against) in a confidence motion Mollet organized over his Algerian policy. It was only in its July 1956 XIVth Congress that the PCF distanced itself from Mollet's Algerian policy. But even after this date, it was too preoccupied with an effort to control the damage created by the Soviet invasion of Hungary and too mindful of its large *pied-noir* constituency to make the war or Algerian independence its main political concern.[26]

The political consensus over the Algerian policy was complemented by a strong bureaucratic commitment to Algeria. The very same ideas and psychological drives that governed the thinking of the political elite dominated the thinking of many army officers. Pro-French Algeria officers perceived the choices France faced in Algeria in black and white terms. For them, France faced a choice between "the permanent interests of France" and capitulation in Algeria, which would inevitably result in France's internal decay and international decline.[27] They were convinced that it was their duty, as guardians of the national interest, to prevent such a development. Finally, to the basic

[25] Quoted in Hamon and Rotman, *Les porteurs*, 235. See also the article of Marc Sadoun, "Les socialistes entre principes, pouvoir et mémoire," in Rioux, *La guerre d'Algérie et les Français*, 225–34.
[26] On the Algerian policy of the French Communist party (PCF), see Jean-Jacques Becker, "L'intérêt bien compris du Parti communiste français," in Rioux, *La guerre d'Algérie et les Français*, 235–44. The PCF rejected repression and advocated more freedom prior to the war, but it still defended integration as a matter of national interest. See Hamon and Rotman, *Les porteurs*, 20–24; and Jean-Pierre Rioux (trans. Godfrey Rogers), *The Fourth Republic, 1944–1958* (NY: Cambridge University Press, 1987), 254–55.
[27] Quoted in Edmond Jouhaud, *Ce que je n'ai pas dit* (Fayard, 1977), 185, from his February 1961 letter to *Le Monde*. General Massu argued that he "sincerely believed" that France "could not leave the southern shores of the Mediterranean without being dangerously weakened and running mortal risks." See *La vraie bataille d'Alger* (Paris: Plon, 1971), 54. See also in Colonel Antoine Argoud, *La décadence, l'imposture et la tragédie* (Paris: Fayard, 1974), 351–53; and the Epilogue of Colonel Roger Trinquier's book, *Le temps perdu* (Paris: Éditions Albin Michel, 1978).

existentialist reasoning army officers (and others) also added their own version of the domino theory, suggesting that the uprising in Algeria was part
of the grand Communist strategy of using the Third World as a means of
chipping at the rims of the Free World.[28]

Of course, French military officers were motivated by a parochial psychological bias as much as by strategic and etatist considerations. The humiliations France suffered in war in Europe and Indochina were first and foremost
humiliations of the army, and as such resonated strongly within its ranks.
Some of the commanders were convinced that "all must converge in order to
restore confidence and dispel the haunting obsession with Indochina,"[29] and
Algeria was supposed to be the elixir for this obsession. In addition, some of
the officers had a strong sentiment toward Muslims who served on their side
or toward those they befriended. Consequently, they developed a powerful
emotional justification for keeping Algeria French. Here again, Indochina
loomed in the background, as officers were reminded of the "nightmare" of
abandoning and betraying their allies there.[30] Finally, the army's commitment was deepened as a consequence of the decisions of the civil authorities
in France. Desperate to win the war, the political leadership made the army
the supreme power in Algeria. In doing so, however, it raised the stakes the
military had in the war. Algeria became the touchstone for the army's ability
to perform its duties.[31]

At least in a narrow military sense, the army had an advantage going into
Algeria after Indochina. First, Algeria was close to home and thus a lesser
logistical problem. Second, with fewer colonies, France could concentrate
its resources and attention in Algeria. Third, the army enjoyed an unusual
operational maturation. The war in Indochina had left it with a cohesive doctrinaire legacy and a cadre of experienced, aggressive, and reform-minded
middle-level officers. These officers closely studied the Maoist version of the
people's war, and based on their studies invented a comprehensive theory
and a doctrine to win such wars – the "revolutionary war" theory. In summary, the mental and doctrinaire legacy of Indochina left the military command with a cadre of highly motivated and competent officers commanding

---

[28] See, for example, General Allard's November 1957 analysis in Girardet, *L'idée coloniale*, 240–
41; and in Peter Paret, *French Revolutionary Warfare from Indochina to Algeria* (Princeton:
Center for International Studies, 1964), 25. See also Jacques Soustelle's analysis in "France
Looks at Her Alliances," *Foreign Affairs*, 35:1 (1956), 116–30; and General Jean Delmas, "A
la recherche des signes de la puissance: l'armée entre l'Algérie et la bombe A, 1956–1962,"
Relations Internationales, 57 (1989), 80–81.

[29] Jouhaud, *Ce que*, 47.

[30] See ibid., 75; and Jean-Marie Domenach, "The French Army in Politics," *Foreign Affairs*,
39:2 (1961), 194.

[31] On the politicization of the army, see George A. Kelly, "The French Army Re-enters Politics," *Political Science Quarterly*, 76:3 (1961), 367–92; and Claude d'Abzac-Epezy, "La
société militaire, de l'ingérence à l'ignorance," in Rioux, *La Guerre d'Algérie et les Français*,
245–48.

battle-tested and well-equipped crack forces determined to destroy the Algerian national revolt.

Finally, France had no particular reason to be concerned with global considerations during the war. The major players of the international system were in no position to either oppose or restrain France. The Americans loathed French colonialism in general and disliked the war in Algeria in particular. However, they did not want to further upset their relations with an already troubled ally whom they had humiliated, yet considered as essential for their European alliance against Communism. In any event, the prospects of success of an American policy that was antagonistic to what the French considered vital national security interest were inherently bleak. Indeed, the American reluctance to take issue with France over Algeria was in complete contrast to the swift and humiliating reaction of the Eisenhower administration to the October 1956 joint British-French-Israeli Suez adventure.[32] The Soviet Union, on the other hand, although an ally of the Algerians by virtue of its rivalry with the West, was in no position to wield power in North Africa. Its power-projection capacity was limited, and consequently it could at best help supply the Algerians with armaments, directly or via proxies, and lend them diplomatic support. Furthermore, the FLN was not a Communist movement and therefore did not offer a particularly strong justification for great commitment.

Still, the conversion of political will, military enthusiasm and maturation, and favorable global conditions were not sufficient to guarantee a military victory in Algeria. Ultimately, the French had to let the essence of Realist thinking – the notion that necessity dictates behavior – trickle down from the level of national-interest formulation, through the level of war-strategy construction, to the level of combat methods. The French political elite understood this necessity very well, and was ready to go a long way in order to let the army operate free of political and moral constraints.

First, the French leadership gradually transferred the governing power from the civil to the military authorities in Algeria. In March 1955, a six-months' state-of-emergency was declared in the more trouble-laden part

---

[32] On the nature of French-American relations during the 1950s and the Algerian war, see Pierre Mélandri, "La France et le 'jeu double' des États-Unis," in Rioux, *La guerre d'Algérie et les Français*, 429–50; Richard H. Immerman, "Perceptions by the United States of its Interests in Indochina," in Lawrence S. Kaplan, Denise Artaud and Mark R. Rubin (eds.), *Dien Bien Phu and the Crisis of Franco-American Relations, 1954–1955* (Wilmington, DE: SR Books, 1990), 1–26; George C. Herring, "Franco-American Conflict in Indochina, 1950–1954," in ibid., 29–48; Herring, Garry R. Hess, and Immerman, "Passage of Empire: The United States, France, and South Vietnam, 1954–1955," in ibid., 171–95; Charles G. Cogan, *Old Allies, Guarded Friends* (Westport, CT: Praeger, 1994), 99–120, and "France, the United States and the Invisible Algeria Outcome," *Journal of Strategic Studies*, 25:2 (2002), 138–58; and Irwin M. Wall, "De Gaulle, the 'Anglo-Saxons' and the Algerian War," *Journal of Strategic Studies*, 25:2 (2002), 128–29, 135.

of Algeria. Less than two months later, the military assumed powers in this zone and took over control of the civil administration and the police force. In August 1955, the zone of the state-of-emergency was extended to cover all of Algeria, and in March 1956, the Mollet government passed a special-powers act for Algeria in the National Assembly. Local authorities were granted extensive powers to pursue any exceptional measures required by circumstances. In January 1957, in order to cope with terror, General Massu was given total police powers in the city. By early 1957, then, Algeria had gone through the last stage of a transformation that turned it into a vast battlefield.

Second, as becomes clear from this discussion, the French leadership let the army run the war as it pleased. The army for its part decided to rely mostly on force and fear. In that respect, the debate over how to rule – benevolently or by intimidation – and how to react to upheavals, had been resolved prior to the Algerian war in places like Madagascar, Indochina, and even in Algeria itself. Thus, for example, the French retaliated with extreme brutal force to the May 8, 1945, massacre in Sétif (which cost the Algerian Moslem population several thousand lives). Moreover, the fact that French reaction was motivated by more fundamental considerations than the heat of the moment became clearer in the aftermath of that massacre. On June 1, General Breuillac reported to the Minister of War on the post-massacre maneuvers he had conducted during the marketday in Tizi-Ouzou, explaining: "The real objectives of this military demonstration, *which are not mentioned in the enclosed document,* were to ... carry out an *operation of intimidation*; strike the imagination of the masses of Kabylie by a visible development of massive military means; constituting a deterrent to the intents of the nationalist agitators."[33]

The last stage of the trickling-down process was achieved when coercive strategy was translated into tactical brutality that emphasized *efficiency* and *effectiveness*. In fact, the process of brutalization, though expeditiously accepted, was made easier by the realities of the war, and in particular the savagery of the countryside and urban terrorism of the FLN and other groups. Revolting sights of blood-letting and mutilation – that regularly included throat-slitting, axing, and disemboweling of the old and the young, women and children, settlers and locals, and occasionally soldiers[34] – helped French commanders justify to themselves and to their soldiers why *necessity, effectiveness,* and *efficiency* were to be emphasized. They vindicated their ruthlessness by evoking biblical notions of "an eye for an eye [and] a tooth for a tooth."[35] In practice, all of that meant that the day-to-day war – the

[33] Quoted in *La guerre d'Algérie par les documents* (Vincennes: Service Historique de l'Armée de Terre, 1990), 84 (italics added).
[34] See Horne, *A Savage War of Peace,* 112, 114, 120–21, 134–35, 153, 171, 186, 192, 209–10.
[35] Massu, *La vraie bataille,* 168, see also p. 165.

intelligence-gathering, the isolation of potential FLN supply sources, the control of the population, and the "seek and destroy" missions – was conducted with little regard for moral considerations. The army reverted, with the full knowledge of the government, to "irregular" and often savage methods. As a routine, it resorted to violent "reprisals," summary executions, and, above all, the systematic use of torture.[36] In addition, the army evacuated a great number of Algerians from territories it considered strategic – perhaps as many as one million (11–12 percent of the Muslim population), and then resettled them, usually under miserable conditions, in new locations (the policy of *regroupement*).

The French Economy and the Algerian War

The question of whether the French war effort in Algeria was subject to deep or crippling economic constraints can be answered with relative ease: It was not. The direct total cost of the Algerian war is assessed at 50–55 billion new francs. Presented otherwise, the war consumed between 50–60 percent of the defense budget, and 10–15 percent of the general budget.[37] In fact, these assessments are perhaps too conservative since they ignore a significant hidden cost – the cost incurred by the loss of the labor of some 200,000 potential young workers who served as soldiers precisely when the labor market was suffering from a shortage that had developed as a result of World War II.

Indeed, as of 1956, the war in Algeria increasingly contributed to the over-burdening of the French budget and economy. As a result, de Gaulle inherited a very serious economic crisis. By 1958, monetary, budgetary, and commercial disequilibria led the French economy into an impasse.[38] The overwhelming budget deficit stood at 1,200 billion old francs (roughly $2.9 billion). The foreign debt stood at over $3 billion (about 7.5 percent of the GNP, half of which was repayable in a year), the trade balance was negative, foreign exchange was depleted, gold reserves stood at only $630 million (equivalent to five weeks of imports), and sources of external credit were vanishing. In short, France's solvency and the government's ability to continue to finance the whole range of policies were all but exhausted.[39]

---

[36] According to General Massu, "the highest civil authorities of the time came on inspection visits. to Algiers. Messrs. Bourgès-Maunoury and Max Lejeune, visited the interrogation centers and enthusiastically supported this formula (that is, torturing in order to obtain information)." Quoted in *La vraie bataille*, 153.

[37] See Horne, *A Savage War of Peace*, 538–39; Rioux, *The Fourth Republic*, 281; and Éveno and Planchais, *Dossier et témoignages*, 322–24.

[38] Jean-Charles Asselain, "'Boulet colonial' et redressement économique (1958–1962)," in Rioux, *La guerre d'Algérie et les Français*, 290.

[39] Charles de Gaulle (trans. Terence Kilmartin), *Memoirs of Hope: Renewal and Endeavor* (NY: Simon & Schuster, 1971), 138.

The impact of the war on the economic deterioration, however, should not be over-estimated. The French economy was brought to an impasse not simply as a result of the rising cost of the war, but rather because of the conjunction between that cost and other ambitious social and economic commitments.[40] Thus, in the mid-1950s, France was improving its international competitiveness by investing in infrastructure renovation, including industrial and agricultural modernization, nuclear energy development, and the expansion of its education system. Indeed, even during the Algerian War, several civil sections of the budget grew more rapidly than the military one.[41] Moreover, on top of these commitments and the increased investment in the war, the Mollet government mandated even greater entitlements. In a nutshell, France's 1958 economic crisis was the result of an irresponsible distribution of resources. French governments, and particularly that led by the Socialist Mollet, put France on the road to bankruptcy because their political calculations and ideological proclivities prevented them from seriously scaling down their ambitious social agenda at a time when they decided to commit resources to war.

In any event, the state of the French economy did not prevent nor hamper the war effort for the simple reason that de Gaulle and his administration managed to heal the economy shortly after he assumed power in mid-1958. In order to put France back on track, de Gaulle, Antoine Pinay, his finance secretary, and a special committee headed by Jacques Rueff introduced an extremely tough and painful economic program.[42] The results of this package, known as the French *economic miracle*, were dramatic, decisive, and quick. As of 1959, the balance of current payments had a surplus. The reserves of the Bank of France grew from 70 million francs on December 31, 1958, to over 7 billion francs a year later, 10 billion in 1960, 15 billion in 1961, and 20 billion at the end of 1962. Within two years of the initiation of the recovery program, France was able to slash its external debt by a third and make the franc an attractive currency. Levels of investment in France increased, and in 1961 it became a creditor of the IMF (whereas previously it had been a major recipient of funds). With an average industrial growth rate of 7.8 percent a

---

[40] Total war expenses comprised about 18 percent of the 1961 GDP. Also note that the French economy had a healthy base, irrespective of the 1957–1958 economic crisis. This was reflected in the growth of various indicators of economic activity in the 1950s, including GNP, average national income, consumption, capital formation, industrial output, and investment. In fact, the French economy did better than other peacetime economies such as the British or Belgian, although it paid for war from 1946 onward. See Rioux, *The Fourth Republic*, 318–35.

[41] The cost for education was 33 percent of military expenses in 1955, 46 percent in 1958, 50 percent in 1961, and 62 percent in 1962. The cost for personnel in the education system exceeded that for military personnel for the first time in 1960. See Asselain, "Boulet colonial," 302; and Edward L. Morse, *Foreign Policy and Interdependence in Gaullist France* (Princeton: Princeton University Press, 1973), 164, figure 4.

[42] See de Gaulle, *Memoirs of Hope: Renewal*, 142–44.

TABLE 5.1 *The Algerian war and the French economy*

Year	All state expenses in Algeria[a] (% of GDP)	Military expenses (% of state expenses)	All military expenses (% of GDP)	All state expenses (% of GDP)
1953		36.3	12.6	
1954		32.7	11.4	
1955	1.0	27.5	9.6	
1956	1.4	28.3	10.1	
1957	1.6	29.7	10.3	34.6
1958	1.7	25.8	8.6	33.3
1959	2.8	26.8	8.6	32.0
1960	2.7	27.9	8.1	29.1
1961	2.5	26.9	8.1	30.0
1962	2.0	24.7	7.5	30.3
1963		21.7	6.9	

[a] Civil and military expenses in Algeria computed from the combined overseas expenses of the different administrations.

Source: Asselain, "Boulet colonial," 292, 296.

year and the growth of exports within and outside the franc zone, France's industry became strong enough to meet the EEC tariff reduction agreed upon in the 1957 Rome Treaty without major disruptions.[43]

Most important for our discussion, the impact of military expenses on the budget and the economy decreased dramatically as a result of the economic upswing. The combined civil and military investment in Algeria continued to grow until 1959, reaching an all time high of 2.8 percent of the GDP (about two-thirds of which were military, though the civil portion grew faster). However, thereafter the GDP grew faster than the expenses incurred by the war in Algeria, and therefore the relative military cost of keeping Algeria French declined.[44] Moreover, after 1958, total military expenditure as a portion of the total state expenditure and as a portion of the GDP, which grew steadily between 1954 and 1958, also declined (see Table 5.1).[45]

The Algerian war, then, did not destroy the French economy nor did it hurt the latter's foundations significantly. In fact, the economy was booming precisely when the military investment in Algeria started to produce the results the French had hoped for. Thus one can safely conclude that the economy

[43] See Asselain, "Boulet colonial," 290–94. The effect of the French economic miracle on individual citizens, at least in the short run, was not as clear cut. See Morse, *Foreign Policy and Interdependence*, 110, and 171 table 6; and Rioux, *The Fourth Republic*, 368–74.

[44] Asselain, "Boulet colonial," 296.

[45] According to Michel Martin, relative military expenditure peaked in 1959 and 1960. However, Martin also argues that military expenditure as a percentage of the GNP declined as of 1957! See "Conscription and the Decline of the Mass Army in France, 1960–1975," *Armed Forces and Society*, 3:3 (1977), 372, table 5.

did not limit the investment in war, nor did the war preclude investment in other political objectives. In fact, one can even go so far as to suggest – on the basis of extrapolation from the economic growth and the military successes to significantly diminish the FLN in 1959–1961 – that had France decided to pursue the war beyond 1962, the relative economic cost of the war would have decreased further. Finally, let us note that the economic recovery was clearly evident before de Gaulle started to drift away from the policy of "French Algeria." In short, there was *no objective reason* to think that the economic demands of the military effort in Algeria prevented France from achieving its other ends.

Conclusion

France was well prepared, highly motivated, and sufficiently endowed to fight in Algeria. The balance of power was clearly in its favor. International conditions did not critically constrain it, and the global rivalry between the superpowers kept them at a safe distance from the war. Having learned the lessons of Indochina, and after the humiliations there and in World War II, the French army was well prepared and highly motivated to win the war. Algeria was part of constitutional France, and French governments and politicians told the public, almost in one voice, that Algeria was French. Indeed, French governments committed society to the war, and let the army fight as it saw fit. Moreover, while the economic burden of war may have initially been debilitating, reforms turned the economy around precisely when the army was starting to "deliver." In fact, the war was close to being militarily won almost as much as can be imagined. Why, then, did the French pull out of Algeria after they had achieved dominance in the battlefield?

6

French Instrumental Dependence and Its Consequences

The major dilemma French governments faced when they were presented with the Algerian upheaval was not whether to fight, but rather how to fight – whether to take the political risk of involving French society at large in the war or remain cautious and insulate society as much as possible from the war. Certain attributes of Algeria made this dilemma more wrenching. Most notably, some of the reasons that pushed French governments toward the resolution of the conflict by force – a long mutual history with Algeria, memories of its significance for Free France, its proximity, and the magnitude and power of the vociferous Algerian lobby – also made harder the isolation of the events in Algeria from the French collective consciousness. The massive presence of Algerian labor in France also promised that the Algerian problem would create echoes at home.[1]

Thus, French politicians maintained a duality in their statements, and a gap between the latter and their actions, in order to accommodate the inherent saliency of Algeria with the desire to avoid the political uncertainty involved in the nationalization of the war. Indeed, French leaders initially boasted with high rhetoric, but avoided backing their words with a real commitment of society to the war. Politicians vowed to preserve French Algeria, but the events in Algeria were depicted to the public as comprising a set of police operations against bands of outlaws.

This duality was not all unreasonable as France was well-equipped to fight a limited war at a relatively safe distance from its society. The French army was quite large at the time, consisting of 2.5 percent of the population, whereas the Algerian nationalist forces were badly organized and not particularly effective. Thus the immediate military manpower needs in Algeria were addressed by relying on existing forces. Units from French colonies

---

[1] The Algerian community in France coordinated demonstrations in different industrial centers, including Paris, Lyon, and Saint-Étienne. See Danielle Tartakowsky, "Les manifestations de rue," in Rioux, *La guerre d'Algérie et les Français*, 131–32.

were sent to Algeria, and the release of conscripts on active duty was delayed. However, while the French were deploying existing forces in the hope
that they would prove sufficient, the Algerian rebels were not standing still.
Rather, they improved the quality of their organization and mobilization
process, and with them increased the scope of their actions. In short, the
war escalated and the French found themselves needing to increase their
investment in order to avoid catastrophic deterioration of their position in
Algeria. Consequently, in 1955, the government started to recall soldiers who
had already been demobilized.

Overall, though, the mobilization initiatives were designed very carefully.
In May 1955, the Faure government announced a delay in the release of
100,000 active-duty conscripts, and perhaps in an effort to test the water
and establish a precedent, it also announced the mobilization of a small number of reservists with technical and professional skills. In August, two days
after the massacre of French settlers in Philippeville (in Algeria), the government took the opportunity created by French public outrage to announce
a delay in the release of another 120,000 active duty soldiers. This time it
also announced the mobilization of some additional 60,000 reservists (see
Table 6.1).

Thus, about a year elapsed from the beginning of the war and until the
situation in Algeria was perceived as problematic enough to warrant the
political risks involved in a full nationalization of the war. But nationalization
was not simply the result of developments in the battlefield. Rather it was

TABLE 6.1 *Growth of French forces in Algeria, 1954–1956*

	Date	Force level	Comment/decision
1954	November	~57,000	Starting level of forces in Algeria
1955	January	~80,000	Transfer from Indochina
	May		Transfer of one division from Nancy; call-up of few reservists and residents of Algeria from 53/2 cohort; transfers from Morocco and Tunisia
	July	~114,000	
	August	~120,000	Transfer from Germany; call-up of those released in April; extension of service of 54/1 cohort
	October	~160,000	
	December	~200,000	
1956	March	~220,000	April decision on massive recall of reservists
	July	~400,000	

Sources: Delarue, "La police en paravent et au rempart," 259; Rioux, *The Fourth Republic,* 239,
250, 268; Alleg et al., *La guerre d'Algérie,* Vol. 1, 573, 574, Vol. 2, 51, 53.

also the result of lessons the French political and military elite had learned from the Indochina war. In Indochina, France conducted a war considerably removed from society. Investment of human and material resources of home origin was kept at a minimum. While the officer corps was French and some of the soldiers were French volunteers, much of the fighting force was composed of foreign nationals. Likewise, most of the financial cost of the war was not paid for by France but rather by the United States.[2] The problem was that Indochina *had been lost*, and the French elite dreaded a repetition of this debacle. Thus, one lesson Indochina taught French leaders was that, at some point, national commitment to the war might be indispensable, as only the latter could guarantee victory. In Algeria, this point became obvious in 1956. The Socialist government of Guy Mollet accepted the demands of the military for a large increase of the contingent in Algeria, and called up a large number of reservists, extended the service of conscripts from 18 to 27 months, and formally nationalized the war.[3] To his constituency, Guy Mollet explained: "The action for Algeria will be effective only with the confident support of *the entire nation, with its total commitment*."[4]

The most obvious result of the elevated level of instrumental dependence was the ascent of Algeria on the public's list of national problems. Certainly, dramatic events in Algeria such as the August 1955 massacre in Philippeville in which Algerian nationalists slaughtered seventy-one Europeans and about a hundred Francophile Moslems, and the May 1956 *Palestro* ambush, in which the FLN virtually annihilated a platoon of French reservists, also contributed to the salience of the war.[5] But the link between mobilization and the nationalization of the war in the minds of the French people was unequivocal (see Table 6.2).

Toward the end of 1955, the war in Algeria was gaining ground as a major issue on the national agenda. As of early 1956, it became the most conspicuous problem, and it remained so until the end of the war.[6] The

---

[2] According to Donald Lancaster, 54,000 French soldiers, out of a 175,000-strong land contingent, served in Indochina in the spring of 1953. 175,000 Vietnamese, Laotian, and Cambodian soldiers also fought for the French. See *The Emancipation of French Indochina* (London: Oxford University Press, 1961), 265, 411 note 45. On the cost of this war and American financing see ibid., 280; Herbert Tint, *French Foreign Policy Since the Second World War* (London: Weidenfeld and Nicolson, 1972), 23–24; Edgar S. Furniss, *France, Troubled Ally* (NY: Harper and Brothers, 1960), 162; R.E.M. Irving, *The First Indochina War* (London: Croom Helm, 1975), 102–04; and Lacouture, *Mendès*, 186.

[3] See Armand Frémont, "Le contingent: témoignage et réflexion," in Rioux, *La guerre d'Algérie et les Français*, 79.

[4] Quoted in Smith, *The French Stake*, 133, from *Le Populaire*, April 16, 1956 (italics added).

[5] The Palestro casualties were particularly significant because they were the first reservists killed in Algeria, and because they were from the Paris region.

[6] See *Sondages: Revue Française de l'Opinion Publique* (Paris: Institut Français de l'Opinion Publique), 1960:3, 39 (henceforth *Sondages*). That Algeria was the top problem for the French was first revealed in *L'Express*, December 16, 1955. See Talbott, *The War Without a Name*, 56.

TABLE 6.2 *Rise in prominence of the Algerian war at home*

Question: "In your opinion, what is the most important problem for France now?"

	Algeria and North Africa (%)	Constitutional– governmental (%)	Economic– personal[a] (%)	Economic– National[b] (%)
September 1954	24	–	24	18
January 1955	27	–	25	12
December 1955	25	–	15	3
April 1956	63	–		2
July 1956	60	5[c]		9
September 1957	51	9[c]	9	27
January 1958	37	10	7	36
August 1958	40	19	11	15

[a] Economic personal – salaries and buying power.
[b] Economic national – financial and (general) economic situation.
[c] Month not mentioned.
Source: Sondages, 1958:4, 5. See also 1956:3, 3.

independent institution that monitored public opinion in France during the war, *Institut Français de l'Opinion Publique* (IFOP), both noted and reflected this change when it conducted its first comprehensive public-opinion poll devoted entirely to the Algerian war in early 1956.[7] The change was also reflected (and encouraged) by the press. Whereas in 1955 the amount of front page news devoted to Algeria in newspapers such as *L'Humanité, Le Monde,* and *France-Soir* was in the lower single digits (2–4.3 percent), in 1956 it more than tripled (11.4–15.5 percent).[8]

It is important to note that besides bringing awareness to the war, the fall 1955 and spring 1956 mobilization waves also sent shock-waves through the body of French society. In September 1955, some 400 reservists refused to board a train in the Gare de Lyon train station. In late September, about 300 reservists, and soldiers whose draft was extended, gathered to pray in the Saint-Séverin Church in Paris, quietly protesting the immorality of the war. In early October, about 600 air defense reservists refused to board trucks that were about to transport them to the train station in Rouen. A crowd of civilians and reservists clashed with police forces in the streets for several hours. Lezignan, Mans, Antibes, Saint-Nazaire, and other cities also became sites of clashes between the CRS (police riot-control units) and militants

[7] See also on the relations between the growing salience of the war and the April 11 decision to send conscripts and reservists, in Charles-Robert Ageron, "L'opinion française à travers les sondages," in Rioux, *La guerre d'Algérie et les Français,* 27.
[8] See Paul C. Sorum, *Intellectuals and Decolonization in France* (Chapel Hill: University of North Carolina Press, 1977), 107.

who tried to block the conscripts' trains heading for Marseilles. All in all, in 1955 the mobilization resulted in demonstrations in nine departments. In 1956, between April and June, mobilization resulted in seventy-seven street manifestations in thirty-six departments.[9]

Still, the reverberations the mobilization action created should not be overstated. Overall, the protests and demonstrations of 1955–1956 were of a limited nature and short-lived. The participants were mostly those directly affected by the war – often-times Algerian workers, those called to serve, their families or peers, and Communist militants.[10] French society at large never challenged the right of the state to fight in Algeria, and the number of deserters remained low throughout the war.[11]

At the same time, one should not draw far-reaching conclusions regarding the general French support for the war from the lethargic reaction of the French public to the war and from the very limited reaction to the mobilizations. An initial, more radical reaction to the mobilization was prevented partly by the well-calculated timing of the latter, and partly because the major commitment to Algeria was taken by a Socialist government. Put simply, the constituencies most likely to protest against the war were also those supporting the parties in power. Thus, these constituencies were rather likely to exercise self-restraint or agree to comply with requests to remain relatively idle.

Finally, the French state did not remain idle in the face of its mounting instrumental dependence. A conspicuously discriminatory draft as a way to control the political cost of mobilization was out of the question. Any such *obvious* inequality would have necessarily led to a social and political crisis. Therefore, the draft was in principle national, cutting across all classes and sectors.[12] However, this picture of civic equity must be amended. The officer corps was socially narrow (almost 60 percent sons of officers and civil servants) and, as such, inherently loyal to the state. Moreover, the government had other ways to control the potential political cost of mobilization. Most of the soldiers who went to Algeria were not only relatively young (between 18 and 30 years of age), but also unmarried. In fact, marriage and fatherhood could protect designated conscripts and reservists from the draft, and students, arguably the most problematic constituency to draft, enjoyed initial deferrals.[13] The latter could delay being drafted until the age of 25, and avoid the draft, under some conditions, for another year or two.

[9] Tartakowsky, "Les manifestations," 131–33; Hamon and Rotman, *Les porteurs*, 17–18, 44–45; and Alleg et al., *La guerre d'Algérie*, Vol. 1, 574–75.

[10] Tartakowsky, "Les manifestations," 134.

[11] See Hamon and Rotman, *Les porteurs*, 212–13. The numbers ran from the Left's (editors of *Vérité pour*) 3,000 to the army's 200 "political" deserters.

[12] Massu argues that around 3 million young Frenchmen served in Algeria. See letter to *Le Monde*, March 22, 1972, quoted in Éveno and Planchais, *Dossier et témoignages*, 136.

[13] Rioux, *The Fourth Republic*, 401; and Armand Frémont, "Le contingent," 81.

According to Michel L. Martin, between 1950 and 1960 the percentage of students benefiting from deferments doubled, from 8 percent to 16 percent of those called to serve.[14] Most important, while the government's ability to control the potential political cost of the war through the conscription process was somewhat limited, it could do so by regulating the distribution of the actual war cost through battle assignments. Thus the army established a division of labor that may have reflected primarily operational considerations, but also happened to be politically convenient.[15] Most of the soldiers in Algeria, some 90 percent, served in the *quadrillage* system. They manned a network of static positions, and were generally assigned defensive tasks and basic intelligence gathering. The lion's share of the most dangerous assignments, the offensive operations, was carried out by 8–10 percent of the French military force in Algeria – a body of some 30,000–40,000 highly mobile and aggressive intervention troops, drawn from the 10th and 25th Paratroop Divisions, the 11th Infantry Division, and smaller detachments of naval and air commando units.[16] These soldiers, many of whom were Foreign Legionnaires, carried the brunt of the war. Indeed, the casualties among these forces were disproportionally high. About one-third of the French fatalities in Algeria were due to accidents. Slightly over one-third perished in ambushes or other kinds of harassment. Thirty percent died in offensive acts of war.[17] Finally, the cost of war was also regulated by reliance on a substantial number of indigenous Moslem troops – according to Peter Paret, more than one-quarter of the ground forces involved in the Algerian war.[18]

There is little doubt that the French policy of committing society but regulating the human cost – through loopholes in the draft, selective battle assignments, reliance on proxy troops, and above all, elimination of restrictions on the use of violence – paid off politically. The stronger kind of motivation leading people to take part in disruptive mass politics, a personal stake, was effectively blunted.

The gap between the "national" and "personal" perceptions of the Algerian problem indicated in Table 6.3 illustrates this success. While Algeria was perceived in 1956 and 1957 as the most important *national* problem, economic problems at a *personal* level outweighed it by a large margin.

---

[14] Martin "Conscription and the Decline of the Mass Army," 360.
[15] See Talbott, *The War Without a Name*, 63–64.
[16] d'Abzac-Epezy, "La société militaire," 248–49; and Paret, *French Revolutionary War*, 35. While some of the paratroop units belonged to the Foreign Legion, paratroop units apparently drew up to 70 percent of their manpower from the conscript population (including reservists). See Éveno and Planchais, *Dossier et témoignages*, 154.
[17] Montagnon, *La guerre d'Algérie*, 174. In periods of intense fighting, such as in the eastern provinces in 1958, front-line units suffered up to 50 percent dead and wounded. See ibid., 249, 250. On the role of the Legion in Algeria, see Porch, *The French Foreign Legion*, 565–618.
[18] Paret, *French Revolutionary War*, 40–41.

TABLE 6.3 *National vs. personal perceptions of the Algerian problem*

Question: *"For you and your family personally,* what are the most important problems now?"

	The repercussions of the situation in Algeria[a] (%)	The pecuniary question (%)
April 1956	12	46
July 1956	7	40
September 1957	6	55
January 1958	3	58
February 1959	14	

[a] *"Rappel des disponibles"* and *"enfants sous les drapeaux."*

Question: *"For France,* what is, in your opinion, the most important problem now?"

	The problem of Algeria[a] (%)	Salaries and buying power[b] (%)	Price stability, and economic and financial situation (%)
April 1956	63	2	
July 1956	60	9	
September 1957	51	9	27
January 1958	37	7	36[c]
September 1959	68	14	7
February 1960	78	5	–
April 1961	78	5	–

[a] Definition varied – for example "North Africa" or "Peace in Algeria" – but essentially Algeria was the issue.
[b] Definition varied – for example, "the standard of living and prices."
[c] In January 1958, this category included the balancing of the budget.
Source: *Sondages,* 1957:2, 4–6; 1958:3, 4–5; 1958:2, 27; 1960:3, 39; 1961:1, 8.

Obviously the gradual and cautious mobilization and the military division of labor in Algeria demonstrated the political acumen of the French governments and the army. However, they also reflected some basic insecurity. In their cautious approach, French governments implied that the war was not all that vital, and if it was not that vital, then the "necessity" argument underlying the army's emphasis of "effectiveness" and "efficiency" lost credibility. This point brings us to the longer-term detrimental consequence of the growth of instrumental dependence. The army expected its soldiers to obediently participate in or silently witness deeds that even the state considered as morally wrong in principle. Yet the ultimate size and nature of the French contingent in Algeria promised that a critical mass of soldiers, mostly reservists, would monitor and refuse to tolerate the army's conduct.

We shall discuss the precise nature of the dilemma the French military conduct created for conscripts, as well as the consequences of that dilemma to the war effort, in the next chapter. Let us note here, however, one indication of the results of the meeting of French soldiers with the ugly face of the dirty French war in Algeria as they appear in a 1960 study of the Catholic magazine *La vie catholique illustrée*.[19] *La vie* asked its readers who had served in Algeria to answer several questions regarding their social characteristics and experience in the war. Admittedly this study cannot be considered as representative of the whole population of conscripts in Algeria since those who responded were not selected in a random manner nor could they be a representative sample. Indeed, the study was biased because the majority of those responding were *practicing* Catholics before their departure for Algeria (85 percent). However, the study-population was representative on other parameters: The respondents came from diverse military units, they were of different vocations and military ranks, they were conscripted in various years, and they served in various parts of Algeria for various periods of time. Thus, in spite of the religious bias of the study, it is worth considering its results. The majority of the 607 veterans who answered the questionnaire felt that Algeria was a foreign land. They also had a better opinion of the Muslims than of the *pieds-noirs*. About 14 percent of them described torture as their worst memory from Algeria, and another 7 percent singled out other excesses during the operations of French forces as their worst memory. Together, these memories *slightly surpassed* other bad memories such as the general "hardships of war" (21 percent) and the "death of comrades" (20 percent).

Indeed, many soldiers (and officers) of the French army were deeply affected by what they saw or experienced, and as their experiences offended their innermost values, they reacted with revulsion. In particular, the reserve troops could not ignore their experiences, and thus they created a critical link between the military front and the civilian rear that became ever more obvious once a surge of criticism of the military conduct reached the Continent following the beginning of the 1957 Battle of Algiers. Indeed, some of the most revealing and effective criticism was initiated or based on the stories of reservists. *Témoignage Chrétien* reproduced *Dossier Jean Muller* – an account of the horrors of the army's conduct in Algeria by a former chief of scouts who was killed in action. A group of Catholic humanists published *Les rappelés témoignent* – a collection of seventy-one letters of soldiers describing, in detail and with revulsion, tales of their experience as soldiers

---

[19] Xavier Grall, *La génération du Djebel* (Paris: Cerf, 1962). Some findings are also summarized in Rita Maran, *Torture: The Role of Ideology in the French-Algerian War* (NY: Praeger, 1989), 76–78.

in Algeria.[20] Pierre-Henri Simon published *Contre la torture*, which included a number of letters and testimonies by soldiers that corroborated his claim that the army was acting immorally in Algeria. And upon returning from reserve service in Algeria, Jean-Jacques Servan-Schreiber, the founding editor of *L'Express*, started to publish a series of half fictional and moderately critical stories on his Algerian experience. Soon these stories came out in the book *Lieutenant en Algérie*.

Understandably, then, the French authorities hesitated when they considered whether to depend on full national mobilization in order to fight in Algeria. Circumstances, however, forced them to ignore their doubts, and act with full force and full national commitment in the Algerian battlefield. In the final analysis, this decision promised a military success, but it also opened the road for political failure. The large-scale mobilization made the war the major item on the national agenda, and at the same time provided the critical nucleus of soldiers who were determined to oppose and expose the army's conduct in Algeria. Furthermore, the critical stories of these soldiers had greater news value precisely because of the salience of the war. Thus, while the state was frantically struggling to reconcile conflicting demands – a need to send many more soldiers to Algeria, engage the Algerian insurgents, and avoid casualties – its solutions only managed to alter and delay the cause of protest and domestic impact of the war. Thus, while the government bought relative domestic peace during the first two years of war, it lost precious time and punch-power that could have thrown the Algerian nationalists off balance, at least for a while. Such a strategy could not eliminate the Algerian problem but it could provide the French with an opportunity to negotiate from a position of strength for a compromise that might have protected some of the interests France eventually had to give up in Algeria.

[20] See Hamon and Rotman, *Les porteurs*, 65–67.

7

The Development of a Normative Difference in France, and Its Consequences

The French opposition to the Algerian war was not focused on a single issue, nor was it consistent in content and emphasis throughout the war. Rather, several ideological camps opposed the war, or some particular aspects of the latter, each for its own reasons. The precise disagreements among the factions within what one can define as the anti-war movement were identified, traced, and classified by a few scholars, most notably Raul Girardet and Pierre Vidal-Naquet.[1] For purposes of convenience, and in order to simplify things for readers who do not find it rewarding to delve into the intricacies of political and ideological cleavages, I have altered somewhat the existing classification of the French literature. Here I refer to the terms Rational-Utilitarian, Marxist, and Moralist when I describe the main camps of the anti-war movement.[2]

The Utilitarian Debate about the Necessity of the War

The debate over the French war effort in Algeria never really focused on the worthiness of the war in terms of battle casualties. It is rather problematic to explain the non-occurrence of this issue, as it is the case with other non-events. One can only speculate and suggest that the relative civil indifference to casualties was the result of the fact that most of the French draftees did not run a great risk when they served in Algeria, for reasons that were discussed in Chapter 6. Still, the Algerian war did not escape a utilitarian debate, though the latter centered around economic considerations.

---

[1] Girardet, *L'idée coloniale*, 211–34; and Pierre Vidal-Naquet, *Face à la raison d'état* (Paris: Éditions La Découverte, 1989), 58–63.

[2] My label "Marxist" is for the camps Pierre Vidal-Naquet defined as "Bolshevik" and *Tiers-mondiste*, and Raul Girardet unified under the term "revolutionary." I kept Girardet's label "Moralist" for the camp Vidal-Naquet defined as *Dreyfusard*. My label "Rational-Utilitarian" is for the camps Girardet called "hexagonal fall-back" and "national grandeur" and others called "Cartierist."

At first, the necessity of fighting in Algeria was debated as part of a general discourse about the economic rationality of maintaining colonies. This debate opened in August-September 1956 in a series of articles by the journalist Raymond Cartier in the weekly *Paris-Match*. Cartier, originally a firm supporter of the French empire, became convinced, following a visit to Africa, that colonialism was a "costly philanthropy." He thus concluded that the cost of maintaining the French colonies outweighed the benefits they generated. They kept France from developing economically, and their eventual emancipation seemed all but certain. This kind of thinking, which the French political scientist Nathalie Ruz aptly defined as "utilitarian anti-colonialism," was quite revolutionary at the time, and with few exceptions, it was either ignored or attacked.[3]

However, in 1957, the utilitarian economic argument registered support from the most important conservative intellectual in France – Raymond Aron. Aron, in what the historian (and staunch supporter of French-Algeria) Raul Girardet described as "preoccupation with the empirical," and "will to approach the debate beyond all ideological or sentimental a priori," courageously analyzed the French stake in Algeria.[4] Aron argued that on the one hand the war split the French national consciousness, threatened traditional French liberties, prevented modernization, and therefore weakened France's position in Europe and in the world. On the other hand, he concluded, the alternative to war – disengagement – did not threaten France's well-being, moral fortitude, or status in the world. Thus, a comparison along these lines, of the cost associated with keeping Algeria French and that of giving it up, convinced Aron that disengagement was the right policy choice for France.

Analysis of the economic relations between Algeria and France leaves little doubt whether the Rational-Utilitarian camp was right. The Algerian economy was but a fraction of that of continental France and, as such, of little value as an export market. Furthermore, Algeria was a competitor of certain French sectors and regions, as its export of cheap wine to France suggested. Even worse, it was a net consumer of continental capital. In 1960, for example, two-thirds of Algeria's imports from France were paid for by France. In fact, oil and natural gas were the only promising commodities Algeria could offer France.[5]

Indeed, utilitarian arguments were particularly well accepted by business circles in France. In 1956, the journalist Jean Daniel pointed out in *L'Express*

---

[3] See Nathalie Ruz, "La force du 'cartiérisme,' " in Rioux, *La guerre d'Algérie et les Français*, 328–36; Girardet, *L'idée coloniale*, 228–30; and Sorum, *Intellectuals*, 201–04.

[4] Girardet, *L'idée coloniale*, 231. Aron's argument appeared in *La tragédie algérienne* (Paris: Plon, 1957), and *L'Algérie et la république* (Paris: Plon, 1958). See also Marie-Christine Granjon, "Raymond Aron, Jean Paul Sartre et le conflict algérien," in Rioux and Sirinelli, *La guerre d'Algérie et les intellectuels français*, 115–38.

[5] See Asselain, "Boulet colonial," 298–300; Jacques Marseille, "La guerre a-t'elle eu lieu? Mythes et réalités du fardeau algérien," in Rioux, *La guerre d'Algérie et les Français*, 285–86.

that the French private sector would "prefer to lend to the Algerian state with interest, rather than invest capital doomed to loss."[6] In September 1957, this position became semi-official when an "unofficial" report of the business community, which was circulated among members of the National Assembly, referred to the idea of economic integration with Algeria as suicide.[7] By June 1959, Michel Debré, explaining why Algeria was not a colony, all but confirmed in parliament that from an economic point of view, Algeria was a bad bargain.[8]

Cartierist ideas reached a large audience if only because *Paris-Match* had a large circulation, but it is not clear how many were convinced by such arguments. Nevertheless, some people clearly came to Cartierist conclusions. For example, in October and November 1957, Serge Adour (under the pseudonym Gérard Belorgey), a reserve officer returning from Algeria, attacked the war in utilitarian terms in a series of articles in *Le Monde*.[9] In fact, toward the end of the war, for reasons not utterly clear, the utilitarian argument was propagated by de Gaulle himself. In April 1961, for example, he declared that "decolonization is in our interest and therefore our policy. Why should we remain caught up in colonizations that are *costly, bloody, and without end, when our country needs to be renewed from top to bottom?*"[10] One has no way of knowing whether de Gaulle adopted utilitarian Cartierist conclusions on his own, or whether he thought they offered him a good means for selling the abandonment of Algeria in France. The important thing is that, for whatever reasons, such ideas still made their way to the top somehow.

Overall, however, the utilitarian argument was shared by a relatively small group of people – mostly members of the business community, journalists, and a few politicians. France, as a state, did not fight, nor did it abandon Algeria, because of economic reasons. Algeria was a net consumer of French capital, but it did not constitute the kind of economic burden that prevented France from developing. At least not in the post-1958 years, as the economic data presented in Chapter 5 suggests.

The Debate about the Morality of the Conduct of the Military in Algeria

As I noted, the wish to balance conflicting battlefield and domestic demands, as well as the nature of the Algerian war, forced the French to emphasize counterinsurgency methods, torture in particular, which were in sharp opposition to the values of a significant portion of French soldiers and citizens.

---

[6] Ruz, "La force du 'cartiérisme,' " 333.
[7] Ibid., 333–34.
[8] Ageron, "L'Algérie, dernière chance," 132.
[9] See Talbott, *The War Without a Name*, 74, 262 note 58.
[10] Horne, *A Savage War*, 444 (italics added).

Indeed, one who wants to get to the bottom of the Algerian puzzle and understand why the war eventually failed must discuss the most divisive issue of the Algerian controversy: the debate about the ethics and legitimacy of the army's conduct in Algeria.

In fact, it is most proper to start the discussion by observing that the ethical questions that were raised in Algeria regarding proper conduct in war were hardly new for the French. The generation that controlled the political system of the Fourth Republic was largely composed of veterans of the Resistance. These veterans, along with the rest of France, had experienced first-hand, the evil involved in counterinsurgency war under the Nazi occupation.[11] At least a few members of this generation were also familiar with the 1949 work of the French lawyer Alec Mellor on the reappearance of torture in modern France, and with revelations on the use of torture during the Indochina war.[12] In the early 1950s, it became clear that the ethics guiding the police operation in North Africa were of questionable nature, and a small group of intellectuals started to pay attention to the conduct of the army and the police there. In December 1951, Claude Bourdet asked, for the first time publicly, whether a French Gestapo was operating in Algeria. In 1953, François Mauriac, Louis Massignon, and others decided to organize in order to defend the law and oppose the repression methods in North Africa. On the one hand, the practice of torture by the French was already endemic in North Africa, and the nature of the war and the political dilemma of mobilization in France led to abuses of power in the battlefield. On the other hand, when the war broke out, the ground for moral dissent was fertile, and thus the seeds of criticism of brutal conduct could sprout, strive, and even prosper.

Thus the expected surge of French abuses of power in Algeria, following the start of the war, immediately generated revelations concerning the ethics of conduct of the French army and police, and almost simultaneously drew sharp criticism. This initial reaction was evident among leading intellectuals and in the left-of-center press. In January 1955, Claude Bourdet, in a *France Observateur* article entitled "Your Gestapo of Algeria," answered the question he posed to his fellow French citizens in 1951. At the same time, François Mauriac started a series of three articles on torture in *L'Express*. The first of these articles appeared under the title "*La question.*"[13] In early February, an Algerian member of the National Assembly raised allegations of torture in parliament, and Mitterrand, the interior minister, all but confirmed the validity of the complaint.[14] These revelations induced Mendès-France and

[11] Rioux, *The Fourth Republic*, 406–07.
[12] See Maran, *Torture*, 5. The use of torture in Indochina was discussed in an article by Jacques Chegary, in *Témoignage Chrétien* in July 1949. See Planchais, *Une histoire politique*, 302–03.
[13] See Hamon and Rotman, *Les porteurs*, 25–26; and Jean-François Sirinelli, "Guerre d'Algérie, guerre des pétitions?," in Rioux and Sirinelli *La guerre d'Algérie et les intellectuels*, 276, 300 note 30.
[14] Lacouture, *Mendès*, 333.

Mitterrand to commission Roger Wuillaume, a high-ranking colonial official, to compile a report on the alleged use of torture in Algeria. In March
1955, Wuillaume delivered his report for limited government consumption.
The report not only acknowledged the use of torture in Algeria, but actually recommended sanctioning it because, on the one hand, Wuillaume
decided it was effective and indispensable, and on the other, concluded that
it could not be concealed. In fact, Wuillaume's only reservation was that torture must be exercised under "controlled conditions" – that is, its practice
should be left only in the hands of professionals so as to prevent "abuse."[15]
In essence, then, after Wuillaume had discovered that torture was widespread
in Algeria, instead of condemning it as unacceptable, he recommended it be
formally institutionalized. Toward the end of 1955, *L'Express*, apparently
trying to embarrass Faure's government before the elections, published pictures of a summary execution in Algeria. In April 1956, the Army's conduct
was condemned again in the press, this time by the historian Henri Marrou
in *Le Monde*.[16] In May 1956, Mendès-France resigned from Mollet's four-
months-old government because of a dispute over the Algerian policy. In his
resignation letter, which immediately became public, he indicted the French
repression policy, predicting that it would cost France Algeria.[17]

   All of the developments I have reviewed occurred within the intellectual
elite and among members of the higher bureaucracy. That, however, does
not reflect fully, or accurately, the scope of the revulsion within France toward the military conduct in Algeria. At least on a small scale, wider social
spheres developed a similar attitude. Catholic and Communist conscripts,
for example, gathered in September 1955 in the Saint-Séverin church to
protest the war. "Our conscience," they declared, "tells us that this war . . . is
a war opposed to all Christian principles, to all the principles of the French
constitution . . . to all the values of a civilization in which our country rightly
takes pride."[18] In fact, in a few notable cases, the moral dilemma that the
army created pushed soldiers to extremes. In August 1956, for example,
a paratrooper sergeant named Noel Favrelière deserted from the French
army, taking with him an Algerian prisoner whom he saved from a summary
execution.[19]

---

[15] The report is reprinted in Pierre Vidal-Naquet, *La raison d'état* (Paris: Les Éditions de Minuit,
    1962), 57–68. See also the summary in Maran, *Torture*, 45–50.
[16] See Éveno and Planchais, *Dossier et témoignages*, 97–101; and Sorum, *Intellectuals*, 113–14.
[17] Lacouture, *Mendès*, 365–66.
[18] Quoted in Patrick Rotman and Yves Rouseau, "La résistance française à la guerre d'Algérie,"
    Doctorat du Troisième Cycle en Sciences Politiques (Université de Paris 8, 1981), 10; and
    Hamon and Rotman, *Les porteurs*, 17.
[19] Or, for example, Lieutenant Jean Le-Meur refused to go on serving in a combat position after
    his commander had instructed his unit "not to take prisoners." See David L. Schalk, "Péché
    organisé par mon pays: Catholic antiwar engagement in France, 1954–62," *The Tocqueville
    Review* 8 (1986/87), 89 note 38.

In spite of these significant publications and reactions, the army's conduct in Algeria did not become a central and highly inflammable topic on the public agenda until the 1957 Battle of Algiers, which became a watershed in the debate about the war. Several factors combined to make this battle so significant. First, Algiers by definition attracted the media's limelight simply because it was the capital of Algeria. Second, the battle in the capital was perceived by both the FLN and the French as critical, and therefore both, trying to project an image to their liking, actively sought media attention through spectacular "successes." Third, a number of attributes of Algiers and of the French forces in the capital operated in synergy to produce a particularly dirty war, which by virtue of being sensational, attracted even more media attention. Thus the sheer size and urban nature of the city required great amounts of fresh intelligence, and that in and of itself made the resort to torture particularly appealing. In addition, the emphasis of the French command on "necessity" and "efficiency," above all else, further encouraged this tendency. Finally, the forces assigned to Algiers – the paratrooper and Foreign Legion units – were those with the least affinity for civic values. As should have been expected, the combination of these factors inevitably produced a critical volume of abuses.[20] Fourth, and most important, the composition of the population in the capital, in conjunction with the scale of the anti-terrorist campaign and the indiscrimination involved, promised that some of the abuse would be inflicted on conspicuous and politically unpalatable targets. In other words, a fraction of the brutality was certain to be inflicted in outrageous ways on people who were "too visible." Such cases included the "suicides" in detention of Larbi Ben M'hidi, *a member of the FLN executive committee*, and Ali Boumendjel, an Algerian *lawyer and graduate of a French law school*; the "disappearance" of the *pied-noir mathematician* Maurice Audin; and the torture and death sentence (that was not carried out) of Djamila Bouhired, a female member of the FLN.

That 1957 was a watershed in terms of public awareness of the mushrooming moral dilemma in Algeria was marked by the tide of press articles and publications dealing with torture, other infringements on civil rights, and the ethics of military conduct in war. The most notable among these publications were *Dossier Jean Muller* in *Témoignage Chrétien*, *Les rappelés témoignent*, Pierre-Henri Simon's *Contre la torture,* and Servan-Schreiber's half-fictional stories in *L'Express*, which were soon assembled in a book under the title *Lieutenant en Algérie*. Moreover, once the issue of inappropriate and criminal military conduct took center stage, it was not going to

---

[20] Paul Teitgen, the police commissioner of Algiers at the time, assesses that more than 3,000 people "disappeared" during the Battle of Algiers. See Alain Maillard de La Morandais, "De la colonisation à la torture," Ph.D. thesis (Université de Paris-Sorbonne, 1983), 397, 604–05; and Vidal-Naquet, *Face,* 21.

disappear. Indeed, other reports, testimonies, and criticisms of the army's conduct continued to be published after 1957. Among the most famous of these were Pierre Vidal-Naquet's *L'Affair Audin*, on the murder of Maurice Audin; Henri Alleg's *La question* (1958) on his own experience of torture at the hands of the paratroopers; and *La gangrène* (1959), an account of Algerian students on their torture at the hands of the internal security service (DST) in Paris.[21]

The Battle of Algiers resulted in another detrimental moral development. It brought into the open the fact that the army's conduct in Algeria was also creating moral resignation at almost all levels of the administration. It was already clear, before 1957, that disapproval of the brutal conduct in Algeria was building up in high circles. But this development remained relatively inconspicuous. In June 1955, Soustelle's *chef du cabinet militaire*, Vincent Mansour Monteil, resigned because of his disagreement over the methods and meaning of the repressive policy in Algeria. Following the October 1956 interception of Ben-Bella and other FLN leaders, both the Secretary of State for Tunisian and Moroccan Affairs (Alain Savary) and the French ambassador to Tunisia also resigned.[22] At the same time, the Director General of the French police, Jean Mairey, who held the position between July 1954 and August 1957, compiled several confidential reports that criticized, in no uncertain terms, police conduct in Algeria. For example, Mairey summed up his December 1955 report by declaring: "[As] the chief in charge of the National Police Force, it is intolerable for me to think that French police officers can evoke by their behavior the methods of the Gestapo. Likewise, [being] a reserve officer, I cannot bear to see French soldiers compared to the sinister SS of the Wehrmacht."[23]

Much as in the case of public attention, and probably as a result not only of the intensity and the scale of brutality involved in the battle of Algiers, but also of the mutual influence between the public and administrative realms, the dissonance and the pressures at the executive level spilled over, and became common knowledge in 1957.[24] Of all such indications, the most notable involved General Jacques Pâris de Bollardière, who resigned from his command in Algeria in March of that year.

[21] Other publications included Colonel Roger Barberot's *Malaventure en Algérie avec le General Paris de Bôllardiére* (1957); Joseph Vialatoox, *La repression et la torture*; Georges Arnaud and Jacques Vergès, *Pour Djamila Bouhired* (1957); Francis and Colette Jeanson, *L'Algérie hors-la-loi* (1957); and Djamal Amrani's, *Le témoin*, (1960). Alleg's *La question* sold some 65,000 copies before it was seized, and altogether some 90,000 copies. See Rotman and Rouseau, *La résistance française*, 137–38 note 1; Vidal-Naquet, *Face*, 65 note 54; and Claude Liauzu, "Intellectuels du Tiers Monde et intellectuels français: Les années algériennes des Éditions Maspero," in Rioux and Sirinelli *La guerre d'Algérie et les intellectuels*, 172.
[22] See Éveno and Planchais, *Dossier et témoignages*, 107–08.
[23] Quoted in Hamon and Rotman, *Les porteurs*, 72–73; and Vidal-Naquet, *La raison d'état*, 89.
[24] See also Rioux, *The Fourth Republic*, 284, 492 note 40.

General Bollardière, who operated under Massu's direct command, was in many respects a model officer. However, he opposed Massu's battle philosophy, which emphasized "instant efficiency [that] overrides all principles and all scruples," because he felt it contradicted "the sacred principles of [Western] civilization."[25] In fact, he had already instructed his soldiers that the "temptation, which the totalitarian countries did not resist, of considering certain procedures as normal methods of obtaining intelligence, must be rejected unequivocally, and *formally condemned*."[26] Thus, Bollardière preferred to put the emphasis of his operation on targeting the hearts and minds of the Algerians in his sector rather than their bodies. In any event, Massu opposed Bollardière's independent approach, and instructed him to act "efficiently" and assign priority to police work. Bollardière, however, did not budge. Rather, he made clear to Massu and to the supreme commander of the French forces in Algeria, General Raoul Salan, that he would not accept their methods. Instead, he asked to be immediately relieved of his responsibilities and put at the disposal of the command *in France*.[27] Upon returning to France, Bollardière, true to his conscience, decided to make his plight public. In a letter to *L'Express* under the title, "In the Name of Respect for Human Beings and Our Civilization," he articulated his moral objection to the methods used in Algeria.[28]

Following Bollardière, several other key officials also "defected" on moral grounds. In September 1957, Paul Teitgen, the police commissioner of Algiers and a former victim of torture at the hands of the Gestapo, resigned because of his inability to end the use of torture in Algiers.[29] In October 1957, retired General Pierre Billotte, the former Minister of Defense in Edgar Faure's government, followed suit. In a letter to Preuves (which *Le Monde* reproduced), General Billotte explained: "On the subject of torture, I am unequivocal: In whatever form, for whatever purpose, it is unacceptable, inadmissible, and to be condemned; it casts a slur on the honor of the army and the country."[30] And for those who argued that the ethics of conduct come second to the ethics of responsibility, he added: "The ideological character of modern war changes nothing in this... a commanding officer must not

[25] Quoted in Maran, *Torture*, 110, 109.
[26] Quoted in Massu, *La vraie bataille*, 222 (italics added).
[27] Ibid., 223–24.
[28] *L'Express* was edited by Servan-Schreiber, who served under Bollardière's command in Algeria and made him the thinly veiled hero of his articles and then his book, *Lieutenant en Algérie*. See also on Bollardière in Maran, *Torture*, 106–17.
[29] Teitgen warned Mollet several times about the army's actions before he resigned. In a March 1957 letter of resignation he sent Lacoste (which was rejected), he warned that the sinister nature of the army's operation in Algiers brought it ever closer to committing war crimes. The letter was published in *Le Monde* on October 1, 1960, after it was presented as evidence in the trial of the Jeanson network. Teitgen was expelled from Algeria in May 1958. See Éveno and Planchais, *Dossier et témoignages*, 144–47.
[30] Quoted in Paret, *French Revolutionary Warfare*, 73–74.

hesitate to expose his men, and even the population under his protection, to greater danger rather than make use of a dishonorable practice."[31]

Meanwhile, in the fall of 1957, three out of twelve members of a public committee the Mollet government created in May 1957 in order to study the state of individual rights in Algeria – Delavignette, Garçon, and Pierre-Gérard – resigned successively because they were frustrated with the inherent weakness of the committee.[32] And in December, *Le Monde* published a report that Garçon had compiled in June, which once again confirmed the use of torture in Algeria.[33] Needless to say, the "defection" of high ranking officials such as Bollardière, Billotte, Teitgen, Mairey, Delavignette, and Garçon were of particular significance. Their revelations and criticism could not be dismissed as the product of politically motivated agitation, defeatism, or opportunism. These were state agents, highly placed, with good access to military information. Moreover, their actions were necessarily associated with a readiness to suffer severe personal consequences – a guaranteed setback to their careers and benefits, certain condemnation from peers, and the probable end of some life-long friendships.

To this important group of critics, one can add two other groups of an almost equal importance whose moral criticism had particular weight: former state servants of significant public and/or national standing, and clerics who enjoyed inherent spiritual authority. Among members of the first group one should note such famous Resistance and World War II heroes as "Vercors" (Jean Bruller) and Jules Roy, former liaison to the intellectual community and the press of premier Mendès-France; Jean-Jacques Servan-Schreiber; and the former mayor of Algiers and state secretary for war in Mendès's government, Jacques Chevallier.

Jules Roy, who was a *pied-noir* by origin, served as a Free French pilot in the RAF, and resigned from the military in 1953 as an air force colonel during the Indochina war because he rejected the ends and methods of the counter-revolutionary war practised there, made his criticism of the military conduct in Algeria public, including two books: *La guerre d'Algérie* (1960) and *Au tour du drame* (1961). Servan-Schreiber, after a tour of duty as a junior reserve officer in Algeria, opened the pages of *L'Express* for a rather unprejudiced coverage of the war, including the reporting of his own wartime impressions of the realities of the Algerian battlefield.[34] Vercors, who returned his *Legion d'honneur* to President Coty in protest over the military conduct in Algeria, called for clandestine anti-war activity, and even supported the movement *Jeune Résistance*, which provided logistical support to draft dodgers and army

[31] Ibid. See Massu's reaction in *La vraie bataille*, 228–34.
[32] See Vidal-Naquet, *Face*, 18; and Alleg et al., *La guerre d'Algérie*, Vol. 2, 515–17.
[33] See Vidal-Naquet, *La raison d'état*, 129–67.
[34] Pierre Leulliette also wrote an account of his Algerian service, which included tales of French brutality, in his 1961 book, *St. Michel et le dragon*.

deserters. And Chevallier, who also happened to be a *pied-noir*, was initially not pleased with the government's "too timid" response to the rebellion, but eventually became one of the toughest critics of the French repression.[35]

The second group included clerics, both in and out of uniform. In Algeria, the Archbishop of Algiers himself, Léon-Etienne Duval, and several other clergymen such as Father Jean Scotto were propagating the rather unpopular idea that the war was based on unjust exploitation and was being conducted in an unjust manner. Whereas in continental France, the leading Catholic establishment, and its leadership in particular, remained largely indifferent to the realities of French conduct in the war, among the rank and file moral issues were raised and debated. For example, the conservative Catholic newspaper *La Croix*, with an average circulation of some 100,000–150,000, opened its pages to critical articles by Jacques Duquesne.[36] Moreover, as priests served in the reserve, often as regular soldiers, they could not avoid seeing, and rejecting, the brutal behavior of French troops. When Massu managed to mobilize the support of his brigade's chaplain for torture, shock waves went through the body of chaplains in Algeria.[37] Catholic disapproval of the military conduct increased again in 1959. On April 9, *Témoignage Chrétien* published a letter from thirty-five priests, who asked their superiors – the highest religious authorities in France – for moral guidance. In the letter, the priests described what disturbed them during their military service in Algeria in the following way:

From a comparison of our experiences, it emerges broadly that methods are used in the conduct of war that our consciences condemn... arbitrary arrests and detention... torture. Summary executions of prisoners, both civil and military, ordered by judicial authorities, but concealed on the plea of 'attempted flight'... are not exceptional. Finally, it is not unusual during operations for the wounded to be finished off...[38]

At the same time that this letter was conceived, the office of the army's Catholic chaplains issued a document entitled *Study of Moral Behavior in a Subversive War,*[39] which marked the departure of the office from the army's, and particularly Massu's, justification of torture on "necessity" and "efficiency" grounds. Instead, the document characterized the practice of torture as inherently wicked and corrupting. Parts of the document became public as they were released to the press.

---

[35] See *L'année politique* (Paris: Presses Universitaires de France, 1956), 233; and Montagnon, *Guerre d'Algérie*, 210–11.

[36] See André Nozière, *Algérie: Les chrétiens dans la guerre* (Paris: CANA, 1979), particularly pp. 47–77, 135.

[37] See discussion in Chapter 8 following note 19, and also see note 20.

[38] Quoted in Paret, *French Revolutionary Warfare*, 67–68; see also Nozière, *Algérie: Les chrétiens,* 135–37.

[39] Paret, *French Revolutionary Warfare*, 67; and Nozière, *Algérie: Les chrétiens,* 138–41.

In retrospect, it seems clear that the readiness of highly placed French officials to criticize the military conduct on ethical grounds made the public more attentive to the moral campaign, and gave the latter much credibility and added strength. The contribution of the two other groups was as significant, because, in the eyes of many, Resistance veterans and clergymen possessed an inherent moral authority. Moreover, the moral agenda found support among the soldiers, both because they experienced or witnessed military conduct that contradicted their values and because the aforementioned figures of authority forced them to think over the legitimacy of the "efficiency" and "effectiveness" justification that the Army promoted.

The proof of the "destructiveness" of these processes (as far as the state was concerned) is in the action of individual soldiers who conveyed the tales of misconduct to the activists who upheld the moral agenda on the continent. But the proof is also in the forceful reactions of the state and the army's command to the Moral problem among the troops. As most of these reactions are discussed in detail in Chapter 8, I shall confine myself here only to the observation that the army's reactions clearly indicated that it feared that the inherent "liberal creed" of its reserve officers and the general discourse on torture within civil society would affect its soldiers' performance. Indeed, such considerations at least partly underpinned the existential rationale army officers developed in order to justify the resort to torture. We also know that toward the end of the Battle of Algiers, the army concluded that the political ripples created by the resort to torture were getting out of control – that is, the command assessed that torture created too much outrage among the soldiers and, consequently, in continental France. In short, and as we shall observe, the army perceived the moral problem among its ranks as significant enough to justify a rather strong cocktail of counter measures.

A last point to note about the moral debate is its genuine nature – that is, it being largely independent of expediency, political or otherwise, although many of those who promoted the Moralist agenda belonged to the political Left. Indeed, many members of the Moralist camp did not ignore the immorality of the FLN conduct. The Catholic intellectual Domenach belonged to the moderate Left, but he had no anti-etatist bias. While he bitterly opposed the army's conduct, and consequently the war, he also observed the FLN's savage conduct with great trepidation.[40] It is also interesting to note that many of those who initially objected to the army's methods, and became opponents of the war by 1960, still maintained a distance from the extreme actions and rhetoric of the radicals within the anti-war movement. Indeed, among prominent Catholic intellectuals, only Barrat and Mandouze signed the "121 manifesto." Other intellectuals who opposed the state's

---

[40] See Jean-Marie Domenach, "Commentaires sur l'article de David L. Schalk," *The Tocqueville Review*, 8 (1986/87), 94; and "Un souvenir très triste," in Rioux and Sirinelli *La guerre d'Algérie et les intellectuels*, 353–57; and Vidal-Naquet, *Face*, 14.

conduct and the war refused to sign the vehemently anti-war manifesto, precisely because of the strong anti-etatist and pro-FLN message it propagated.[41]

In fact, the "purity" of the moral argument was tested and vindicated several times when the interests of Moralists conflicted with their principles. For example, in March 1960, *Esprit* declared that in case Soustelle – the most effective figure to promote the idea of French Algeria – were to be incarcerated, it would protest such action. Similarly, *Esprit* and the members of the Audin committee denounced the torture of OAS (*Organisation Armée Secrète*) members, although the OAS constituted a major threat to the physical well-being of anti-war intellectuals and organizations, including their own.[42] Indeed, because of its genuine nature, the moral debate managed to pull together unlikely political partners: Catholics, Communists and even a few conservatives. Moreover, the moral questions raised in Algeria were apparently so disturbing that people who were not committed to Algerian independence, did not oppose a policy of integration, or even supported the concept of French Algeria in the beginning found themselves in opposition to the state and the army as far as the conduct of the latter was concerned. Neither Mauriac, who condemned the methods of repression from early on, nor Pierre-Henri Simon, who ignited the moral debate in 1957 with *Contre la torture*, were originally partisans of Algerian independence. Jean-Jacques Servan-Schreiber was a Mendèsist, and like Mendès-France himself opposed the methods of repression because he thought that the brutality involved doomed the prospects of Algerian integration. Edmond Michelet and Robert Delavignette both turned against the use of torture and other abuses of power, and Michelet, while serving as justice minister under Debré, even sided with the Moralist camp to such an extent that he was forced out of the government.[43] Yet they were both among the founding members of the "Union for the Salvation and the Renewal of French Algeria" (USRAF) in 1956.

The Debate about the Identity of the State

It is obvious that the objections to the military conduct in Algeria were deeply connected to the concern that the ultimate consequence of events

---

[41] See Étienne Fouilloux, "Intellectuels catholiques et guerre d'Algérie (1954–1962)," in Rioux and Sirinelli *La guerre d'Algérie et les intellectuels*, 105. Many, however – including Daniel Mayer (*La ligue des droits de l'homme*), Pierre Gaudez (National Union of French Students – UNEF), Jean Dresch, Paul Ricoeur, Jean-Marie Domenach, Jean Effel, and Jacques Prévert – signed another anti-war declaration that did not call for disobedience. See Jean-François Sirinelli, "Les intellectuels dans la mêlée," in Rioux *La guerre d'Algérie et les Français*, 128.

[42] See Schalk, "Péché organisé," 76. An *Esprit* article and testimonies that were published in May 1962 are reprinted in Vidal-Naquet, *Face*, 170–86.

[43] See Joseph Rovan, "Témoignage sur Edmond Michelet, garde des Sceaux," in Rioux, *La guerre d'Algérie et les Français*, 276–78.

in Algeria would be the irreversible distortion of France's identity and the French political order. In the final analysis, the protests against the military conduct in Algeria were not only expressions of universal humanist values, but also reflections of an honest national wish to save the identity of France, the nation that had aspired, from the time of the Revolution, to lead the world into better times.

Thus, while both Marxists and Moralists were nurtured by ideologies that rejected the military conduct in Algeria for different reasons, and associated it with a threat to their own preferable domestic order, many of them felt proud being French, and as such cared much about their country's identity. Robert Bonnaud, a Communist militant from Marseilles who published the first reservist's book that criticized the army's conduct in Algeria, explained the complexity he and his friends encountered as a result of harboring Marxist, liberal, and national values, at the same time:

As intellectuals, at the time, three problems engaged [our minds] a lot. First, as French intellectuals convinced of the *injustice of the colonial situation* and the legitimacy of the struggle of the Algerians, *the problem of the national belonging*, of national solidarity (the solidarity with the French of Algeria in this instance). *The national values, are they absolute? Should they come first, before all other [values], under all circumstances? Should national belonging prevail always?*[44]

Of the two groups, the Moralists and the Marxists, the former focused more on the state's identity than did the latter. For the Moralists, the issue boiled down to how they could save the state, which tolerated if not actively encouraged what they considered as behavioral aberrations, from itself. Indeed, it is this dark cloud of immorality hovering over the identity of the French state that Pierre Henri-Simon raised in March 1957 in *Contre la torture* and that Philosopher Paul Ricoeur discussed again, a year later, in *Esprit.*[45] Abbot de Cosse-Brisac, a descendent of an old and distinguished military family, explained the very same point when he wrote in *L'Express* in July 1960: "We are responsible, as Frenchmen, for the torture whose use is spreading like gangrene. It is the very archetype of collective sin. We are all torturers if by a collective silence we allow it to happen..."[46] And Jérôme Lindon, whose publishing house, *Minuit*, published many of the most damning books on the army's and state's conduct, explained the point in no uncertain terms: "[W]hat I may have done, I have done for France, not for Algeria."[47]

---

[44] Quoted in Rioux and Sirinelli *La guerre d'Algérie et les intellectuels*, 350–51 (italics added).
[45] La Morandais, *De la colonisation*, 825; and Schalk, "Péché organisé," 82.
[46] Quoted in Paret, *French Revolutionary Warfare*, 74–75. See also ibid., 150 note 34.
[47] Quoted in Vidal-Naquet, *Face*, 59.

8

The French Struggle to Contain the Growth of the Normative Gap, and the Rise of the "Democratic Agenda"

Presumably the French state started the war in Algeria from an advantageous position in the marketplace of ideas. Algeria was a part of sovereign France, and therefore the war could be convincingly portrayed as involving core national security interests. Furthermore, the war could be framed in the larger context of the Cold War, and thus be presented as a part of the effort to defend the free world from "the Communist threat." Finally, FLN savagery provided a moral justification of the war and the resort to "extreme" measures. For the last two reasons, France could also present the war not only as a matter of national security, but also as embodying a struggle between the forces of light against those of darkness.[1]

At the same time, however, the Algerian situation also involved some complexities that made the marketing of the war less easy and the state less enthusiastic to engage in such an endeavor in the marketplace of ideas. First, although legally part of sovereign France, Algeria nevertheless remained somewhat remote from continental consciousness.[2] Second, the war in Algeria had to be fought in the shadow of a long and disastrous involvement in Indochina (which eventually was ill-received by the French) and thus not likely to create much enthusiasm or willingness to sacrifice among residents of continental France. Moreover, the temptation to play down the situation in Algeria was further strengthened by the fact that initial FLN activity was somewhat sporadic and therefore seemed to require only a limited investment. This, in turn, had several advantages. It gave the government the opportunity to rely on existing conscription and professional soldiers

---

[1] French residents of Algeria, including Catholic clergymen, propagated such ideas very forcefully. See Nozière, *Algérie: Les Chrétiens*, 192–216, particularly p. 193.

[2] In early 1958, when the war was already a salient public issue, over 40 percent of the French did not know how many *pieds-noirs* lived in Algeria or what was the overall size of the Algerian population. Only about one-third estimated these numbers within a margin of 20–50 percent error. See *Sondages*, 1958:3, 39.

and permitted the army to relax the rules of engagement and thereby also increase military effectiveness. Consequently, the need to send in reinforcements was reduced, and the army could control the information coming out of Algeria rather well. In short, the limited investment prevented the war from becoming a major source of controversy in France.

Indeed, the fears of the political consequences of either too timid or too vigorous a reaction were well reflected in the initial declarative position of the government and its members. In an effort to keep the actual involvement of the continental citizenry minimal and avoid the pitfall of being blamed for impotence, French politicians vowed to preserve Algeria and repress the uprising. At the same time, they also played down the gravity of the situation and described the army's task as "maintaining" order and peace in Algeria rather than fighting a war. For about a year, this duplicity helped French governments keep their citizens at a safe distance from the war and themselves away from a serious battle in the marketplace of ideas. However, the scope of FLN activity and consequent army requests for reinforcements convinced the government of Edgar Faure, and even more so that of Guy Mollet, to abandon the cautious muddle-through policy, nationalize the war, and flood Algeria with conscripts. The President, René Coty, could passionately preach from Verdun, the powerful symbol of national devotion and sacrifice:

There [in Algeria], the fatherland is in danger. The fatherland struggles. The duty [of the soldiers] is simple and clear. Those who are not subject to the military discipline, must at least submit to the civic discipline that forbids any act, and even any utterance liable to upset the spirit of the children of the fatherland whom the Republic calls to arms. . . .[3]

This final acknowledgment that Algeria necessitated a collective effort may not have thrilled the French, but neither was it opposed by most of them. As I have noted, several basic factors – such as historical ties, national pride, vested interests of the army, and the substantial political power of the vocal "Algerian" lobby – made sure that a war would be depicted in the public's mind as a national imperative. At least, in this respect, the government enjoyed some extended period of grace before the war started to attract some "undesired" attention, and consequently called for an ever-growing governmental effort to maintain and defend the legitimacy of the war.

The Domestic Reaction of the Government and the State

We have already noted that the 1957 Battle of Algiers was one of the most critical turning points of the Algerian war. While the forces of Massu crushed the FLN network in the capital, inadvertently they also made the issue of the ethics of military conduct in Algeria the linchpin of the struggle against

---

[3] June 17, 1956. Quoted in *L'année politique*, 1956, 63.

the war. Whereas the 1956 mobilization made Algeria the major issue in the continental marketplace of ideas, the Battle of Algiers turned this marketplace into a defining battleground. By that time, however, the presence of the French government in the marketplace of ideas was already established. Indeed, the French administration proved, right from the start of the war, that it had little patience for critics of the war and the army's conduct in Algeria. Journalists and other critics of the army or the state discovered that criticism soon ended in harassment.[4] Invariably such critics were accused of "demoralizing the army" or "damaging the security of the state." Journalist Robert Barrat (formerly the secretary general of the Catholic Center of French Intellectuals) was arrested for publishing in *France-observateur* an account of a meeting with FLN leaders and warriors as early as September 1955. In March 1956, the DST visited the residence of five members of the editorial committee of *France-observateur* and detained the editor, Claude Bourdet, for a day in retaliation for an article that opposed the dispatch of more troops to Algeria. Julien Rouzier, the director of the Communist daily *L'Écho du Centre* was sentenced to a year in prison for "undertaking to demoralize the army," and his newspaper was suspended for fifteen days. In May 1956, Claude Gérard was arrested briefly for similar deeds. In November, it was André Mandouze's turn (co-founder of *Témoignage Chrétien*). He was detained for five weeks. In early April 1956, the DST ransacked Henri Marrou's apartment after he criticized, in *Le Monde*, the resort to torture and collective punishment and the creation of concentration camps in Algeria.[5] Houses of contributors to *Le Monde* were "searched" as a way to threaten them and thereby the economic stability of the newspaper. State censorship of radio broadcasts increased.[6]

These rather resolute reactions to criticism may give the reader a false sense of drama. Overall, the criticism of the war did not attract too much attention, and thus could be handled with relative ease. All that however, changed in 1957, as the gushing revelations and protests that came in the wake of the Battle of Algiers started to cut right into the legitimacy of the war effort and the image of the government, its members, and state officials. Thus only in 1957 did the government start to feel what it was like to be on the defensive.

Indeed, the pressure of publications and protest convinced the government to make some tactical concessions. Thus, to previous denials of abuses of power by the army and police in Algeria, officials added "in principle"

---

[4] Servan-Schreiber was recalled to serve in Algeria in 1956, possibly because he opened up *L'Express* for criticism of the army's conduct. The draft deferment of another critic, Maurice Maschino, was cancelled. See Sorum, *Intellectuals*, 155.

[5] Hamon and Rotman, *Les porteurs*, 43, 48, 60–61, 64. See on despotic acts during de Gaulle's reign in ibid., 235, 306.

[6] Smith, *The French Stake*, 150.

condemnations of any misconduct, and promises to investigate allegations of wrongdoing. For example, in early March 1957, the National Assembly established a committee to study alleged cases of torture in the Oran region. To steer the committee, however, Mollet appointed a Socialist friend, and not surprisingly, the committee – excluding a Mendèsist member who happened to be a physician and as such the only one qualified to judge the merit of the evidence – concluded that the allegations could not be corroborated.[7] Similarly, in late March, the government was forced to investigate allegations of torture, following the pressure of law professor René Capitant to explain the mysterious "suicide" of his former student, the Algerian lawyer Ali Boumendjel. At no point, however, was the government's public stand against abuse of power – a stand that was the result of public and intra-party pressure – genuine.[8] Rather it was part of an effort to minimize the significance and damage of revelations. Thus from the floor of the National Assembly Mollet could promise that "if it is true that [prisoners] were the object of torture, I declare that no excuse would be valid."[9] But at the same time he tried to discredit the critics: "I am sure . . . that none of you would do the injustice of thinking that the government, the army, or the administration might want to organize torture . . . Cases have been cited. I must say that it is rare, too rare alas, that those who bring such accusations against us, agree to provide sufficient evidence . . ."[10]

When pressures mounted further, however, Mollet had to "make good" on his word. In April, he finally announced the creation of the "Commission for the Defense of Individual Rights and Liberties" (the Garçon committee) – an investigative body that was instructed to report to the government about possible abuses of power in Algeria. Its public image aside, the commission was not intended to serve its declared purpose. Indeed, it included public figures that the government considered safe, and its powers were severely limited. For example, it could report only to the Prime Minister and the Resident Minister in Algeria, it had no power to subpoena documents or record testimony under oath, and it was authorized to deal only with individual cases rather than the phenomenon of torture as a whole.[11]

Meanwhile, the government balanced its "concessions" with threats that "unpatriotic" accusations would not be tolerated. On March 15, 1957, the defense ministry issued a press release, taken up by *Le Monde*, in which it was argued that the military command never tolerated the few improper actions that were brought to its attention. However, the communiqué also

---

[7] See Hamon and Rotman, *Les porteurs*, 66–67.
[8] Mollet faced opposition also from within the SFIO, though it did not threaten party integrity or his leadership. See Talbott, *The War Without a Name*, 74–75.
[9] Quoted in Smith, *The French Stake*, 149.
[10] Ibid.
[11] See Vidal-Naquet, *La raison d'état*, 129–33.

attacked, in no uncertain terms, a "certain press," condemned the use of "faulty, grossly exaggerated and distorted facts" in the campaign against the army, and added a dire warning that the government would act against all those associated with the anti-military campaign.[12] In Algeria, of course, the tone (and deeds) of the government were more resolute. In July 1957, for example, the Resident Minister in Algeria, Robert Lacoste, talking to the War Veterans of Algeria, put the blame for the "resurgence of terrorism" squarely on "the exhibitionists of heart and mind who initiated the campaign against torture...."[13]

But what was the state to do when it turned out that increasingly the "exhibitionists of heart and mind" popped up in its own yard, at all levels, and – as was revealed in the cases of Bollardière, Teitgen, Mairey, Delavignette, Garçon, and Billotte – even among the top echelon of its military and bureaucratic establishments? And how were the state and the army to react to the "defections" that shattered the cohesion of the state apparatus, reduced public confidence in official reports, undercut the esprit de corps of the army, and cast a long shadow over the legitimacy of the war?

At least as far as military personnel were concerned, the answer seemed simple. The authorities possessed inherent powers of the first degree – the powers to refine and tailor policy, to try to mold the mindset of soldiers, to demand obedience and loyalty, to sanction undesired conduct, and to discredit dissenters – and they could use them. Indeed, the army put all of its powers to work. First, it punished dissenters and "defectors" in order to deter further erosion within its ranks. Thus, the dissenting General Bollardière was sentenced to two months' arrest in a fortress for publicly deploring the army's conduct in Algeria (whereas General Jacques Faure, who plotted a coup in Algeria in 1956 was sentenced to only thirty days of arrest). And Noel Favrelière, who had deserted in 1956 with an Algerian prisoner he had saved from summary execution, was sentenced to death in absentia in 1958.

Second, the army set out to modify French civil law in an effort to gain legitimacy for its methods and diminish the scruples its soldiers experienced when faced with the realities of the battlefield in Algeria. In the army's sanitized language, it was looking for a way to adapt the law "to the particular conditions of the Algerian conflict."[14] The type of modification the army had in mind was revealed when Massu, ever less subtle than his superiors, went to the attorney general in Algeria, Paul Reliquet, and asked him to find a way to enable his soldiers to "carry out efficaciously, but on legal foundations,

[12] Ibid., 107–08. Today – particularly after Paul Aussaresses' book, *Service spéciaux Algerie 1955–1957: Mon témoignage sur la torture* (Perrin, 2001) – the allegations are obviously vindicated.

[13] Quoted in Sirinelli, "Guerre d'Algérie, guerre des pétitions," 286, from *Le Monde*, 9 July 1957, 4.

[14] Quoted from a letter of General Allard. See Massu, *La vraie bataille*, 376, or Maran, *Torture*, 104.

the task assigned to them."[15] In short, having failed to control the normative gap by other means, the army was looking for a way to eliminate what it perceived as the seminal factor that set its soldiers apart from its policy – the laws that banned torture.[16]

Third, the army never stopped trying to win the hearts and minds of its own soldiers and convince them that France must win the war, and that it could do so only by resorting to unconventional practices.[17] In this campaign, the army enjoyed the important support of its chaplains. For example, French soldiers who were sent to Algeria from Germany received a note from their chaplain explaining that "the nation sends you to Algeria to re-establish order. Fulfill your duty as a soldier in the here and now. This is the will of God." Similarly, the chief army chaplain, Cardinal Feltin, told soldiers that "the orders of pacification that you receive cannot raise conscientious objection. . . . They are not anti-Christian because first and foremost they aim at establishing peace . . . They rest on the preferential love you owe to your country."[18]

This effort to condition the soldiers to the "demands" of the battlefield, and get them to accept the need to use "unusual" measures, only increased once they had been dispatched to Algeria. Furthermore, once there, the soldiers were exposed to a second, existential rationale, that specifically justified the resort to unusual measures such as torture. As recalled, the problem of "mis-adjustment" of soldiers to the "necessities" of counterinsurgency war was particularly acute in the period of the intensive Battle of Algiers. Indeed, in this period, concern with the "moral problem" of the forces in Algiers was so serious that Massu decided to lobby personally for the use of torture. In a secret memorandum to his officers, he urged them not to be discouraged by a "certain press" and to accept the extreme methods as necessary and morally valid.[19] Moreover, with Colonel Roger Trinquier he approached Father Louis Delarue, the chaplain of his 10th Paratroops Division and asked him (presumably as an effective moral authority) to sanction torture. Delarue, in what may be seen as an almost medieval point of view, told the soldiers that their Christian conscience commanded them, as civilized people, to toughen themselves and put effectiveness and efficiency above moral hesitations.

We need to find, without hesitation, efficient means, even if they are irregular, and to apply them without weakness . . . truth to tell, here it is no longer a matter of waging war, but of annihilating an enterprise of generalized, organized murder . . . What then

---

[15] Quoted in Massu, *La vraie bataille*, 100 (and Maran, *Torture*, 104).
[16] Massu was never given the legal latitude he was seeking. However, he simply denied the supremacy of civil law in Algeria by declaring that Algiers was under military authority. See Massu, *La vraie bataille*, 360, 367–68.
[17] See also Maran, *Torture*, 16–17.
[18] Quotes are from Nozière, *Algérie: Les Chrétiens*, 126.
[19] See Ambler, *The French Army*, 321–22; and Vidal-Naquet, *La raison d'état*, 110.

is required? That on the one hand you protect *efficiently* the innocent whose existence depends on the *manner in which you will have carried out your mission,* and that, on the other hand, you avoid all *arbitrary acts* ... Between two evils: making a bandit, caught in the act – and who actually deserves to die – suffer temporarily, and [letting the innocent die] ... it is necessary to chose without hesitation the lesser [evil]: an interrogation without sadism yet efficacious.[20]

In fact, this rationale, in one way or another, was presented to the officers who served elsewhere in Algeria. Furthermore, the army eventually even created, in late 1959, a special preparation center for its officers, whom it apparently found "mentally unprepared" for the "requirements" of the battlefield, in Arzew (Algeria).[21]

Meanwhile, the army also established in July 1957 the *Détachments Operationnels de Protection* (DOP), a special centralized military agency that was designed, according to Massu, to interrogate "suspects who would tell nothing."[22] The idea behind the DOP was to let "professionals" deal with "tough" cases and thus avoid excessive and possibly fatal damage "amateurs" could inflict on suspects, which the French command suspected could play into the hands of the Moralists.

As events heated up further in Algeria, the position of the government became ever more tenuous. When all has been said and done, the government was still trying to defend a position full of contradictions. Indeed, its efforts to deny, its promises to investigate and punish, and its harassment campaign proved of limited value, if not wholly counterproductive. Under such conditions, the search for a change was all but certain, and the temptation to neglect competitive strategies in the marketplace of ideas in favor of increased resort to coercion was particularly irresistible. In fact, only more so because the nature of the French state – its vast executive powers and its role as a major employer – made a more despotic intervention in the marketplace of ideas relatively easy. Thus, while the government started to use coercive measures at home early on, once it became clear that the cover-ups and brazen lies only weakened its position, it increasingly turned to blunt executive instruments.

The growing resort to despotic powers was evident in several ways. During the first six years of the war (1954–1960), twenty-one issues of *L'Express,* sixteen of *France-observateur,* four of *Témoignage Chrétien,* and several issues of *L'Humanité, Esprit,* and other periodicals were seized.[23] Similarly, between 1957 and 1962, whole editions of books that documented torture and other excesses and criticized the army and the state were seized on the pretext that

[20] Massu, *La vraie bataille,* 160–62.
[21] On Arzew, see Heggoy, *Insurgency,* 176–82, particularly p. 179.
[22] Quoted in Massu, *La vraie bataille,* 165.
[23] Sorum, *Intellectuals,* 147 (see also p. 259 note 78).

they injured the morale of the army or encouraged military disobedience.[24] Apparently these seizures were motivated by more sinister, economic calculations than the simple wish to prevent any specific publication. After all, any book-seizure was almost certain to be circumvented by repeated publication elsewhere or even by the same publisher whom the authorities targeted. However, the seizures threatened to push the publishing houses, particularly the small ones, into serious financial trouble, perhaps even to the point of forcing them out of business.

At the same time, the metropolitan and Algerian forces of order – the CRS, the DST, the police, and the army – were increasingly willing to treat French and Algerians on the continent, and French citizens in Algeria, in ways similar to those practiced by the army in Algeria against the rebels. In Algeria, opponents of the war of European origins, mostly Communists, were not spared torture, and in rare cases even summary execution. In France, the DST, albeit on a much smaller scale, was also torturing – initially Algerians, but eventually also French suspects (including, during the final stage of the war, members of the OAS). In fact, in June 1959, evidence regarding the use of torture against Algerians in France was published in Alleg's book, *La gangrène*.

The antagonism between the state and the press (and intellectuals) reached a climax in 1960. In April, George Arnaud, a freelance journalist, published in *Paris Presse* an account of a clandestine press conference with Francis Jeanson, the fugitive leader of a network that smuggled money it collected in France to support the FLN.[25] For state authorities, this particular affair must have involved great embarrassment and frustration. It seemed to indicate that authority was slipping right through the fingers of state bureaucrats. The network operated undetected for over three years, its leader eluded the police thereafter, and now, as if to add insult to injury, Jeanson ridiculed the agents of order by convening a clandestine press conference right under their noses. Furthermore, efforts to control the damage that was done by the clandestine press conference only reinforced the impression of administrative ineptitude. The authorities tried to stop the publication of *Paris Presse*, but failed. Then they tried to block its distribution, and failed again. Thus, not surprisingly, once Arnaud's story was published, he became an intolerable reminder of the incompetence of the state. Arnaud was also all that was left for the frustrated state-agents. The DST arrested him and demanded the names of those who participated in the press conference. Arnaud refused

---

[24] Apparently, twenty-three books were seized, twenty-one of which were published by Minuit and Maspero. See Vidal-Naquet, *Face*, 24, 25. Among the more famous are Alleg's *La Question?* (1957), Minuit's *La gangrène* (1959), Fanon's *L'an V de la révolution algérienne* (1959) and *Les damnés de la terre* (1961), Favrelière's *Le Désert à l'aube* (1960), Hurst's (pseudonym Maurienne), *Le Déserteur* (1960), Maschino's *Le refus* (1960), Jeanson's *Notre guerre* (1960), Charby's *L'Algérie en prison* (1961), Leulliette's *St. Michel et le dragon* (1961), and Mandouze's, *La révolution algérienne par les textes* (1961).

[25] See Hamon and Rotman, *Les porteurs*, 205–10.

to disclose any names, and was therefore put on trial and sentenced to two years imprisonment.

In September 1960, the state faced a new challenge. A day after the beginning of the proceedings of the trial of Jeanson's network, a group of intellectuals published a proclamation that became known as the "121 Manifesto," in which they called for military disobedience and support for the FLN. Reacting to this absolute defiance of French authority, the police brought in some of the signatories for questioning, raided the publishing house *Seuil*, and detained for a short while those it found there, among them Lindon, Vidal-Naquet, Barrat, and Domenach (the latter had not even signed the manifesto). Meanwhile, Debré's government increased the penalties for instigation of disobedience, threatening the signatories of the manifesto with imprisonment for up to three years and fines of 100,000 new francs. Entertainers who signed the manifesto were excluded from programs of the state-run TV. The government also issued a disciplinary ordinance facilitating the suspension of state employees guilty of condoning desertion, and then put it into effect. Pierre Vidal-Naquet was suspended from his teaching position in Caen University (for one year as of October 1960), as was Laurent Schwartz – a world-class mathematician who taught in the *École polytechnique* (controlled by the Defense Ministry) – by the minister of the armed forces, Pierre Messmer.[26]

Finally, as the war progressed, it became evident that the police were ever more ready to become brutal and repressive inside France. Much of the brutality was clandestine, and its more sinister and violent expressions may not have been ordered explicitly by higher authorities. However, on several occasions, particularly during demonstrations, police violence was intense and perhaps pre-planned. This evident brutalization came to an ugly climax during demonstrations in Paris. On October 27, 1960, the police brutally assaulted demonstrators of the Left, in the process beating journalists and prominent public figures who participated in the manifestation. On October 17, 1961, the police killed scores of Algerian demonstrators, wounding many more in the process. The police also arrested over half of the demonstrators (more than 11,500 people!), and eventually some 1,000 of them were deported. Finally, on February 8, 1962, police forces were responsible for the death of eight French protesters, including a fifteen-years-old boy and three women, during a demonstration of the French Left against the terror campaign of the OAS.[27]

[26] See Hamon and Rotman, *Les porteurs*, 303–08; and Rotman and Rouseau, *La résistance*, 505–13. Professor Peyrega, the dean of the Faculty of Law in Algiers, complained to the Minister of Defense about several killings of Algerians in March 1957. His letter was published in *France Observateur*, and then was discussed in *Le Monde* and *France-Soir*. The Minister of Education revoked his deanship, presumably in response to the request of Peyrega's peers.

[27] See Tartakowsky, "Les manifestations," 138, 139, 141; Alleg et al., *La guerre d'Algérie*, Vol. 3, 365–71, 381–85; and Hamon and Rotman, *Les porteurs*, 316.

As we have noted, the attitude of the government and its executive agencies during the time of the Fifth Republic was essentially similar to its attitude during the Fourth Republic. On the record, the new regime was more committed to opposing unscrupulous conduct, particularly in Algeria. André Malraux, a person to the liking of many Moralists, was appointed Minister of Information. He was a long-time Gaullist and could provide the government with a veneer of respectability. Indeed, one of Malraux's first actions as minister was to propose that three French literature Nobel Laureates (Martin-du-Gard, Camus, and Mauriac) investigate the problem of torture in Algeria. Malraux' superiors, de Gaulle and Debré, apparently sincerely, instructed the army to avoid torture, and threatened to severely punish violators of this policy. However, Malraux's offer was never taken up, while de Gaulle and Debré never enforced their anti-torture policy, although they were well aware that it was not taken seriously in Algeria.[28] Moreover, as has been documented and noted, the official forces of order continued to struggle inside France ever more energetically against those who criticized the war.

The Secondary Expansion of the Normative Gap

Considering the position and actions of state officials toward protest against the war and the conduct of the army – in Algeria and in France, in the marketplace of ideas, in the interrogation rooms, on the streets of Paris and other cities, in court, and in the workplace – it is obvious why the anti-war agenda soon developed a third dimension that linked the war to a threat to the democratic order in France. And considering the seditious acts of important parts of the army command in Algeria, and their collusion with the most extreme circles among the settlers, it is no wonder that this dimension became of the utmost significance among the rest of the issues of protest.

As emerges clearly from Chapter 7 and the discussion up this point, there was simply no democratic way by which the state could overcome the discrepancy between its actual and declared policy, or overcome its opponents in a free competition in the marketplace of ideas. Consequently, French governments adopted a strategy that exploited a mixture of benign and despotic measures in defensive and offensive ways. On the one hand, the administration denied wrongdoing, argued that torture and other excesses were unacceptable, and tried to discredit critics. On the other hand, the forces of order intimidated and punished critics. The contradictions in this Janus-faced approach, as well as the coercion involved, corroborated

---

[28] See La Morandais, *De la colonisation*, 414. See also ibid., 406–17; Maran, *Torture*, 57; and Jean Lacouture, (trans. Alan Sheridan), *De Gaulle: The Ruler 1945–1970*, Vol 2 (London: Harvill, 1991), 242–43.

some of the claims against the government, and fueled fears among growing sectors that democratic order was in jeopardy.

Indeed, it did not take too long for the intellectuals to figure out that the political order in France had become a hostage of the war. Certainly there was no need to stretch the imagination in order to reach such conclusions. A political "disaster" seemed to have become ever more likely, in one of two ways, either through a swift coup d'état, or through a slow and incremental process of degeneration.

As with other aspects of the Algerian war, the threat of sedition had roots in the history of the two decades preceding the war. When the war started, the army's confidence in civil institutions was already weakened as a result of the events in World War II and the war in Indochina, which left its officers with a particularly bitter feeling of betrayal. In addition, the long years officers had spent in the colonies – that is, away from civic and democratic life – did little to promote their appreciation for democratic order. Finally, the Algerian war only strengthened a conviction, which the defeat in Indochina had created among French officers, that democracy was too soft and indecisive to win "revolutionary wars."[29] Indeed, the anti-insurgency doctrine that middle-level officers in Algeria had developed in Indochina was inherently anti-democratic. They advocated countering the total mobilization of insurgents with equal ideological and material commitment of the "incumbent," and included in this adaptation of sort a restructuring of the normative foundations of its society.[30] Worse still, several officers professed their anti-democratic bias, arguing that "it is time to realize that the democratic ideology has become powerless in the world today."[31] In this sense, the army's collusion with the Algerian "ultras," as well as the latter's repeated challenges to civil supremacy in France, were natural. This also demonstrated that the war was breeding a real threat to democratic life in France. Jean-Marie Domenach captured the putrefied fruits of this process in one of the most insightful analyses of the relations between the war and French democracy. In an article he published in *Foreign Affairs* in September 1958, he explained:

A fascist orientation in Algeria cannot long coexist with the practice of democracy in France ... Hence I am convinced that Algerian fascism, when it finds itself pitted against the force of things as they are and the resistance of human beings, will turn on the metropolitan country and seek to crush the liberal forces remaining there ... The war in Algeria will last as long as Frenchmen refuse to satisfy the aspirations of the

[29] Indeed, in all conspiracies (excluding Faure's), and later in the ranks of the OAS, high-middle-level officers, many who developed and/or practiced "revolutionary war" were prominent.

[30] See, for example, Colonel Lacheroy's opinion in Kelly, "The French Army Re-enters Politics," 384–85.

[31] Major Hogard, quoted in Paret, *French Revolutionary Warfare*, 28. See also ibid., 27–28, 115–16.

Algerian people; and as long as the war lasts the Algerian situation will continue to breed fascism ... fascism will inevitably spread in the army as long as the war continues. We are in a race against time.[32]

While the threat of sedition and coups d'état was obvious, and as such frightening enough to serve as a cause for counteraction, the threat of incremental decay of democracy was, at least in theory, less arousing. Incremental deterioration is by definition more subtle and harder to discern, expose, and organize against than a one-strike revolt. This was not the case during the Algerian war.

As soon as the state of emergency was announced, and special powers granted to the Algerian administration and the army, civil liberties and rights in Algeria – that of the press, individuals, and groups – became precarious.[33] Moreover, the army readily admitted that there was no room for democratic procedures in Algeria. In September 1958, Colonel Trinquier, a leading thinker of the army in Algeria and Massu's right hand, explained that "what we have to do [in Algeria] is to organize the population from top to bottom. You may call me fascist, but we have to make the population docile and everyone's actions must be controlled."[34] Even more important, although Algeria was geographically separated from France and, unlike France, governed by emergency regulations and de facto military rule, there were clear signs that the anti-democratic developments in Algeria were spilling over to metropolitan France. In late 1957, after Delavignette resigned from the Mollet "human rights" committee he explained that "[w]hat is true of Algeria may very quickly [become true] of France."[35] Indeed, the process of the "Algerianization" of France was evident, particularly in the violations of proper judicial procedure and individual rights and in the efforts to curb freedom of speech. Many of the reactions of the administration in the marketplace of ideas – be it the brazen denials in the face of incontrovertible facts, the deceit, the efforts to discredit critics, their harassment, the brutal use of force, and the state vendettas – carried a strong odor of despotism. Thus, not surprisingly, Jérôme Lindon, responded in July 1959 to the seizure of books *Minuit* had published by writing the following words to Edmond Michelet, the Minister of Justice: "It remains for us to disassociate ourselves ... from these dreadful tortures and these arbitrary measures *which tend to establish in France the kind of regime, against which we fought together not so long ago.*"[36]

[32] Jean-Marie Domenach, "Democratic Paralysis in France," *Foreign Affairs*, 37:1 (1958), 38–39.
[33] See about the content of the "state-of-emergency" law in Maran, *Torture*, 40.
[34] Quoted in Domenach, "The French Army in Politics," 187.
[35] Vidal-Naquet, *Face*, 18 (see also *La raison*, 170).
[36] Quoted in Anne Simonin, "Les Éditions de Minuit et les Éditions du Seuil: Deux stratégies éditoriales face à la guerre d'Algérie," in Rioux and Sirinelli, *La guerre d'Algérie et les intellectuels*, 237 (italics added).

It was also becoming ever clearer that the judicial system was losing its impartiality, particularly, though not exclusively, in Algeria. Moreover, the good name of the Continental system was bound to be tarnished in any event because *Continental* magistrates perpetrated much of the judicial abuse in Algeria.[37]

Above all, I suspect, the fascist threat was quickly identified because it directly threatened the interests of those who benefit most from the democratic order – intellectuals, the press, individual politicians, and political parties. Obviously, fascism could not cohabit with the freedoms these groups and institutions considered non-negotiable because these freedoms provided the basis for their power, status, and material and spiritual gains. Indeed, while the threat to democracy resonated well among French citizens, it resonated much more forcefully, as Charles-Robert Ageron noted, among the educated elites.[38] In early 1959, after de Gaulle returned to power but before the army became seditious again, 65 percent of those who had received higher education (*instruction supérieure*) believed that the army had a tendency to overstep the scope of its normal functions. By comparison, only 52 percent of the male population and 42 percent of the total French population believed so.

The first major challenge to democracy after this poll, the January 1960 *Barricades Week* helped transform such opinions into actions. Those most likely to suffer from the consequences of sedition – the labor unions and parties of the Left, but also the Radicals and members of the MRP – rallied for negotiations over the future of Algeria because they became convinced that the continuation of the war was turning into a real threat to democracy.[39] Furthermore, the threat to democracy resonated so well that it became an attractive mobilization theme. For example, in October 1960, the labor unions organized a national day of action, not only for "peace in Algeria through negotiations" but also "for the safeguarding of democracy and its fundamental principles."[40] Indeed, Danielle Tartakowsky, who studied the mass protests during the Algerian war, concluded that most of the demonstrations were organized around the quest for peace and, above all, around the threat the war presented to life in France and to French democracy.[41]

Finally, one must note that the Algerian war led to a deep rift, even if limited in scope, in the relations between the French state and portions of

[37] See Vidal-Naquet *Face*, 115–31. On the signs of the spread of torture to France, see Hamon and Rotman, *Les porteurs*, 118, 127–38, 156–59.

[38] Ageron, "L'Opinion Française," 34.

[39] Jacques Julliard describes the "Mendèsists" as utilitarian, and as concerned with army insurrection, the vanishing of liberties, and the potential collapse of parliamentary rule rather than with the immoral conduct of the army. See "La réparation des clercs" in Rioux and Sirinelli, *La guerre d'Algérie et les intellectuels*, 391.

[40] Tartakowsky, "Les manifestations," 137, 139.

[41] Ibid., 143. Some suspect that de Gaulle purposely exploited the fear of fascism in order to break the French Algeria camp. See Montagnon, *Guerre d'Algérie*, 341.

its society.[42] As noted, the Moralists became convinced that, at bottom,
the immoral behavior in Algeria was not spontaneous but rather organized
under the aegis of the state. Thus they decided that the problem should
be addressed at that level. Pierre Vidal-Naquet epitomized this logic when
he wrote in the summer of 1960 in *Vérité-Liberté* that "it is the Mollets, the
Bourgès-Maunourys, the Lacostes, and the Cotys who create the Massus and
the Charbonniers."[43] In retrospect, he articulated his logic of anti-etatist activism (rather than simply moral or legal activism) even better. He and others
struggled against torture, "not only because the use of torture deprives the
victim of his fundamental right ... but [also] because ... it was a question of
showing that the responsibility of the French State was involved at all levels: that
of the army, of the police, of the justice [system], of government."[44] From
such a position, particularly when one believed that the French struggle was
fundamentally illegitimate, the decision to switch sides was not that far. Indeed, several anti-war activists, including Chevallier (who initially supported
the national position) and several clergymen, were willing to provide shelter
to FLN fugitives, while others, Communists as well as Catholic-progressives,
were even ready to participate in the operations of the insurgent forces.[45]
Most notably, as of mid-1956 and for three years thereafter, the Jeanson network was ready to transfer money that the FLN had raised in France into
bank accounts in neighboring countries.[46] Members of this network, and
others, also provided logistical support for army deserters and FLN opera-
tives, including those who tried in the early fall of 1958 to assassinate Jacques
Soustelle. These kinds of support activities are said to have been performed
by some 4,000 French citizens! That was, of course, just a fraction of French
society. However, the number still seems enormous when one considers the
known number of citizens of other countries who supported their homeland's
enemies during such wars. Finally, as the war progressed, some intellectuals and others became radicalized, and were ever more ready to criticize
the state itself rather than only its conduct. Most famous among such acts

[42] For example, the isolation of the army seems to have resulted in a drop in the number of
applicants to St. Cyr. See Ambler, *The French Army*, 101, 132–34.
[43] Quoted in Vidal-Naquet, *Face*, 157. Charbonnier was an intelligence officer who was allegedly involved in Audin's "disappearance."
[44] In ibid., 17 (italics added).
[45] On the activity of progressive clergymen such as Scotto, Bérenguer, and Kerlan, see Nozière,
Algérie: Les Chrétiens, 204–28. In April 1956, Henri Maillot, a Communist reservist and
officer-cadet, defected with a truckload of light weapons. He was killed in a battle against
French forces two month later. Another Communist, Fernand Iveton, placed a bomb in his
factory (he was executed in February 1957). Several young European women also participated in the Algiers terrorist campaign. For a discussion of cases, see Hamon and Rotman,
Les porteurs, 196–201, particularly p. 198; and Alleg et al., *La Guerre d'Algérie*, Vol. 2,
470–76.
[46] This money, according to the FLN head in France, "nourished decisively the war treasury of
the GPRA." Quoted in Montagnon, *Guerre d'Algérie*, 219.

was the September 1960 "Declaration on the right of insubordination in the Algerian war," which was better known as the "121 manifesto."[47] Perhaps the readiness of some communal pillars, such as Cardinal Duval and retired mayor Chevallier, to eventually become Algerian nationals, also attests to the depth of state bankruptcy.

[47] See Hamon and Rotman, *Les porteurs*, 303–18, 391–94.

9

Political Relevance and Its Consequences in France

The French people at large were not captive of the *"Algérie française"* theme. But neither were they convinced that Algeria must be abandoned because of moral or economic considerations. Most French people simply chose to exclude themselves from the debate over Algeria. Of course, some regions and sectors tended to be more opposed to the war than others. Certainly the war was not too popular in Left-dominated regions or in regions whose inhabitants' interests were threatened by the Algerian economy. Thus, Marseilles, where a solid one-third of the vote was Communist, produced greater activity against the war than other cities. And among the least favorable to the plight of the *pieds-noirs* were regions in the south-east of France, where the population was inherently separatist and regarded the Algerian settlers as unwelcome competitors in the market of lesser wines.[1]

The general French apathy was above all shown by the small scale of anti-war activity. The first demonstrations against mobilization and against the war usually attracted no more than several hundred people. Later, demonstrations grew larger. At times, they consisted of a few thousand people. But, with the exception of three mass demonstrations that had either to do with the future of democracy or were organized after Algeria's future was all but decided, they rarely consisted of more than 10,000 people.[2] In fact, at least initially, the pro-French Algeria forces were able to counter-rally against the anti-war protestors. For example, in early 1956, nationalist students led by

---

[1] Ageron, "L'Opinion française," 36.
[2] See Tartakowsky, "Les manifestations," 132, 136, 138; Horne, *A Savage War*, 297; and Vidal-Naquet, *Face*, 57–58. The first big demonstration occurred on May 28, 1958. The second mass demonstration occurred in October 1960, long after de Gaulle's self-determination announcement, and his March 1960 reference to "Algerian Algeria." The third 1962 mass demonstration took place after the fate of the war was essentially sealed. It was triggered by the earlier death of eight demonstrators in a much smaller demonstration against the OAS.

a past president of the National Union of French Students (UNEF) demonstrated in several major cities in response to anti-war demonstrations.[3]

Equally important, the success of the anti-war campaign, even within educated sectors of the population, was limited, at best. Those who benefited
most from the liberal-democratic order in France, and did particularly well
in the 1950s – people of administrative and managerial vocations or the self-
employed – were the most sympathetic to the political Right and the "French
Algeria" idea.[4] In fact, even the academic community was deeply split over
the war. Many academics, including some highly respected non-conservative
scholars such as Jacques Soustelle, Paul Rivet, and Albert Bayet, firmly supported the cause of French Algeria and the war.[5] Perhaps it is not surprising
that academics publicly expressed support for the government, the country,
and the military effort in Algeria (as they did, for example, in the pages of
the May 23, 1956 issue of *Le Monde*) at a relatively early stage of the war.
Yet this support did not collapse even long after the army's abuses in Algeria
and the state's arbitrary acts at home became common knowledge. For example, in October 1960, more than 300 intellectuals, artists, and others of
liberal vocations condemned the September "121 manifesto" and condoned
the state and the war effort. In a public declaration, they described the role
of the French army in Algeria as "social and humane," the 121 manifesto as
"one of the most cowardly forms of treason," and its signatories as a "fifth
column."[6] Of course, to such voices of support for the war and the state one
should add those of loose cannons and politicians-of-hate from the vulgar
far Right. These considered anyone who opposed the war – in particular, the
intellectuals whom they regarded as degenerate, sexually and otherwise – to
be traitors. For example, Jean-Marie Le Pen, at the time a young Poujadist,
declared in December 1955 that "each time one gets a kick in the ass one
needs to brush the trousers afterward. France is governed by fagots: Sartre,
Camus, Mauriac."[7]

Finally, while it is clear that the anti-war campaign was led by forces
from the political Left, these forces operated outside the institutions that
funnelled their political demands in normal times. Thus, while Communists
and Catholics played the key role in the protest, and from among them
came the militant anti-war activists, the Church and the Communist party
remained ambivalent for a long time. The PCF, as already explained supported the Socialists' Algerian policy until its XIVth Congress in July 1956,

---

[3] Tartakowsky, "Les manifestations," 132–33.

[4] Ageron, "L'opinion française," 33–34; and *Sondages*, 1958:4, 51. This sector, however, appears
also to have been the most polarized.

[5] See also Berstein, "La peau de chagrin," 213; and Hamon and Rotman, *Les porteurs*, 305.

[6] See Éveno and Planchais, *Dossier et témoignages*, 277–80; and Sirinelli, "Guerre d'Algérie,"
282–84, 290–93. See also, on the "manifestos' war" of Fall 1960, Sirinelli, "Les intellectuels
dans la mêlée," in Rioux, *La Guerre d'Algérie et les Français*, 124–28, particularly p. 125.

[7] Quoted in Sirinelli, "Les intellectuels dans la mêlée," 123.

while shortly thereafter it was distracted by the Soviet suppression of the Hungarian uprising. At the same time, it did not make much sense for the Church to endorse an agenda that its bitter ideological enemies, the Communists, promoted, particularly as the pro-French Algeria Right wing propagated a family-patrimony-church message.[8] In fact, paradoxically, the Communist and Catholic establishments had another common reason for discouraging anti-war commitment: Both had a large *pieds-noirs* following. Indeed, within the Catholic community, opinions varied as much as they did within other circles. Whereas for every anti-war "resister" such as Jean Bruller (Vercors) there was another such as Rémy Roure, for every clergy-man that condemned the injustice of occupation and repression there was one that sided with the state, and many more that kept silent on issues that begged for their moral intervention. Likewise, for every anti-war view published in *Esprit* or *Témoignage Chrétien* there was an opposing view in *France Catholique* or *Verbe*.

In short, the struggle over the future of Algeria in France, as the record of public protest shows, was one between committed minorities.[9] Within France, the anti-war minority was clearly gaining power as time passed. Yet it remained a minority, enlisting members mostly from narrow, educated, and class-conscious bands of society – from among university staff and professors, progressive clergy, trade unions (particularly the unions of the teaching professions), the higher administration and the justice system, Communists, and the press.[10] Indeed, it is precisely because the results of the war seem to have depended on such a minority, that the nature of this minority deserves further discussion.

The Intellectuals and the Press

It is hardly possible for anybody who studies the Algerian war to ignore the French intellectuals. Few will contest that they formed the spearhead of the anti-torture and anti-war campaign, kept the moral critique alive on the national agenda, and articulated the relations between the war and the threat to democratic order in France. Yet it is hard to fathom the depth of the impact intellectuals had on the war without first understanding the unique place French intellectuals traditionally occupied in French political life. Stanley Hoffmann's insight on the issue is of great value:

The intellectual was heir of the defeated aristocracy, since prowess as a source of prestige and prestige as a source of power continued to dominate the values and human relations of 'bourgeois' society. He was also the heir of the Church, since

---

[8] For example, the Catholic establishment could hardly stomach the "worker-priests" mission, which had a proletarian commitment. See Rioux, *The Fourth Republic*, 428–29. See also, on the position of the Church, Girardet, *L'Idée coloniale*, 266–74; and Nozière, *Les Chrétiens*.

[9] See also Rioux's assessment in *The Fourth Republic*, 291.

[10] See also Domenach's, "Democratic Paralysis in France," 44.

the Enlightenment displaced religion and put lay science in its stead, and since the intellectual was called upon to provide the spiritual guidance once associated only with the Church... entry into intellectual or quasi-intellectual professions was for centuries the privileged method of social ascent... This privileged position was an honor and a risk. In times of purges the intellectuals were hit hardest, *precisely because their pretense at being the conscience of society was taken seriously.*[11]

Indeed, the history of the Algerian war vindicates Hoffmann's view. Clearly the state had good reason to fear that the leaders of the anti-war campaign, who had access to different levels of French society, would serve as a moral compass for many others. Even a partial list of the academics, literary icons, journalists, and other celebrities who opposed the war explains why they were perceived by the institutions they criticized, as well as by themselves as having critical weight. How could any government of a country obsessed with the notion of being at the forefront of Western civilization remain aloof in the face of an indictment by its best and brightest?[12] Indeed, the intellectuals were well aware of their inherent powers. Jean-Marie Domenach, a leading anti-war Catholic intellectual, wrote in a 1958 *Foreign Affairs* article I quoted in Chapter 8, that "it will be difficult to lead France in a direction they (the anti-militarist and anti-fascist forces) oppose."[13]

Yet it would be a mistake to limit our understanding of the capacity of French intellectuals to influence society and policy to structural causes only. Circumstantial factors also played an important role, increasing the probability that the opinion of intellectuals would matter. Above all, the war years were also years of expansion of the French education system. The population of students, and consequently that of teachers, increased dramatically between 1954 and 1962.[14] The number of students grew by 80 percent (from 140,000 to 252,000) and that of teachers and members of the literary and scientific professions by 55 percent (from 77,000 to 120,000). Furthermore, the change was also qualitative – the number of students in the humanities and sciences for the first time exceeded that of students of

---

[11] Stanley Hoffmann, *Decline or Renewal* (NY: The Viking Press, 1974), 128 (italics added).
[12] Consider for example the following partial list: André Mandouze, Henri Marrou, Louis Massignon, Raymond Aron, Jean-Marie Domenach, Paul Ricoeur, Maurice Duverger, Jean-Paul Sartre, Germaine Tillion, Madeleine Rebérioux, Pierre Vidal-Naquet, Laurent Schwartz, Jacques Madaule, Jacques Berque, Jean Dresch, René Capitant, François Mauriac, Pierre-Henri Simon, Claude Bourdet, Robert Barrat, Jean-Jacques Servan-Schreiber, Jean Daniel, Roger Stéphan, Claude Gérard, Jean Bruller (Vercors), Simone de Beauvoir, Florence Malraux, Françoise Sagan, Simone Signoret, Tim (the cartoonist), François Truffaut, Jean Effel, and Jacques Prévert. See also Étienne Fouilloux, "Intellectuels catholiques et guerre d'Algérie (1954–1962)," in Rioux and Sirinelli *La guerre d'Algérie et les intellectuels*, 93–109. International figures included Frederico Fellini, Alberto Moravia, Heinrich Böll, Gertrude von Lefort, Norman Mailer, C. Wright Mills, John Osborne, Sean O'Casey, and Max Frisch.
[13] Domenach, "Democratic Paralysis in France," 44.
[14] See Rioux *The Fourth Republic*, 400, 420 table 26; Morse, *Foreign Policy and Interdependence in Gaullist France*, 164–66; and Alain Monchablon, "Syndicalisme étudiant et génération algérienne," in Rioux and Sirinelli, *La guerre d'Algérie et les intellectuels*, 176.

medicine and law. Such changes gave the anti-war camp a larger basis for activity, a better ability to disseminate its messages, and the means to achieve visibility. Teachers served as a pipeline for the circulation of ideas, and both they and the students provided a base for mass action and protest at a later stage of the war.

The second important tier of the anti-war movement included Communist and Catholic militants. These "foot soldiers" were responsible for the early anti-war demonstrations and the supply of information about army conduct in Algeria to the intellectual leadership.[15] The more radical elements in this tier participated in the networks of support for the FLN, draft dodgers, and deserters. This entire tier – its mainstream as well as its radical wing – controlled the agenda, first by deeds and only then by polemic: those in the mainstream by pushing intellectuals into action through letters and by elevating the level of public consciousness through demonstrations, and those in the radical extreme by "outrageous" acts of support for the FLN. Such activity, being illegal, often ended up in court, where lawyers took the lead. Indeed, these lawyers – Gisèle Halimi, Roland Dumas, Jacques Vergès, and others – were not simply agents of the individual defendants, but rather promoters of the anti-war political agenda. Indeed, they skilfully used the courts as launching pads for attacks against the state, the army, and the war.[16]

In the final analysis, however, both the intellectuals and the militants would probably have remained marginal had it not been for the third element of the anti-war movement – the free press and the publishing houses. Without this intermediary, the intellectuals and other anti-war groups would have had trouble conveying their ideas and affecting the rest of society as well as the political world. One does not need much imagination to speculate what would have been the effect of the anti-war campaign had the French state controlled the written press and publishing houses to the same degree as it did the means of electronic communications. In 1955, the first year of the war, about half the French population claimed to have obtained most of its information about world events from the press, and about half claimed to have obtained most of it from the radio.[17] Without the free press, then, the

[15] Examples include the letters of Catholic soldiers that were published in March 1957 in *Les rappelés témoignent*, the letters and testimonies published in Pierre-Henri Simon's *Contre la torture* (Paris: Seuil, 1957), 72–93, and accounts of individuals such as Bonnaud and Mattei. See Hamon and Rotman, *Les porteurs*, 65–66, 71–72.

[16] See Hamon and Rotman, *Les porteurs*, 281–300; and Vidal-Naquet, *Face*, 31–34.

[17] See *Sondages*, 1955:3, 26–27, 30–31. On the electronic and written media, see Rioux, *The Fourth Republic*, 441–44; and Ruz, "La force du 'Cartiérisme,'" 330. *L'Express* and *Paris-Match* did particularly well in the early 1950s. *Paris-Match* had a circulation of 2 million and an estimated readership of 8 million (some 40 percent of its readers lived in Africa). In 1955, *France-Soir* was the most popular daily, with a circulation of over 1,183,000, and a substantially larger readership. On circulation trends 1945–1958, see *L'année politique*, 1957, 550–55, and 1958, 572–73.

state would have easily monopolized the marketplace of ideas and avoided much of the friction that ultimately doomed the war effort.

Looking back on events in France during the war, it is clear that the different players of the anti-war movement had a remarkable ability to control the national agenda and shape it to their liking, often by collusion with each other.[18] Several examples can demonstrate how each of the tiers of the anti-war movement achieved saliency in the marketplace of ideas, be it through publications, the work of action committees, articles and letters to the press, or a deliberate strategy of provocation by deeds, words, manifestos, and through the courts.[19]

The Audin affair is a classic example. Maurice Audin's wife, Josette, convinced Pierre Vidal-Naquet to investigate his "disappearance." With others, Vidal-Naquet established a committee bearing Audin's name. Meanwhile, Jacques Duclos and Pierre Mendès-France publicly blamed the paratroopers for his murder, and the Teachers' Union and the League of the Rights of Man demanded an investigation into his disappearance. Audin's tragic "disappearance," then, provided the anti-war campaign with a powerful and lasting symbol. The "Audin Committee" managed to mobilize support from among members of an important layer of society – the educators – and became an effective vehicle in the anti-torture campaign throughout the war.[20] For example, five and a half months after Audin's disappearance, the committee managed to create news by arranging for the defense of Audin's mathematics thesis in absentia.

Other examples are as revealing. As noted in Chapter 7, a group of reservist priests who were shocked by the army's conduct in Algeria sought in March 1959, by way of a public letter, moral advice from the Church leadership in France. Obviously this was a rather clever way to force the Catholic Church to take a stand on an issue its leadership had painfully tried to avoid until then.[21] In April 1960, Jeanson masterfully seized the headlines, when as a fugitive he managed to organize a press conference under the noses of the authorities. In June of the same year, Georges Arnaud was able to use his trial

---

[18] The success of book publications was in part due to this collusion between intellectuals, publishing houses, and the media. *La question* was given a boost by a review article in *Le Monde*. Once it was seized, a battery of the most distinguished literary figures in France protested to President Coty, calling for an investigation of the allegations in the book. *La gangrène* was reviewed in Radio Europe No. 1 and in *Le Monde*. When it was seized, its publisher, Jérôme Lindon, maximized the exposure by addressing his response to Malraux (de Gaulle's Minister of information at the time) from the pages of *Le Monde*, *France Observateur*, *Libération*, *L'Express*, *Témoignage Chrétien*, and *Tribune du Peuple*. See Simonin, "Les Éditions," 227–30.

[19] On the importance of meetings, conferences, and other activity of the *Esprit* team see Domenach "Commentaires," 93–95. On the provocation strategy, see Bonnaud, "Le refus," 349; Simonin, "Les Éditions," 226–27; and Hamon and Rotman, *Les porteurs*, 280.

[20] See *Face*, 15–16.

[21] Nozière, *Les Chrétiens*, 135–41.

as a means to reveal to the public his observations on the DST's use of torture (later he published these observations in the book *Mon procès*), and Simone de Beauvoir raised a new storm with her publication on the sexual abuse of (a second) female FLN member, Djamila Boupacha, during torture.[22] During the criminal procedures against the Jeanson network, defense lawyers Roland Dumas and Jacques Vergès stunned officials when they called Paul Teitgen, the former Police Chief of Algiers, to testify for the defense, and mobilized Sartre's persona by reading a letter in which "he" declared that he would have been proud to participate in the network's activity.[23] Finally the press and others were able not only to expose the evils of the system but also the names of individual executors. For example, Vidal-Naquet and the Audin Committee singled out, by name, those – such as Colonels Trinquier and Bigeard, Majors Faulques and Devis, Lieutenants Prez, Jean, Erulin and Charbonnier, and the policeman Llorca – who were responsible for ordering and practicing torture.[24]

Until now, our discussion has mostly been concerned with the question of how moral values shaped the nature of the anti-war campaign. However, such analysis is incomplete. Left alone, it would omit other, less altruistic motivations, which probably contributed as much to the shaping of the anti-war campaign as did moral considerations. I have already noted the general anxiety over the loss of democratic order and the understanding of intellectuals that they, as the leaders of the anti-war campaign, would be the first to be targeted by a fascist regime. Such fears necessarily played a role in the readiness of intellectuals and others (including political parties) to take part in the anti-war campaign. Similarly, fears of the consequences of the war, and particularly the draft, played a role in the decision of students to take a stand against the war. Jacques Julliard, for example, observes that until the 1955 mobilization of the *disponibles* (those available in this case for mobilization), the UNEF was unwilling to distance itself from the masses of students and was therefore cautious in its reaction to the war. However, with the 1955 mobilization, the student leadership felt more confident that there was sufficient cause to lead the students into opposition to the war, and therefore decided to join the anti-war campaign.[25]

[22] The use of torture gained further publicity once Picasso's portrait of Boupacha appeared in magazines.

[23] See Hamon and Rotman, *Les porteurs*, 281–302. Marcel Péju and Claude Lanzmann wrote the "Sartre" letter. Sartre merely gave them permission to use his name while he was overseas. In fact, he did not even see the draft.

[24] Published in *Témoignages et documents* 6, October 1959, and reproduced in Vidal-Naquet, *Face*, 141–56 (see particularly pp. 151–52). Similarly, *Le Monde* cleverly cited in its February 6 and 7, 1959 issues judicial proceedings that discussed torture and mentioned the (unintentionally distorted) name of a security agent, Inspector Beloeil, who was allegedly involved in torture. See also Hamon and Rotman, *Les porteurs*, 136–37, 169.

[25] Jacques Julliard, "Une base de masse pour l'anticolonialisme," in Rioux and Sirinelli, *La guerre d'Algérie et les intellectuels*, 359–64, particularly pp. 360–61.

Most interesting by far among such utilitarian motivations are those of the press. As already explained, the Left and Liberal newspapers dedicated themselves to moral criticism of the war, only after Algeria was saturated with conscripts and reservists and Massu's paratroopers started to pacify Algiers. Similarly, the publishing houses joined the anti-war campaign only gradually: *Le Seuil* and *Minuit* as of 1957, the more radical *Maspero* as of 1959, *Julliard* and *Gallimard* in 1960.[26] This timetable, as well as the spread of criticism within the press, should be understood not only in terms of the developments in Algeria, but also in terms of the nature of information as a consumer commodity in a capitalist society.

As is the case with other leaders of the anti-war campaign, there is little doubt that individual journalists joined the campaign against the conduct in Algeria against the war, against the fascist threat at home, and against the state out of moral and ideological convictions. However, the newspapers and publishing houses and some of the journalists were also motivated by other less virtuous considerations – namely, economic, personal, and institutional interests.

First, the press was "caught" in a dynamic of publication that pushed it to deal more with the army's conduct in Algeria. As the volume of news about Algeria, and consequently public interest, grew, so did the interest of individual news producers seeking to find ways to capture the attention of more readers. A more daring coverage and criticism were assured ways to out-compete rivals. Within the progressive press and publishing houses, such a need to probe more deeply into events in Algeria and be more critical in the reporting thereof were inevitable. At the same time, the editors and journalists of the conservative press, who undoubtedly knew about the nature and scope of violence in Algeria as much as their left-wing peers, were forced to reconsider their uncritical support for the army and the war. Indeed, parts of the conservative press gradually joined the criticism of the army's conduct when it became abundantly clear that the unpleasant realities and ramifications of the Algeria war were of great news value. Therefore, as of 1957, the avant-garde anti-war press, which included *L'Humanité*, *Libération*, *Le Monde*, *France-observateur*, *Le Canard Enchaîné*, *Témoignage Chrétien*, *Esprit*, and *Les Temps Modernes*, was occasionally joined by more conservative newspapers such as *Le Figaro*, *L'Aurore*, *Paris-Presse*, *France-Soir*, *Le Parisien Libéré*, and the Catholic *La Croix*.[27]

The case of the publication of Georges Arnaud's report on Jeanson's clandestine press conference is a classical example of the impact of commercial

---

[26] See Vidal-Naquet, *Face*, 23. See also the discussion in Liauzu, "Intellectuels du Tiers Monde et intellectuels français," 155–74; and Simonin, "Les Éditions," 219–45.

[27] Conservative newspapers started to publish critical articles only against torture and the *regroupment* policy (not the war). See Hamon and Rotman, *Les porteurs*, 75–76; Horne, *A Savage War*, 221, 339; and Rioux, *The Fourth Republic*, 292.

competition on the conservative press. Arnaud's objective was to reach the largest number of readers. Knowing that he had a scoop, he turned to *France-Soir*, the daily with the largest circulation in Paris. However, *France-Soir* had no particular reason to risk antagonizing its readership or the authorities, and thus declined the offer. Arnaud then went to *Paris-Presse*. *Paris-Presse* faced a dilemma. It was quite conservative, but its circulation suggested that it could hardly afford being too selective when presented with such a news gem. At one time – in 1946 – the *Presse* was as widely circulated as *France-Soir* (almost 500,000 copies). However, in the late 1940s and 1950s, the readership of *France Soir* tripled, whereas that of *Paris-Presse* was halved. *Paris-Presse* solved its dilemma by way of duplicity. Its editor accepted Arnaud's article, but at the same time distanced the newspaper from Arnaud in no uncertain terms, noting that "the document we publish on the opposite page reached us through the agency of Georges Arnaud ... [w]ho is personally 'engaged' in a political struggle and has espoused, especially in the Algerian affair, positions very remote from ours."[28]

Somewhat similar competitive commercial considerations were present also within the progressive publications, which were essentially motivated by genuine moral and ideological considerations. For example, it seems that a "mixed bag" of moral and commercial calculations guided Jérôme Lindon of *Editions de Minuit*, to publish Henri Alleg's documentary book *La question*. Lindon well knew the risks involved in publishing such a controversial and provocative book. However, he decided to accept the book, in part because, as Anne Simonin argues, he was painfully aware of *Minuit*'s "marginal position in the publishing field" and in part because of the opportunity the publication of a controversial book offered his business.[29]

The second non-ideological interest that made the press particularly stubborn and cohesive, in spite of great ideological divides, had to do with the state's reaction to the dynamic of publications and the position of the press vis-à-vis the latter. As I explained earlier, the state decided to curb press freedoms, harass journalists, and target publishers economically after its agents realized that they were unable to compete in the free marketplace of ideas. In doing so, however, the state threatened the collective institutional interest of the press, and therefore pushed even newspapers that did not oppose the objective of French Algeria to resist the authorities. Indeed, the news media, excluding those of the far Right, closed ranks and defended their right to publish without state intervention. This kind of institutional fraternization was displayed when *La gangrène* was seized, and even more pronouncedly, when Arnaud was detained, tried, and jailed following his participation in

---

[28] Quoted in Hamon and Rotman, *Les porteurs*, 208.
[29] Simonin, "Les éditions," 227. Lindon also decided to publish Arnaud's and Vergès' *Pour Djamila Bouhired*, after Julliard hesitated, saying: "*C'est un paquet de merde ... mais si personne n'en veut, je le prendrai.*" See Hamon and Rotman, *Les porteurs*, 75.

the Jeanson clandestine press conference.[30] Some 200 journalists signed a petition protesting Arnaud's arrest, and at his trial editors and journalists from newspapers of diverse political affiliations, including the editor of *France-Soir*, who had refused to publish his story, testified on his behalf.

The Consequences of Political Relevance

While it is clear that as the war progressed, and particularly as of the Battle of Algiers, the anti-war movement gained numerical strength and established firm control over the agenda, it is equally clear that it failed to achieve its immediate objectives. The army did not change its battle philosophy and combat methods, and the majority of the French remained indifferent to the brutality exercised by French soldiers in Algeria. Nevertheless, the criticism of the intellectuals and the press was at the root of the French disengagement from Algeria. It fractured the state system, drove the loyalists of French Algeria into desperation and sedition, created the irreparable rift between continental France and the army in Algeria, and ultimately made Algeria both the cause and the price of the divorce between France and itself.

In fact, it is inconsequential whether or not the campaign really fractured the cohesion of the state, converted key actors, or shifted public opinion, though it can be shown to have done just that. What mattered most was how the loyalists of French Algeria perceived the impact of the anti-war movement. This perception was the precursor that pushed the extremists to act in ways that presented France with a narrow choice between fighting in Algeria and preserving the democratic order at home, rather than one between defeat and victory. Indeed, these claims are strongly supported by the rhetoric and actions of army officers and other bureaucrats, as well as by political developments during the war.

The type of divisive impact the anti-war campaign had at the bureaucratic level can best be demonstrated by reviewing the tension between police officers of continental origin and their *pieds-noirs* and army counterparts.[31] The "continental" officers had complained several times to their superiors, first Mairey and then Verdier, about the investigative practices widespread in Algeria. Some of these complaints were probably motivated by sincere revulsion. However, as a collective, the continental officers complained because they felt vulnerable rather than because they were ethically concerned. In a letter they submitted to Verdier in the fall of 1957, the "continentals" expressed their opposition to being subjected to the army's authority, particularly in the DOPs, because they felt they had to pay an unreasonable and

---

[30] Hamon and Rotman, *Les porteurs*, 157, 249.

[31] Toward the end of 1956, Lacoste even refused police reinforcements from the continent. See Jacques Delarue, "La police en paravent et au rempart," in Rioux, *La Guerre d'Algérie et les Français*, 258, 262.

unjust price for this mixed authority structure. They pointed out that while military officers were protected by their superiors and by anonymity, they were being held responsible for acts they did not commit, or were not even aware of. Most important from our current perspective, they articulated clearly the reason that made them complain:

French public opinion, on the whole, and even that of those citizens least informed, does not ignore that there is in Algeria a serious problem of 'special' methods that are occasionally applied...The habitual hostility of all, the defamatory rumors spread at all times against our corps, rumors all too rarely officially refuted, the suspicion cast, even officially, upon the 'police methods'...all that shows too well what peril threatens us.[32]

At the same time that continental police officers concluded that the campaign against the authorities' lawlessness in Algeria threatened them personally, the military commanders in Algeria concluded that it threatened the very future of Algeria. They were certain, as their rhetoric and actions indicated, that the press and the intellectuals had actually succeeded in undercutting the effort to preserve Algeria.[33]

Frustration with the meddling of the anti-war forces and with being castigated built up with every wave of revelations of army misconduct. For example, in the wake of one of these waves of criticism against the army, General Allard, the Commander of the Army Corps in Algeria, found it necessary to complain to his superior about the domination of the continental marketplace of ideas by the anti-war camp:

Be it the big daily or weekly press, or the publications of publishing houses, hardly a week goes by without articles denouncing to readers the army's attitude in this affair (of Djamila Bouhired) being published...two particular texts catch one's attention: [in] *L'Express*...Mauriac...[and in] a brochure edited by the *Editions de Minuit*, Georges Arnaud and Jacques Vergès...It seems to me impossible to leave without response to such grave attacks, whose persistent renewal jeopardizes the army's morale.[34]

In fact, the army in Algeria had good reasons to fear the political relevance of the press and the intellectuals and the consequent growing likelihood that France would negotiate with the FLN and perhaps give up Algeria. In February 1956, Mollet offered to negotiate with the Algerians (though only on reforms and not with the FLN), and between April and September he conducted probing negotiations with Mohammed Khider, a spokesman for

---

[32] Delarue, "La police," 267.

[33] In 1957, Soustelle had singled out *France-Observateur*, *L'Express*, *Témoignage Chrétien*, and *Le Monde* as "the big four of anti-French propaganda." In retrospect, Massu arrived at the conclusion that Servan-Schreiber's series of articles in *L'Express*, *Lieutenant en Algérie*, constituted "the starting point of the campaign against the army in Algeria." Quoted in Massu, *La vraie bataille*, 227.

[34] A December 4, 1957 letter to General Salan, quoted in Massu, *La vraie bataille*, 258–59 (italics added).

the FLN. Soon after the beginning of the Battle of Algiers, signs multiplied that Mollet was caving in to the pressure created by the attacks on the army. In April 1957, he agreed to create a committee (the Garçon committee) to study alleged cases of torture in Algeria against the wishes of the three top ministers in charge of Algeria: Lacoste, Bourgès-Maunoury, and Lejeune.[35] Admittedly the army had little to fear from the work of the committee (which was nominated in May) since its members were presumably "safe" and its powers very limited. But the fact that the government was ready to bow to pressure and that the committee's findings were eventually leaked to the press were real causes for concern. Furthermore, the rhetoric coming from officials seemed to indicate that the Moralist campaign was effective, as the government started to distance itself from the Army. Indeed, the occasional criticism within Mollet's cabinet – particularly by Gaston Defferre, Alain Savary, and even François Mitterrand – did little to alleviate fears that the anti-war movement was getting the upper hand.

In February 1958, such fears within army circles seem to have been vindicated. After repeated skirmishes along the Tunisian border, the French air force decided to strike FLN installations in the Tunisian village of Sakiet-Sidi-Youssuf. Unfortunately, scores of civilians, including children, were killed in the bombing, and as a result a public outcry and a political storm broke out in France. The political and press pressure, and the army's feeling of being victimized, were vividly described in the memoirs of General Jouhaud, the air force commander in Algeria: "In France, Robert Buron would speak of 'the cynical and placid unawareness of certain military or civil authorities' . . . It was no longer the military people, whose heads were sought, but *it was Robert Lacoste or Felix Gaillard, who were fiercely attacked . . . Happily, not all the French press inveighed against us.*"[36]

Shortly after the bombing, Jacques Chaban-Delmas, a former brigadier general and the Gaullist Minister of Defense in Félix Gaillard's cabinet, visited Algeria. Upon his arrival, General Salan offered to show him the air force's Corsair airplanes. Chaban-Delmas, clearly concerned with the potential ramifications of having his picture taken with the airplanes, snapped at Salan: "Hide these planes so that I should not see them."[37] Thus, by 1958, the senior officers in Algeria (and other loyalists of French Algeria) were convinced that unless the military acted, the position of the French press, sympathy among Communists and Catholics for the FLN cause, divisions over Algeria – and what they considered a faint political response to these "threats" – would doom the French future of Algeria.[38] Four days before the May 13 coup, this sense of desperation was conveyed to President

---

[35] Hamon and Rotman, *Les porteurs*, 72.
[36] Jouhaud, *Ce que*, 58–59 (italics added).
[37] Ibid., 62.
[38] See also ibid., 75–77.

Coty in a telegram sent to him by General Salan and other senior officers in Algeria via the French Chief of Staff, General Ely. In the telegram, the officers explained:

The contemporary crisis shows that the political parties are profoundly divided over the Algerian question. The press gives the impression that the abandoning of Algeria was to be considered by a diplomatic process, which would start with negotiations in view of a cease-fire... The army in Algeria is troubled... The entire French army would feel that the abandonment of this national heritage [Algeria] would be an outrage...[39]

Under de Gaulle's presidency and Debré's premiership, things did not look any better for the loyalists of French Algeria. The press and intellectuals continued to attack the army, and their support for negotiations with the FLN and Algerian independence increased steadily. The labor unions and the students' and teachers' organizations added their voices to the anti-war campaign, and de Gaulle was gradually abandoning the tough French position on Algeria that governments before him dared not challenge. Finally, forces within the government seemed to collude with the opposition to French Algeria. Officials in Michelet's Ministry of Justice, for example, made members of the anti-war camp privy to documents such as the Michel Rocard report on the conditions in the internment camps and the classified preliminary report of the International Red Cross on torture and the *regroupement* camps.[40]

We have no direct measurement to tell us whether, or to what extent, the moral campaign influenced the general public in France, though we know that after the Algerian war began to be debated in the press, public opinion increasingly shifted toward negotiations and compromise in Algeria. However, we have indications that in spite of years of state instigation against intellectuals and the press, the French did not consider both, as one might have expected, as obstacles to peace, not even in the high days of "treason" following the Jeanson trial, the "121 manifesto," and "Sartre's" letter. In response to a November 1960 IFOP poll, only 5 percent of the French blamed the intellectuals of the Left for preventing a quick resolution of the war.[41] In contrast, the army was blamed by 19 percent and the colonialists by 36 percent. Even de Gaulle was blamed by 6 percent – 1 percent more than the intellectuals.

---

[39] Quoted in Montagnon, *La Guerre d'Algérie*, 252.

[40] *Le Monde* published the Rocard report on April 18, 1959, and the Red Cross report on January 5, 1960 (also published by *Témoignage Chrétien*). The Rocard report is reprinted in Éveno and Planchais, *Dossier et témoignages*, 223–28. On the relations of *Le Monde* with Michelet's 'lieutenants,' see Rovan, "Témoignage sur Edmond Michelet," in Rioux, *La guerre d'Algérie et les Français*, 277–78; Vidal-Naquet, *Face*, 22 note 38, 53; Alleg et al., *La Guerre d'Algérie*, Vol. III, 217–19; and Hamon and Rotman, *Les porteurs*, 159, 171.

[41] *Sondages*, 1961:1, 12.

Finally, there are strong reasons to suggest that the moral campaign contributed considerably to the shaping of the conscripts' and reservists' opinion. I have already discussed the opinion of Massu, the army command, and the Defense Ministry that the anti-war campaign "demoralized" the soldiers in Algeria. In April 1961, the belief system of the soldiers was put to a test when four generals and numerous colonels staged a coup d'état in Algeria. The conscripts were faced with a personal choice of loyalty and an ideological choice between military supremacy, in order to keep Algeria French, and the preservation of democracy on the continent. After de Gaulle appealed through the electronic media for support for his legitimate rule, the coup quickly collapsed. Many units in Algeria refused to join the conspirators. Even if one accepts the idea that de Gaulle's firm broadcast kept the balance from tilting in favor of the putschists, we are still left with the question as to why many soldiers were receptive to de Gaulle's message in the first place. Some would attribute this receptiveness to de Gaulle's rhetorical skills, stature, and passion. Admittedly, de Gaulle was a skilful communicator. However, soldiers hesitated to join the insurrection prior to his speech, and in any case his success was predicated on the soldiers' readiness to accept the massage he delivered rather than his authority alone. At least partially, this mental readiness to be turned away from the insurrection was the result of the embedded values the conscripts held, as well as years of exposure to the struggle against the war and the fascist threat in France.[42]

Whatever de Gaulle's precise objectives or thoughts on the future of Algeria were, the fact remains that he gave up Algeria entirely, only after he failed to secure a different outcome.[43] Indeed, he sought to defend as many French interests in Algeria as possible, and thus tried as of 1958–1959 to bring the nationalist forces to the negotiation table from a position of strength and on his own terms. He repeatedly failed, and it is plausible that these failures were at least partly due to the domestic struggle. By 1960, the role the domestic struggle in France played in the final outcomes of war was no more an issue to speculate about. In less than two years, it led France to capitulate on every major issue it had considered as vital national interests.

In January 1960, de Gaulle lost control of events in Algeria. On the 19th he ordered Massu back to France after Massu had criticized him and his policy in a public interview. This resulted in a rebellion of the Ultras in Algiers. A French mob took control of key sites in the city, shot and killed

---

[42] See Planchais, *Une histoire politique*, 359–64. Domenach argues that articles in *Esprit* influenced a significant number of junior officers in Algeria. See "Commentaires," 93. Julliard makes a similar argument, as a former soldier in Algeria, about the impact of the intellectuals. See "Une base," 363.

[43] See Gil Merom, "A Grand Design? Charles de Gaulle and the End of the Algerian War," *Armed Forces and Society* 25:2 (1999), 267–88. See Lacouture's opinion of de Gaulle's fundamental objective in Algeria in *De Gaulle*, 199.

fourteen gendarmes (wounding over 120), and barricaded itself for a week in what became known as the Barricades' Week. It was by the end of this week that de Gaulle must have finally realized that he could never have the upper hand in Algeria. The Algerian problem needed to be solved not only before it dragged France into a full-fledged civil war, but also before it consumed him. On the one hand, the anti-war intellectuals and the press dominated the marketplace of ideas, and the continental public was convinced that Algeria had to, and would be, given up. On the other hand, the French Algeria camp, fearing that Algeria was about to be lost, was on the verge of explosion. In addition, the confidence of the political elite in its own ability to control events was deteriorating rapidly.[44] De Gaulle simply ran out of options. He could only join one camp, and if he planned on remaining significant, it was necessarily the continental one. He wavered for a while, trying to please both the continental public and the Ultras in Algeria.[45] But it did not matter any more. On the continent, Algeria was already considered a lost cause, and, as the ferment against de Gaulle and attempts on his life in Algeria suggested, he enjoyed no credibility among the members of the French Algeria camp.

Regardless of de Gaulle's thoughts, the FLN leaders, who were attentive to developments within France,[46] seem to have realized that he needed a settlement in Algeria more desperately, and sooner, than they did. Thus, the fact that the FLN was badly beaten mattered little. Once de Gaulle embarked on the road of compromise, the French political readiness to continue to pay the same price for lesser objectives was, by definition, doubted. This indeed was indicated in early June 1960, when the French were approached by the commanders of the FLN's Wilaya IV with a presumed offer for separate peace negotiations. De Gaulle declined the offer, and instead publicly invited the "leaders of the insurrection" to start negotiations with France without preconditions. The second pillar of the French Algeria policy – the refusal to recognize the FLN as a partner to negotiations – was buried, and as a result, a delegation of the GPRA met with French representatives in Melun. However, because de Gaulle was not ready to capitulate on core issues, no significant tangible achievements were accomplished in Melun.

In the fall of 1960, de Gaulle buried the last pillar of the French Algeria policy – issuing several declarations conceding that Algeria would become independent – and thus the Muslim population, realizing that it would shortly

---

[44] See Lacouture, *De Gaulle*, 258.

[45] On January 29, during the upheaval in Algeria, de Gaulle appeared on TV, restated his support for the self-determination plan, but also told the French that it did not involve an abandonment of Algeria. During a March 1960 tour of French bases in Algeria, he told his public: "There shall not be a diplomatic Dien-Bien-Phu.... The insurrection will not throw us out of this country ... France must not leave. She has the right to be in Algeria. She will stay." Quoted in Montagnon, *La Guerre d'Algérie*, 312. See also ibid., 308.

[46] The FLN targeting of French public opinion was revealed already in the 1956 text of the Soummam conference. See Tripier, *Autopsie*, 580, 583, 597.

be governed by the FLN, flocked to its side.[47] As of January 1961, France had a new player on the scene, the OAS, the terrorist organization of the extreme loyalists of French Algeria. Algeria, in short, was getting out of control, and as OAS terror spread to Paris, France seemed to follow suit. Thus, when the FLN came to negotiate in Evian, it was well aware of France's predicament. Still, as if unmoved by events, the French tried again in May to start negotiating from a position of strength. The FLN, however, confident of its bargaining power, promptly refused to start negotiations until France had caved in on its demand to partition Algeria so as to preserve the oil-rich Sahara. Still, as de Gaulle clung to the last vestiges of the French national interest, and refused to bargain over the future of the Sahara, the talks soon collapsed. Only in September 1961, after a failed July round of talks in Lugrin, did de Gaulle accept the loss of the Sahara. A few additional rounds of talks ensued, resulting in the final negotiations in March 1962 and the agreement of Evian. France gave up its last substantial demand concerning the status and nationality of the European community in Algeria.

---

[47] When de Gaulle addressed the French people on September 5, 1960, he declared that "the Algerian Algeria is in the making." On November 4, he referred on television to "Algeria [that] would have its government, its institutions, and its laws," and to "the Algerian republic that would exist some day." Finally, on April 11, 1961, he stated in a press conference that "this state [Algeria] will be what the Algerians would wish it to be. As far as I am concerned, I am persuaded that it will be sovereign, inside and out... France would not obstruct." Quoted in Montagnon, *La guerre d'Algérie*, 326, 340; Jouhaud, *Ce que*, 189–91; and Ageron "L'Opinion française," 36.

PART III

The Israeli War in Lebanon

A Strategic, Political, and Economic Overview

In June 1982, Israel invaded Lebanon on a massive scale. The three most important reasons driving Israel's decision involved its basic conventions of national security, the perception that future confrontation with the PLO in Lebanon was inevitable, and the personal composition of the pinnacle of the defense establishment during the second Likud government.[1] On the one hand, the political alternative to using military power – negotiations with the PLO over the resolution of the Israeli-Palestinian conflict – was rejected a priori by both the right-wing government of Israel, its main opposition party, Labor, and most of the leaders of the various PLO factions. On the other hand, the two main strategic alternatives to a deep invasion – continuation of the ceasefire in the theater of operations and the maintenance policy that combined measured retaliations and limited operations – were rejected as insufficient by the central leadership of the Israeli defense establishment. This last point was critical. While Israeli leaders up until 1981 preferred to exercise restraint and treat the PLO threat from Lebanon as a maintenance problem, once Sharon became Defense Minister, the probability that Israel would chose a 'once and for all' strategy against the PLO presence in Lebanon dramatically increased. In that respect, the June 1982 invasion was predetermined, though its exact date was left for the "next" random yet certain Middle East spark – in this case, the May 31, 1982, attempt to assassinate Shlomo Argov, Israel's ambassador to the United Kingdom.

---

[1] On the international, strategic, and decision-making aspects of the war, see Richard A. Gabriel, *Operation Peace for Galilee* (NY: Hill and Wang, 1984); Zeev Schiff and Ehud Yaari, *Milhemet Sholal* (Jerusalem: Schocken, 1984) [published in English as *Israel's War in Lebanon* (NY: Simon and Schuster, 1984)]; and Avner Yaniv, *Dilemmas of Security* (NY: Oxford University Press, 1987).

The Strategic Dimensions of the War in Lebanon

Israel's war effort in Lebanon was not a single homogeneous effort. Rather, Israel conducted three wars in Lebanon, in two phases. In the first phase, it fought two wars of territorial acquisition: one against the PLO and one against the Syrians. In the second phase, Israel was involved in a protracted guerrilla war against various groups in Lebanon. Irrespective of the criticism one may hold against the decision to go to war, the quality of the performance of the Israel Defense Forces (IDF), or the cost of the war, from a military point of view Israel did quite well in Lebanon.[2]

In the first phase of the war, the IDF succeeded in achieving the territorial objectives that the Israeli government instructed it to accomplish. Within a relatively short time, superior Israeli forces – consisting of some 80,000 soldiers, organized in six to seven divisions with strong armored power and protected by the superior Israeli Air Force (IAF) – compelled the Syrians and the PLO to take extremely unpleasant decisions.[3]

The Syrians had to withdraw from their positions in the southeastern part of Lebanon and in Beirut. Their air defense system in Lebanon was virtually eliminated. They lost about a quarter of their first-line combat jets and many of their best pilots. A considerable number of their armored vehicles were likewise destroyed.

The PLO suffered even more disastrously. By all accounts and against the Israeli expectations, its warriors fought with courage and discipline.[4] But bravery is little comfort for losers. Whatever the PLO had built in southern Lebanon for over a decade, it lost in a matter of days. The PLO leadership tried to put on the best face in its defeat, but being unable to match the Israeli power, the PLO was compelled to choose between the lesser of two evils – a heroic extinction or a humiliating deportation. Because it chose the latter, it had to eventually accept, as Abu Iyyad, Arafat's deputy, promptly defined it, "surrender terms."[5] Indeed, Issam Sirtawi, another prominent PLO member, summed up the war by saying: "Lebanon was a disaster.... If Beirut was such a great victory, then all we need is a series of such

[2] On the military aspects of the war, see Gabriel, *Operation Peace for Galilee*; Trevor Dupuy and Paul Martell, *Flawed Victory* (Fairfax, VA: Hero Books, 1986); Anthony H. Cordesman, *The Arab-Israeli Military Balance and the Art of Operations* (Lanham, MD: University Press of America, 1987); and Yair Evron, *War and Intervention in Lebanon* (Baltimore: Johns Hopkins University Press, 1987), 129–42. For criticism of the military operation in Lebanon, see interview with Dov Tamari, *Monitin*, 62 (October 1983), 78–80; and Yaniv, *Dilemmas*, 135.

[3] For the orders of battle, see Gabriel, *Operation Peace for Galilee*, 21, 50–53, 231–33; and Dupuy and Martell, *Flawed Victory*, 86–89, 91–94.

[4] Refael Eitan with Dov Goldstein, *Sippur Shel Hayal* [A Story of a Soldier] (Tel Aviv: Maariv, 1985), 263.

[5] Quoted in Rashid Khalidi, *Under Siege: PLO Decision-Making During the 1982 War* (NY: Columbia University Press, 1986), 84.

victories, and we will be holding our next National Council meeting in Fiji."[6]

Moreover, the initial Israeli achievements were obtained at a reasonable cost. Richard Gabriel assesses the casualties-to-forces ratio among the adversaries in the first phase of the war as follows: The PLO lost 12 percent of its forces committed to war, the Syrians 2.5 percent, and Israel 0.5 percent. The kill-ratio is assessed by Gabriel to have been 1 Israeli fatality for every 6.5 PLO fatalities and 1 Israeli fatality for every 4 Syrian fatalities.[7] Indeed, the losses, as much as the territorial results, indicated the magnitude of the initial success of the Israeli military.

The war in Lebanon, however, did not end once its dynamic phase was over, because the Israeli leadership never intended to confine itself to its single declared objective of pushing back the PLO artillery forty kilometers from the Israeli border. Rather the war continued in a different manner since Israel intended to achieve four additional objectives. First, Israel wanted Syria to withdraw its forces from Lebanon and terminate its political domination there. Second, the Israelis wanted to restructure the domestic Lebanese balance of power and crown their young Christian Maronite ally, Bashir Gemayel, as the President of Lebanon. Third, they wanted to sign a peace agreement with Lebanon. Last, they thought that Palestinian national demands in the territories would be reduced as a byproduct of victory in Lebanon, and in particular of the destruction and humiliation of the PLO (also hoping that Jordan would come a step closer to becoming the Palestinian state).[8]

On all four counts Israel failed miserably. The Syrians did not relinquish their grip on Lebanon, but rather returned to Beirut after the initial setback, only to become once again the power brokers in Lebanon. On September 14, 1982, a Lebanese agent of the Syrians blew up Bashir Gemayel, in one of the headquarters he visited, and with him the prospects for a strong Christian regime in Lebanon. If anything, the war contributed to the decline of Christian power in Lebanon, increasing instead the power of the far less manageable Shiite fundamentalists. In May 1983, Israel signed an agreement with Lebanon, but the document was a far cry from a peace agreement, and turned out to be utterly worthless.[9] Finally, the Palestinians in the occupied territories turned less docile, and it appears that Israel's general deterrence posture (not only vis-à-vis the Palestinians) deteriorated as a result of the

---

[6] Quoted in Yaniv, *Dilemmas*, 168.

[7] Gabriel, *Operation Peace for Galilee*, 182.

[8] For an excellent analysis of Israel's goals, see Evron, *War and Intervention in Lebanon*, 105–18. See also Shai Feldman and Heda Rechnitz-Kijner, *Deception, Consensus, and War: Israel in Lebanon*, JCSS paper no. 27 (Tel Aviv: Tel Aviv University, distributed by Westview, 1984), 10–24; Aryeh Naor, *Memshala Be'milhama* [Cabinet at War] (Tel Aviv: Lahav, 1986), 32, 74; Yaniv, *Dilemmas*, 100–07; and Eitan, *Sippur*, 256, 266, 286, 290.

[9] The agreement is quoted in Yaniv, *Dilemmas*, 327–38.

war. Thus, in spite of its indisputable military superiority and relatively high degree of international autonomy,[10] Israel failed to achieve any of its political objectives in Lebanon, aside from the expulsion and degradation of the PLO.

The Israeli Political System, the State, the Army, and the War

The Lebanon war was to a large degree the outcome of an extreme version of Realpolitik ideology, which is not much of a surprise considering the historical record of the Jewish people. A long history of persecution, an almost successful Nazi attempt at genocide, and a struggle from inception against neighbors who rejected Israel's legitimacy and sought its destruction instilled among Israelis in general and their political elite in particular a psychological climate that was necessarily conducive to power politics. Indeed, the fundamental security creed of the Israeli leadership sprang directly from the judgment that the historical misery of Jews was due to an anarchic and indifferent world and the Jews' weakness and defenselessness.[11] Hence, the Israeli operational conclusion that self-help was the best solution to the bitter Jewish predicament and the belief that the state, being the guarantor of national survival, should be supreme. Indeed, the latter conviction was well reflected in Israel's institutional structure and budgetary priorities.

Still, although Israel was disposed by nature to act in a Realpolitik manner when threatened, it was not until certain political changes occurred in Israel that Lebanon was likely to become the site of a *large-scale* Israeli invasion. These changes were on two levels. First, in 1977, the center-right Likud party replaced the center-left Labor party in power. Second, the personal composition of the Likud leadership changed in 1981 in the second Likud government.

Traditionally, the leaders of Herut, the dominant faction within the Likud, had supported an activist, irredentist, and opportunity-driven state-centered

[10] Israel's major international constraint was United States policy. Initially the Reagan administration was ambivalent about the war. In June 1982, it even suspended the delivery of aircraft to the IAF, and in July it suspended the delivery of cluster bombs. However, the Secretary of State, Alexander Haig, supported Israel until his dismissal (June 25, 1982). In early-mid 1983, the administration reversed its policy, and indeed, when Israel planned its first withdrawal, in the summer, the United States tried to delay it. See Yaniv, *Dilemmas*, 206–15; and Barry Rubin, "The Reagan Administration and the Middle East," in Kenneth A. Oye, Robert J. Lieber, Donald Rothchild (eds.), *Eagle Resurgent?* (Boston: Little Brown, 1987), 445–46.

[11] See Asher Arian, Ilan Talmud, and Tamar Hermann, *National Security and Public Opinion in Israel*, JCSS study no. 9 (Jerusalem: The Jerusalem Post, distributed by Westview Press, 1988), 16–30; Yigal Elam, *Memalei Ha'pkudot* [The Orders' Executors] (Jerusalem: Keter, 1990), 48–56, 75–84; and Gil Merom, "Israel's National Security and the Myth of Exceptionalism," *Political Science Quarterly* 114:3 (1999), 410–17.

policy.[12] However, not all Likud leaders, nor all ministers in Begin's first government were ready to match the party rhetoric and ideology with deeds. Indeed, as long as moderates such as Defense Minister Ezer Weitzman and Foreign Minister Moshe Dayan surrounded Herut's leader, Menahem Begin, Israel's foreign policy was rather restrained. Unfortunately, the standing of Dayan and Weitzman was weakened, and they both left the government. When that happened, the government was still left with a few moderate ministers, but the latter were far less influential and could not stop the far more "activist" policy line that was forming among the state leadership. Indeed, when Begin decided, in a fait accompli manner, to annex Jerusalem and subject the Golan Heights to Israeli law, his ministers dared not challenge him seriously. Rather, they granted him support with little, if any, objection.

Yet, of all the reconfigurations of the power structure within the Likud government, none was of greater significance than the nomination in August 1981 of Ariel Sharon to the position of Defense Minister in Begin's second government. By mid-1981, then, the top three positions in the Israeli defense hierarchy were in the hands of hard-line power politicians – Prime Minister Begin, the IDF Chief of the General Staff (CGS) Lieutenant General Rafael Eitan, and Defense Minister Sharon. Moreover, Sharon was a most effective political entrepreneur, and Begin was not restrained anymore by powerful moderate figures such as Dayan and Weitzman. Thus, in rather short order, Israel's security orientation assumed a more offensive nature,[13] and its leadership was ready to execute policies that critics previously dismissed as rhetorical reverberations from the Likud's years in opposition.

Indeed, it is important to briefly analyze the special nature of the combination of Sharon, Eitan, and Begin if one is to really understand the shift in policy toward the PLO in Lebanon. Furthermore, it is worthwhile to start with Sharon, as he was the chief catalyst of the policy transformation, and as two of his attributes in particular – his strong etatist conviction and the sheer force of his personality – shaped Israeli war policy more than anything else.[14] These personality traits were revealed long before Sharon became Defense Minister, in his quest for bureaucratic imperialism – swift and irritating

[12] See Gad Barzilai, "Democracy in War," 41–45, 63, and "A Jewish Democracy at War: Attitudes of Secular Jewish Political Parties in Israel toward the Question of War (1949–1988)," *Comparative Strategy* 9 (1990), 179–94; and Ilan Peleg, *Begin's Foreign Policy, 1977–1983* (NY: Greenwood Press, 1987), 53–54.
[13] See Dan Horowitz, "Israel's War in Lebanon: New Patterns of Strategic Thinking and Civilian Military Relations," in Moshe Lissak (ed.), *Israeli Society and its Defense Establishment* (London: Frank Cass, 1984), 91–95. Clearly, Sharon's and Eitan's *minimal* objective in Lebanon was total, in the sense that they sought to *eradicate, once and for all,* the PLO presence there. See Ariel Sharon with David Chanoff, *Warrior: The Autobiography of Ariel Sharon* (NY: Simon & Schuster, 1989), 431–32; and Avraham Tamir, *A Soldier in Search of Peace* (Tel-Aviv: Edanim, 1988), 146.
[14] Sharon's etatist ideology resembled most the Weltanschauung of a nineteenth-century German Realist. See for example *Warrior*, 531.

decision-making style, sour relations with peers in general and the press in particular, and in his periodic slips of the tongue or when he articulated his opinions.[15] They also left a clear mark on his potential critics and sub-ordinates, as was indicated by their submissive reactions to his demeanor.[16] Sharon's essential attributes, actions, and impact continued to have the same effect once he became Defense Minister. He centralized the defense ministry, increased his control over the military, and isolated the latter from the government and the press. Once in office, he created the Unit for National Security (UNS), an agency that General Eitan, his main associate in the Lebanon war "had no doubts" was intended "to create ... a kind of a small 'General Staff,' that would give [Sharon] greater independence in military-security activity and free him from dependence on the [IDF's] General Staff."[17] To head the UNS, Sharon nominated General Avraham Tamir, an officer on active duty with extraordinary organizational skills who was not too choosy about his political masters (or, alternatively, was eager to serve the "state").

The second key personality in the decision to go to war, Lieutenant General Rafael Eitan, shared many views with Sharon. Their opinions were not identical, but both believed in a rather extreme version of Realism. Like Sharon, Eitan supported the supremacy of the state and of military consid-erations, and distrusted the staying power of affluent societies to the same degree. He was also almost as ready to use military power, though he was clearly more patient with the ideas of his peers. Finally, much like Sharon, he distrusted the media, as he believed that the latter must support the state and its interests (as Realists would define them), yet it cannot be trusted to do so.

The third key personality in the decision to launch the Lebanon war, Prime Minister Menachem Begin, was of a different quality, but neverthe-less matched Sharon's and Eitan's grand plans almost perfectly. He was a charismatic leader, he had a penchant for power politics, and he could carry

[15] After the Yom Kippur War, Sharon explained his order of loyalties as follows: "When it comes to fulfilling instructions ... I rank [my duties] thus: my duty to the country; my duty to the soldier; my duty to my commander." In June 1980, while frustrated with his relentless yet futile efforts to secure the Defense Ministry, for himself, he burst out: "Security is above the constitution." Occasionally, Sharon referred to journalists as "creeps" and "traitors." In April 1980, for example, Sharon snapped at a TV crew that caught him chatting with picketing settlers across from Begin's office: "You are a band of saboteurs ... you ruined the country. Look how it is being destroyed. It is your fault." See Uzi Benziman, *Lo Otzer Be'Adom* [*Would not Stop at Red Light*] (Tel Aviv: Adam Publishers, 1985), 268, 229, 232. For other accounts of Sharon's mode of behavior, see ibid., 206–10, 217, 224–25, 226–27; and Schiff and Yaari, *Milhemet*, 114.

[16] Reserve Brigadier General Tamari, an officer of great integrity, described the impact of Sharon on the IDF command as follows: "The General Staff got scared and became terribly frightened. The fear was so great that some of the people did not greet me – and those were people I had worked with for years." See *Monitin* 62 (October 1983), 78, 68.

[17] Eitan, *Sippur*, 196.

the government almost anywhere. In fact, he had already proved quite fearless (or reckless, depending on one's point of view) in the face of international pressures. For example, he was ready to take resolute fait accompli measures, as revealed in his decisions on the Israeli status of Jerusalem and the Golan Heights. Indeed, it was precisely because he "never dreaded the Gentiles" that General Eitan respected him.[18] Finally, what made Begin a good partner for Sharon's and Eitan's plans was his attitude toward military power. On the one hand, he was deeply ignorant of military matters, and on the other, he was highly impressed by military power and greatly admired Jewish military heroes. In this last respect, Sharon and Eitan offered him much. Each had an impressive combat record and each was considered to have been instrumental in Israel's success in fending off and turning around the joint Egyptian-Syrian attack in October, 1973. If Begin had any "vice," from his partners' perspective, it was his true commitment to democracy (though his demagoguery and conduct often brought out the dark sides of populism).

While Begin, Sharon, and Eitan brought about the war, it was not simply the combined attributes of each of them that made the war possible, but rather also the fertile political ground on which these attributes operated. These three leaders brought together a volatile combination of attributes that was reminiscent of that which Wilhelm II, the younger Moltke, and Bethmann-Hollweg displayed on the eve of World War I. Among the three of them there were strong rhetoric, impatience with diplomacy, affinity for radical solutions, militaristic simplism, adventurism, and a civil complacency that resulted in the abdication of authority. However, while this combination produced the overall concept of the Lebanon war, the war could not have been carried out without the consent of the majority of political players and the press in Israel, and a fundamental conformity of Israelis. That, of course, does not mean that Israelis universally or enthusiastically endorsed the war, let alone its ambitious objectives. Rather, what I argue is that the widespread belief in power politics in Israel, and the deep trust of Israelis in the government whenever security matters were concerned, provided the foundations upon which the war leadership could risk launching a bold strategic gamble in Lebanon.

Indeed, the uncritical endorsement of the etatist ideas that underlie power politics, and in particular the idea that the government has the freedom to define the national interest as it pleases and act upon that definition, were expressed rather clearly. Most notably they were indicated by the rather timid reaction of the press and the major opposition party, the Labor Alignment, to the Lebanese initiative. In fairness, one must also note that Labor did not blindly support the government, but instead was convinced to do so, or perhaps maneuvered into giving the government the freedom of action Begin

---

[18] Ibid., 192.

sought.[19] Begin invited the Labor leaders Itzhak Rabin, Shimon Peres, and Haim Bar-Lev *twice* (in April and May 1982) for briefings about the imminent war. According to the Labor leaders, in these meetings they were presented with plans for an expanded but limited version of the 1978 Litani operation. That, however, does not mean that the Labor leadership believed the limited plans Begin and Sharon presented to them. The three Labor representatives were security experts of the first order – leaders with extensive military and security background, vast political experience, and open communications with the IDF command. Moreover, they knew Sharon well and they got wind of the big, up-to-Beirut, war plans. Considering these facts, it is somewhat odd that in spite of their sensing that Sharon was misleading them, they chose to listen and seek reassurances about the limited nature of the war, rather than vigorously oppose the concept of the extended war.[20]

The precise motivations and calculations of both parties at these meetings remain a matter for speculation. Clearly, at a minimum, Begin wanted to build a political consensus on the eve of the war. However, he may have had additional objectives. For example, he may have wanted to get a first-hand impression of just how far politically he could drag Labor. Or perhaps he sought to limit Labor's freedom to maneuver by making its leaders into partners by their acquiescence to his plans. The leaders of Labor, on the other hand, may have thought that Begin opposed a large operation, that he was strong enough to contain Sharon, or that, being a man of honor, he could be trusted when he suggested that he intended to keep the war limited. Or they may have been negligent because they assessed their position as a no-win situation, fearing that strong opposition to the war would provide the Likud with the opportunity to smear Labor and depict it as treacherous and soft on security matters. Whatever the case, it is hard to see how the leaders of Labor could not have been suspicious of the nature of the coming war. After all, why were they called to Begin twice for discussions, and why was their approval so important if the operation was to be of a limited nature?

In any event, the significant point about the meetings between Likud and Labor, and what followed, is that the road for war was paved with a political consensus. Thus while during the first days of battle the Labor Party issued some cautious statements against the expansion of the war beyond the declared 40 kilometers, as the IDF pushed further into Lebanon and defeated the PLO and the Syrians, its warnings increasingly sounded as polite advice and "for-the-record" notes. In fact, if anything, the initial Labor ambivalence was soon replaced by sporadic support. For example, Haim Hertzog, a senior Labor member with a long record in military and state service (who eventually became President), argued that the military victory in Lebanon

---

[19] On the pre-war deliberations and cabinet decisions, see Evron, *War and Intervention in Lebanon*, 118–26.

[20] See in Schiff and Yaari, *Milhemet*, 111–12, 118–19.

opened the road for "new political opportunities" – a euphemism for embracing the ambitious objectives of the extended war.[21] Indeed, in early July 1982, Hertzog explained that "those opposing the exploitation of the newly created political situation [in Lebanon], because we had progressed beyond the 40 kilometers, are unrealistic (naive)."[22]

Hertzog may have been the least subtle, but he certainly was not the only Labor member to support the war policy. Other members of Labor felt that once Israel was engaged in war, national considerations should outweigh all other expedient calculations. Itzhak Rabin, for one, who as Prime Minister avoided the trap of committing Israel to the Christians in Lebanon, refused to invade that country on a massive scale, and did not see much wisdom in the war's ambitious goals, nevertheless refused to attack the war policy at the time. Rather, he took an etatist position, and even volunteered to help to salvage as much as possible from what he believed was a mess the Likud had brought upon Israel. "During war" Rabin argued, "there is no room for public debate."[23] Moreover, he added that he was not "impressed ... [by] statements such as 'misleading' 'lying' and [by] blaming the government [for deceiving], because [he did] not know of an initiated [military] operation in which the government could tell the whole truth with respect to the war goals."[24] Indeed, Rabin faithfully put aside partisan considerations, and went as far as to support Sharon's siege of Beirut because he was concerned that an IDF failure to expel the PLO from Beirut would be a political and psychological failure not just for the Likud, but for Israel as well.[25]

In fact, Labor was not the only opposition party to initially support the war and advise the government to exploit the IDF's achievements. Some journalists in the liberal press, and even left-of-Labor politicians, proved no less supportive of the war than their right-wing counterparts. Among these supporters were those from the kibbutz movement and the Mapam party, most notably Member of the Knesset Imri Ron and the Mapam Jerusalem branch head, Hillel Ashkenazi.[26]

Finally, the war leaders took advantage of what they perceived as the likely widespread popular support for power politics, which sprang from the fundamental etatist proclivities of the Israeli public at large. It was not a secret that Israel's demographic and territorial inferiority vis-à-vis its Arab foes compelled Israel to rely on extremely high levels of resource-mobilization from its citizenry, and that these high levels required the latter to

---

[21] *Maariv*, June 25, 1982.
[22] *Maariv*, July 9, 1982.
[23] Itzhak Rabin, *Ha'milhama Be'levanon* [The War in Lebanon] (Tel-Aviv: Am Oved, 1983), 29.
[24] Ibid., 30. Rabin referred mainly to international constraints on the government's ability to be candid.
[25] Ibid., 35.
[26] See Ran Edelist and Ron Maiberg, *Malon Palestina* [Palestine Hotel] (Israel: Modan, 1986), 307. See also *Ha'aretz*, June 29, 1982.

accept Realist thinking and the supremacy of state power uncritically. Indeed, Israel's Jewish citizens were born and had grown up in an atmosphere that emphasizes the ideas of self-sufficiency, self-defense, and sacrifice. In short, as a result of outside pressure and domestic efforts, the Israeli public was fundamentally supportive of power politics. By the time Israelis became teenagers they had already assimilated basic Realist concepts, and by the time they had become adults they supported state autonomy in national security matters, with little reservations. Indeed, a post-Lebanon-war research study on security perceptions of Israelis (that is, after public confidence in the government was already damaged) demonstrated with great clarity that most Israelis tended to trust and support their government's judgment in matters of national security, were ready in principle to sacrifice even more, and believed that in any event they could do little to change state policy.[27] Furthermore, the study also indicated that most Israelis believed that during times of war their duty was to support the government and state policy.[28] As this study concluded, the Israeli public's "religion of security" provided Israel's leaders with "enormous leverage" in shaping defense policies.[29]

The Israeli Economy and the Lebanon War

The Israeli economy was already ailing when the Lebanon war started. The growth rates of real wages in the public sector and of public and private consumption were higher than the growth rate of the GNP. In fact, the GNP was actually in decline. These ominous developments were accompanied by a large budgetary deficit and import surplus, and consequently by soaring domestic and foreign debt, whose composition was becoming increasingly inconvenient (the share of short-term credits with higher interest payments was on the rise). At the same time, the rate of national savings diminished, and the economy suffered from steep inflation (over 115 percent as of 1980).[30] In large measure, all of these problems occurred because of the economic incompetence of the Likud leadership.

Under such conditions, one could reasonably expect the war to have a negative impact on the economy, and hence on the political position of the government. The direct cost of the Lebanon war was approximately $1 billion in 1982, according to the Israeli Treasury, and closer to $2 billion,

[27] Arian et al., *National Security*, particularly pp. 80–88.
[28] Ibid., 40.
[29] Ibid., 83, 87.
[30] On the state of the Israeli economy at the war's beginning, see Yoram Ben-Porath, "Introduction," in Ben-Porath (ed.), *The Israeli Economy* (Cambridge: Harvard University Press, 1986), 1–23; Moshe Sanbar, "The Political Economy of Israel 1948–1982," in Sanbar (ed.), *Economic and Social Policy in Israel* (NY: University Press of America, 1990), 1–23, particularly pp. 17–20; and Yair Aharoni, *The Israeli Economy* (NY: Routledge, 1991).

according to the defense establishment.[31] As a result of the loss of work-
ing days and commercial international deals and so on, the war probably
generated an additional indirect cost of some $500 million. By November
1982, only five months into the war, it was also becoming clear that Israel's
economy was heading for trouble. In 1982, Israel's net foreign debt increased
by some $2 billion (from $13.437 to $15.473 billion, compared with about
$1.4 billion increase in 1981), inflation reached a new record high, private
consumption and government expenditures increased, yet the GNP declined
by 0.4 percent (further decline was avoided because of defense-related orders).[32] Theoretically, then, the war could potentially burden the stalling
economy, which would then conceivably constrain the conduct of war. Moreover, the combination of war and ailing economy could presumably contribute to a political backlash.

However, none of these effects occurred. While the economic deterioration accelerated during the war, the war did not have a major impact on
the economy, nor did the economy constrain the war effort. The economic
cost of the war and its potential political consequences were effectively kept
at bay for several reasons. First, the cost of maintaining the IDF forces in
Lebanon after the dynamic phases of war and the siege of Beirut declined
sharply. This cost is estimated to have been some $200 million a year, as of
the first year of war.[33] Second, none of the expenses of the war, including
the cost of losses and of replenishing the IDF's stocks, reached unmanageable proportions. Indeed, the overall defense cost as part of net resources
in 1982–1983, after factoring in the massive American aid (which declined
somewhat from its high 1981 level), was smaller than in the 1970s. Moreover, that cost/resources ratio continued to decline in 1984–1985.[34] Third,
the Treasury took effective steps to meet the new budgetary requirements
and mitigate the potential political consequences of the growing economic
burden the war had created. Soon after the war started, the Treasury declared that Israel would need external help, and that therefore borrowing
on international markets would increase.[35] At the same time, the Treasury
also increased its taxation at home, though it did so very carefully, guided
by three principles: diversification of extractive measures, indirect taxation,
and the securing of the cooperation of the Histadrut – the powerful Labor-
controlled umbrella organization of Israeli Labor unions. Thus, the Treasury raised the Value Added Tax (VAT) from 12 percent to 15 percent, while

---

[31] The difference is partially due to the 'replacement costs' the Defense Office adds to the direct
   cost.
[32] Data from Ben-Porath, "Introduction," 20–21, table 1.1.
[33] Yaniv, *Dilemmas*, 256.
[34] See Aharoni, *The Israeli Economy*, 86 table 2.3, and 253 table 6.1. On American aid to
   Israel in the early 1980s, see Kochav, "The Influence of Defense Expenditure on the Israeli
   Economy," in Sanbar, *Economic and Social Policy*, 33–36, particularly p. 35 table 2.3.
[35] *Ha'aretz*, June 17, 1982.

neutralizing 3 percent of the cost-of-living bonus. Then it reached an agreement with the Histadrut on subsidy reduction and on a progressive compulsory state "borrowing" of 2–5 percent from all breadwinners – the *Peace for
Galilee* loan. In addition, the government levied a 2 percent tax on sales on
the stock exchange, added a surtax on imports, and came up with a new tax
on travel abroad. The Treasury also took the opportunity of further cutting
subsidies, and in the autumn of 1982, as a measure to control inflation, set
the increase in the government controlled prices of basic foodstuffs and other
goods and services at a constant 5 percent a month (these included energy,
public transportation, and communications).

This economic analysis indeed reveals that the politically destabilizing effects of the first months of the war were offset by parceling out the additional
burden of war and by the structural artifacts of the war itself on the economy. The war was financed by borrowing, enhanced extraction, and feeding
on reserves, and defense-related economic activity and a reduction of un-
employment resulting from mobilization promised relative peace at home.[36]
Indeed, the initial economic reaction was politically successful. In spite of the
dramatic tax hikes and growing inflation, Israeli society at large remained
socially docile and politically non-vengeful.[37]

Containment of economic deterioration and the isolation of the political
system from the consequences of the former became more difficult with the
passing of time. The unhampered private consumption of 1982 and 1983
was paid for by altering the "saving and investment" behavior of Israelis.
In 1983, the rate of increase in Gross Domestic Investment (GDI) was cut
almost by half compared with the previous year. In 1984, the rate of GDI
turned negative (see Table 10.1). Israelis were desperately trying to maintain
their consumption level by turning their backs on saving and investment.
In early 1983, the stock market collapsed, and in October of the same year,
banking shares crashed, depriving industry of a principal source of capital
and robbing ordinary Israelis of what they considered their savings. Israel's
economy was running out of two healthy sources of capital – investment
and consumption – and was left with a very hazardous one – the printing of
money. In 1984, as the gloomy Treasury assessment presented in Table 10.1
suggests, Israel's economy, and not just the individual affluence of its citizens,
was in deep trouble, exhibiting symptoms of stagflation.

---

[36] At least in one sense, the war "helped" the economy and the social order, as it created a
temporary shift toward domestic consumption. Thus, in 1982 the *domestic component* in
the domestic defense consumption – including labor costs, purchases of goods and services,
and construction – rose by some 9 percent. However, total domestic defense consumption
actually declined by 9.8 percent, because its *import component* declined sharply. See Kochav,
"The Influence of Defense Expenditure," 44 table 2.5.

[37] The overall tax increase up to April 1983 was assessed at 16.4 percent. In the same period,
the GNP remained roughly unchanged. See *Ha'aretz*, April 8, 1983.

TABLE 10.1 *The Israeli Treasury's perspective, 1984/85*

	GDP (%)	PPCC (%)	PC (%)	GDI (%)	Unemployment (%)
1982	1.0	5.1	4.8	14.6	5
1983	1.8	5.1	4	8.0	4.5
1984 (estimated)	1.6	−7.5	−3.9	−9.2	5.9
1985 (projected)	0.8	−2.5	−1.4	−15.1	7.4

GDP: Gross domestic product
PPCC: Private per capita consumption
PC: Public consumption (excluding defense imports)
GDI: Gross domestic investment
Source: Bank of Israel, *National Budget for 1985.*

Presumably, the intensifying economic crisis that unfolded during the war, the tax hike, and the threat to the standard of living were strong factors that could be expected to eventually undermine the legitimacy of the war and the government. In reality, however, no such relations between the war, the ailing economy, and the political situation were established. This may have happened for several reasons. For example, the acute economic deterioration came only after the most intensive period of state-society struggle – after the biggest demonstration against the war, after the greatest number of casualties, and after the eviction of Sharon from the defense ministry. Thus, the economic deterioration came after the anti-war coalition had already exhausted much of its steam and after the outcome of the war had been decided.

However, my argument is more radical. I contend that paradoxically, although the Lebanon quagmire helped to consolidate an economic mess, the war actually shielded the government from the expected backlash of the economic deterioration. The political campaign following 1982, and the elections of July 1984 (that produced the National Unity government), epitomize this paradox. The partisan struggle of Labor and Likud in 1983 and 1984 revolved around the withdrawal from Lebanon and the state of the economy. However, it did not revolve around responsibility for the war or its consequences, as Labor felt that it could only hurt its political standing. Moreover, while the election campaign focused largely on the state of the economy, the war itself was shunned, and the results suggested that not much could have been gained from the monumental failures of the government. Likud lost seven seats (down from forty-eight to forty-one), yet Labor also lost seats, though only three (down from forty-seven to forty-four). Following a deadlocked war and spinning economic crisis, the main opposition party was able to change the parliamentary balance by merely four seats. Considering the events prior to the election – the two 1983 stock-market crashes (that in relative terms were of the magnitude of the 1980s American S&L crisis), the 400 percent rampant inflation of late 1983, the general economic deterioration in 1983–1984, and the departure of the charismatic leader Begin – this was a rather meager achievement, if at all, for the opposition. If anything, Likud did surprisingly well. Indeed, one is tempted to conclude that Likud did well because, rather than in spite of, its flawed war in Lebanon.

Conclusion

Of all Israeli wars, the Lebanon war was launched under the most favorable conditions. The balance of power in the battlefield was overwhelmingly in Israel's favor. Regional and international conditions opened a window of opportunity for Israel that gave it unprecedented latitude (so much so that its forces could lay siege to an Arab capital). Most Israeli citizens considered the PLO a mortal enemy and thus rallied round the flag once the war started, with their civil and political institutions following suit. The economic cost of the war was bearable, the combat fatalities limited, and both were on the decline as time passed. Still, "it is possible to say ... without hesitation," as did Lieutenant General Eitan himself, "that eventually, the results of the war were *radically different* from Israel's wishes and intentions."[38] Why, then, in spite of such favorable conditions and initial successes, did Israel end the Lebanon war on the defensive and without achieving most of its objectives?

[38] Eitan, *Sippur*, 207 (italics added).

11

Israeli Instrumental Dependence and Its Consequences

The most obvious characteristic of Israel's instrumental dependence, in the early 1980s, was its wide scope and depth. Structural features over which Israel had little control – most notably, the magnitude of the security threats Israel faced and its geostrategic vulnerability and demographic inferiority vis-à-vis its Arab enemies – simply called for a comprehensive mobilization of its human and economic resources. Thus, as a result of external threats, the boundaries between Israeli society and the state were blurred. Israeli Jewish society, one could add, was tightly meshed with its military.[1] Hence, Israel's comprehensive conscription policy, large defense budget, and high tax rate. Indeed, the majority of Israeli Jewish males served three years of compulsory military service, conscripts who became officers served additional periods, and both groups continued to serve on active reserve service for several weeks each year for up to four decades following their mandatory conscription service. Similarly, Israel's defense expenditures were always high, and in 1978–1980, for example, they reached 17 percent of GNP (excluding U.S. grants).[2] In 1981, a year before the Lebanon war, the tax burden on Israelis – that is, tax revenues as a percentage of the GNP – amounted to 46 percent.[3]

Considering the skills and commitment of Israeli society at large and the comprehensive extraction and conscription system, it is not much of

---

[1] As David Ben-Gurion explained: "Could we survive without such a big defense force?", answering, "This dilemma obligates [one] to think whether or not to blur the boundaries between army and citizens..." Quoted in Elam, *Memalei Ha'pkudot*, 77.

[2] See Aharoni, *The Israeli Economy*, 86 table 2.3, 245–73; Berglas, "Defense and the Economy," in Ben-Porath, *The Israeli Economy*, 173–91; and Kochav, "The Influence of Defense Expenditure," 25–45.

[3] American support to Israel increased dramatically after the 1973 war, yet the level of domestic "extraction" remained extremely high. In 1980/81, 31 percent of government expenditures were for defense. This equalled 66 percent of the tax revenue (though taxes comprise only 46.5 percent of government expenses). See Yosef Gabbay, "Israel's Fiscal Policy, 1948–1982," in Sanbar, *Economic and Social Policy in Israel*, 86–87 tables 4.1 and 4.2.

a surprise that the Israeli state has at its disposal a particularly powerful military. The deep commitment and involvement of Israeli society in the military, however, also added a rather inconvenient dimension to Israel's instrumental dependence. The fact that Israel was democratic and relied on widespread instrumental dependence limited its autonomy, at least under certain circumstances.

Throughout the years, Israel was admittedly successful in using a large portion of its society efficiently for military purposes, and still retained relatively high degrees of both autonomy and social cohesion. That, in and of itself, however, does not negate the observation about the essentially problematic nature of its vast reliance on society. First, one should recall that the state's autonomy in Israel and the social cohesion it enjoyed owed much to the hostility and pressure of the Arab nations in general, and to their belligerent rhetoric, occasional aggression, and diplomatic harassment of Israel in particular. These, perhaps more than anything else, conditioned Israelis to identify rather unanimously with the state and think in terms of power politics.

Second, social cohesion was achieved as a result of the traditional Jewish emphasis on the community, or in less flattering terms, the Jewish self-imposed "communal intimacy." Moreover, the Israeli state was able to exploit this cohesion by portraying itself as a benevolent mediator between the individual and the community, and thus gained highly motivated and obedient citizens who tend to confuse their subordination to the state with a contribution to the community. Indeed, there is little wonder that the Socialist founders of Israel, and all governments since, have cultivated this communal sense as a measure of promoting state power.

Third, Israeli governments succeeded in deriving from their citizens much for military purposes because combat military service and sacrifice were rewarded materially and socially, while the opposite behavior, even when involuntary, was penalized socially and otherwise. Because military service had become a gauge of individual quality and a determinant of one's position on the social merit scale, the IDF attracted the best people and managed to derive the most from them. This system was so effective that Israel was able to recruit the backbone and spearhead of the IDF – its officer corps, top line combat units, and special operations forces – from classes that in other societies often avoid military service altogether. Similarly, Israel enhanced its citizens' readiness to sacrifice by veneration of battle casualties. Tales of heroic sacrifice were common in popular stories and songs, every combat casualty received considerable coverage in the daily newspapers, and heroic death was commemorated throughout Israel. Indeed, individual and collective battle casualties are immortalized in the names of streets, hills, waterways, public parks, and other sites almost everywhere in Israel.

The fourth, and perhaps most crucial, reason for Israel's remarkable success in extracting and using its social resources effectively was the result of self-restraint. Successive Israeli governments succeeded in exploiting effectively Israel's "excessive" instrumental dependence because they were clever enough not to test the inherent limits of the fragile structure of their reliance on society. Israel's leaders, although adept at power politics, shaped their security policies while considering carefully the state's particular dependence on society. Indeed, the scope of different Israeli operations was often decided on the principle of avoiding casualties as much as possible. For example, during the 1978 Litani Operation, which constituted the largest Israeli military intervention in Lebanon until the 1982 war, Israel's defense leadership decided not to conquer the city of Tyre because of concerns that this would involve a large number of IDF and Lebanese casualties.[4]

In 1982, Sharon, Begin, and Eitan took the gamble of acting beyond the limits that their predecessors had respected. They apparently perceived society more as a surmountable obstacle than as a body that sets limits that must be reckoned with. Indeed, it is not that they ignored the consequences of Israel's instrumental dependence. Rather, Sharon, Begin, and Eitan underestimated the significance of the latter. This underestimation resulted in part from their ideological bias, and perhaps in part from some of their personal attributes. However, it was also a result of objective factors – namely, the fact that they enjoyed initial conditions that truly reduced the significance of Israel's instrumental dependence, and thus made their gamble domestically less risky.

As a result of the 1973 Yom Kippur War surprise, the IDF went through a series of expansions that left the state with a larger conscript and professional army than ever before. The army was thus composed of a larger proportion of young soldiers and of older (professional) soldiers who were by definition in a better alignment with the state. This socially more convenient army permitted Sharon and Eitan to rely less on the reserve forces, which on average included older, less adventurous, and intellectually more independent people.[5] It seems that this change in the IDF composition encouraged the war leadership to believe that it could control the potential problems associated with the structure of Israel's instrumental dependence. Thus, Eitan tried to minimize the use of reserve forces in Lebanon from the very first day of the war, counting on a high turnover rate and short recall periods.[6] According to General Moshe Nativ, the head of the IDF manpower division at the time, 50 percent of the reserve soldiers who were mobilized at the war outset were demobilized within less than a month into the war.[7]

Still, in spite of the efforts to reduce the burden on reservists, Sharon and Eitan quickly realized that they could not easily overcome the consequences

---

[4] Eitan, *Sippur*, 161.
[5] This idea was originally raised by *Peace Now* activist Dr. Avishai Margalit. See Edelist and Maiberg, *Malon*, 324.
[6] Eitan, *Sippur*, 240, 278–80.
[7] See *Ha'aretz*, July 1, 1982.

of Israel's instrumental dependence. Moreover, as the war continued, their problems grew worse. In retrospect, at least, Eitan seems to have fully understood the real magnitude of the problem. In his memoirs he wrote:

When the IDF is rightly said to be the people's army, naturally one refers to the advantages implicit in the definition ... under the conditions created in the Lebanon war, this had also a *less positive expression* ... the division in the people naturally had also an impact on the IDF, its soldiers and officers. It cannot be otherwise. *The debates whether the war was necessary or avoidable, whether it had achieved its objectives or missed [them], whether it was 'worth' the cost of its casualties or not*, could not remain in the civil street. The IDF has no insulation material and it does not ordain its soldiers to remain silent.[8]

The consequences of Sharon's and Eitan's underestimation of the "counterproductive" potential of instrumental dependence cannot be overstated. First, the officers and soldiers in reserve combat units were those to turn most quickly against the war and give the anti-war agenda its critical initial legitimacy. Indeed, reservists were continuously demonstrating before and after their service across from Begin's state residence. In early July 1982, hardly a month into the war, a group of reservists calling itself "Soldiers Against Silence" (SAS) organized, expressing its "total non-confidence in the policy of Defense Minister Sharon, [who was] responsible for the cynical use of the IDF without national consensus." The group demanded an immediate and total halt of the war, the resignation of Sharon, and the disclosure to the public of the whole truth regarding the war's initiation.[9] In early August 1982, SAS delivered a 2,000–signature petition to Begin, asking the government to refrain from further firing on Beirut. In April 1983, reservists organized the "No to (the war) Medal" movement, refusing to accept (and returning) the badge of honor the state issued for those who participated in the war.

Second, the upheaval within the army drove the media into taking a firm stand against the war. Reservists demanded directly that journalists tell the public about the real purpose of the war and how they, the reservists, felt about it. In fact, it is almost certain that the journalists would not have taken a critical position, at least not that early in the war, in the absence of the persistent demands of soldiers that they do so. Hirsh Goodman, one of the military correspondents who raised the first doubts about the war in the press, vividly describes his own turning point from conformity to criticism:

What caused me to change so drastically? One day, I, Yaakov Erez [from *Maariv*] and Eitan Habber [from *Yediot Aharonot*] arrived at a paratroopers' elite unit. We looked for the battalion commander. This guy caught us and said: 'What are you doing here?

---

[8] Eitan, *Sippur*, 221 (italics added).
[9] Advertisement in *Ha'aretz*, July 6, 1982.

Drive to Jerusalem and shout! Our lives are in your hands!' So said the man. Then I understood unambiguously [what was] my destiny as a military correspondent in this war.[10]

The conclusions of Eitan Habber of *Yediot Aharonot* (hereafter *Yediot*) were similar: "We did not, God forbid, create the atmosphere [against the war]. We conveyed it from the front to the rear..."[11] And Ehud Ya'ari, the Israeli TV commentator on Arab affairs and a co-author of the book *Israel's Lebanon War*, also described the origins of press criticism of the war as originating at the front:

The reserve paratroopers' battalions constituted perhaps the major factor that *corrected* the nature of reports.... certain segments among the combat forces and among the senior commanders 'used' journalists they stumbled upon, in order to transmit to the rear a different picture from that emerging from the official announcements.[12]

Third, and of particular significance from the state's perspective, open defiance of the reserve recall – traditionally marginal – became significant and salient. By March 1983, 1,470 Israeli reservists, including 228 officers, had signed the *Yesh Gvul* pledge not to be sent into Lebanon. In addition, by that time the twenty-eighth reservist had been jailed for refusing to comply with the recall. Half a year later, in October 1983, the number of soldiers jailed for refusing to serve in Lebanon had mushroomed to 100. Admittedly these were small numbers, but they were unheard of in Israel as far as open draft resistance in times of war was concerned. At the same time, the number of reservists who evaded mobilization without breaking the law and making defiant declarations (thus avoiding sanctions and/or social censure) also increased substantially. In fact, the number of these "gray dodgers" was apparently so high as to alarm IDF officers, who had to cope with a shrinking manpower base in their units and with the bitterness the latter created among soldiers who reported for service and became enraged over the manifestly uneven burden-sharing.[13]

Fourth, as resentment within the army grew, the structure of instrumental dependence began to affect operational consideration. No other case demonstrates this development as vividly as that of "the brigade that was not mobilized." This case involved a reserve paratroop brigade that fought in the early stages of the war and was released shortly thereafter in line with the policy of relying as little as possible on reserve forces. The brigade was needed again toward mid-late July, a few weeks after its initial release, because the IDF was short of quality infantry units for its planned assault on Beirut. After a short debate, Sharon and Eitan decided to postpone the

---

[10] Quoted in *The Journalists' Yearbook, 1983*, 13 (italics added).
[11] Ibid., 16.
[12] *The Journalists' Yearbook, 1985*, 73–75 (italics added).
[13] *Ha'aretz*, September 27, 1983.

mobilization of the brigade as late as possible.[14] The first details of this internal debate about whether to recall the brigade were revealed by Sharon himself in a TV interview on September 24, 1982. "The entire war," Sharon argued in the interview,

> was waged in the midst of a deep public debate . . . the likes of which we have never witnessed before . . . things have reached such a point, and [there was such an] accumulation of pressure, of anti-war propaganda, of incitement, of preaching, of uncontrolled writing . . . that when we had to mobilize an IDF reserve brigade, the chief of staff and I came to the conclusion that we could not mobilize this brigade . . . because there was a very bad atmosphere . . . [15]

Months later, Sharon revealed more about the reason for his decision not to mobilize the brigade. Although it is often hard to distinguish between what really motivated Sharon, what he believed motivated him, and what he said for political purposes, some of the interview seems revealing.

> [As a result of] information brought to me, including the conversation held between the chief of staff and the brigade commander . . . not only because the unit was released only a short while before, but also because of serious worries about the brigade's morale *and even with respect to negative influence that [the brigade] might have on other units* . . . the decision was taken . . . not to mobilize . . . [16]

If, in the first disclosure, Sharon emphasized an imaginary path of "bad" influence from the public and the media to the soldiers, in his second disclosure, he came closer to the truth. Eventually, about a year into the war, *Al Hamishmar* revealed a more accurate version of the story. Indeed, it was the "mood" among the soldiers of this brigade that deterred the architects of war from mobilizing the brigade prematurely. But it was not the fear that this mood would hamper the soldiers' battle performance that deterred Sharon and Eitan, but rather the fear that soldiers would "overreact" to a recall and that their anger would be contagious. The following excerpts from the discussion between Sharon and Eitan reveal how the reliance on reserve forces, while executing a disputed policy, influenced the considerations of decision-makers. The dialogue started with Sharon's inquiry as to whether there was any other brigade available for recall:

Eitan: "All the other (brigades) have just been released."

Sharon: "It would be bad [if] you mobilized a brigade that suffered a hard blow to its morale."

---

[14] See Schiff and Yaari, *Milhemet*, 276. Apparently, once ordered, the brigade mobilized at an unusually slow pace. See ibid., 276.

[15] Quoted in *Ha'aretz*, September. 26, 1982; and in Feldman and Rechnitz-Kijner, *Deception*, 58.

[16] Quoted in Feldman and Rechnitz-Kijner, *Deception*, 59 (italics added).

Eitan: "I spoke with . . . (the brigade commander). He told me that the people would take it very hard. It is possible to manage without them. But I think in case something happens, and people would say that we were left without [reserve] forces."

Sharon: "It is better to have such a brigade as a general reserve."

Eitan: "I am most reluctant to mobilize it . . . "

Sharon: "There are other [units]. I would think many times whether to mobilize such a brigade."

Eitan: "I am most reluctant to mobilize them. Let's mobilize them on "D" day. I would mobilize them when the shooting starts. Then, there would be no problem."

Sharon: "Don't mobilize them. It would not be good. Better to have one less block of houses. We would take them by air . . . First, let's conclude that you do not mobilize them before that. *It could poison there anybody who participates in the war and returns home . . .* "

Eitan: "I share your opinion in this matter."

Sharon: "What is important is that *other regular forces are brought in. Forces that have not (yet) been in the war.* As far as I am concerned, it is better to mobilize them after that, if [we] mobilize them at all . . . "[17]

In conclusion, the case of "the brigade that was not mobilized" both indicated the kind of problems Israel's inconvenient instrumental dependence created, and how it eventually influenced military policy-making. Moreover, the case is also significant because it happened during the very early stages of the war, when the war was still quite popular. Indeed, the considerations that later led Israel to choose a defensive course in Lebanon had much to do with the reality that was exposed in this case. Israel chose a policy that required a relatively small deployment of forces and that was designed to save casualties because it felt constrained by the certainty that any other policy would increase the political friction within Israel to an unacceptable degree. Such a choice, one might want to note, ultimately prevented Israel from being able to back up its ambitious goals in Lebanon.

The last effect of the nature of Israeli instrumental dependence was on operational, and even strategic, plans, as the IDF's first priority was to reduce its own casualties. Put otherwise, the army traded fire for blood – that is, the command preferred to increase the use of firepower to achieve operational objectives whenever the casualty forecast for direct ground combat was high. This preference reduced Israeli casualties, but had the drawback of involving inevitable collateral damage that undermined the legitimacy of the whole operation. For example, most commanders, including Eitan (who apparently had a change of heart) and Amir Drori, Officer in Command (OC), Northern Command, opposed the idea of entering Beirut because of

[17] Quoted in Naor, *Memshala*, 133–34 (from *Al Hamishmar* daily, June 17, 1983) (italics added).

casualty considerations.[18] However, they still wanted to achieve their military objectives of expelling the Syrians and the PLO. So, they searched for alternative means of leverage.

Indeed, a review of the military decisions demonstrates that the IDF sought alternative ways of deploying its soldiers in casualty-demanding situations, such as urban infantry warfare. Most notably, and as becomes clear from Sharon's words in the case of the "brigade that was not mobilized" – "Don't mobilize them. . . . [it is] better to have one less block of houses" – the war leadership was inclined to search for other methods such as reliance on higher levels of artillery and air power, and assignment of certain missions to proxies, in order to control the number of IDF casualties.

These methods, and in particular the second, had a far-reaching influence on the fate of the war, as was revealed most clearly when Israel's proxies, the Phalangist forces, were sent into the Sabra and Shatilla refugee camps instead of IDF soldiers. As may be recalled, Israel tried several times to convince the Phalangists to join the battle in Beirut but to no avail. Then, in mid-September 1982, Israel's trump card, the Maronite Christian leader Bashir Gemayel, was literally blown up. Suddenly, the Phalangists, crazed by the assassination of their leader and emboldened by Israel's massive presence in Beirut, were ready to act. The refugee camps were precisely the kind of battle environment that promised large numbers of military casualties in case the offender insisted on sparing innocent civilians. This, as we know, led to Israel's disastrous decision to send the Phalangists into the Palestinian refugee camps of Sabra and Shatilla.

Clearly, the ensuing massacre was pivotal in destroying the whole war initiative. In the final analysis, then, Israel's instrumental dependence undercut the war effort in a dialectic manner. It "forced" the war leadership to look for alternative casualty-thrifty ways to achieve its military objectives, of which less-discriminating battle methods and the use of proxies were first. The proxies, however, may have served the anti-war cause no less than the number of casualties, and at least as well as they served the pursuit of the military objectives. Indeed, had the war been kept "cleaner" and more "surgical," Sharon and Eitan would have stood a much better chance of marketing the war in Israel and rendering the opposition to the war ineffective.

---

[18] See Schiff and Yaari, *Milhemet*, 263–64; and Eitan, *Sippur*, 279. Beirut apparently evoked, in the generals' minds, memories of the heavy casualties of the 1967 battle in East Jerusalem and the 1973 battle in Suez. Telephone interview with Brigadier General Giyora Furman, October 24, 1991. See also Eitan, *Sippur*, 279–80; and Schiff and Yaari, *Milhemet*, 263. Brigadier General Amos Yaron, the commander of the Beirut theater of operations, argued that as time passed, consideration of casualties became increasingly constraining. Personal interview with Yaron, October 3, 1991, Tel-Aviv.

12

The Development of a Normative Difference in Israel, and Its Consequences

The Israeli opposition to the Lebanon war did not revolve around a single issue, nor was it consistent in content and emphasis throughout the war. As was the case with the French war in Algeria – and as is perhaps often the case with significant social protest against war – several ideological camps opposed the war, each guided by its own agenda. Still, the Israeli opposition to the Lebanon war was rather homogeneous, embracing mostly people from what can be defined as a non-radical soft left. The majority of the opposition to the Lebanon war was led by, or identified most clearly with, the agenda of the Peace Now movement.[1] Moreover, as the agenda of this constituency was rather consistent, and the time frame of the events rather short, it is easy to follow the development of protest, and identify, define, and classify its key themes with relative clarity.

The Utilitarian Debate about the Human Cost of War

It is clear that the single most important theme of the anti-war campaign, which was responsible for the initial mobilization of Israelis against the Lebanon war, was the rate of IDF casualties. The fact that casualties had a great social impact is far from trivial. Before the Lebanon war, the human cost of Israeli wars hardly ever divided the Israeli public or raised the question of whether a war itself was worthwhile.[2] It therefore remains to be understood

---

[1] See Tamar Hermann's analysis of *Peace Now* in "From the Peace Covenant to Peace Now: The Pragmatic Pacifism of the Israeli Peace Camp," Ph.D. dissertation (Tel-Aviv: Tel-Aviv University, 1989), 357–64.

[2] The high number of casualties in the Yom Kippur War stirred much protest. However, because the war was forced on Israel, it was not the operational need for sacrifice that was disputed, but rather the responsibility of the military and political elite for the complacency that led to this need. During the War of Attrition (1969–1971) that cost Israel casualties weekly, very few Israelis protested. A notable exception was a small group of high school seniors facing conscription. See Barzilai, "Democracy in War," 121.

TABLE 12.1 *Israeli fatalities in Lebanon, 1982–1985*
(as disclosed during the war)[a]

12 days into the war (June 18, 1982)	214 (32%)
$2\frac{1}{2}$ months into the war (August 23, 1982)	332 (50%)
One year into the war	500 (75%)
Total Israeli fatalities in the Lebanon war	**664 (100%)**
Of whom, officers	145 (22%)
All other IDF fatalities in the same period	490 (74%)

[a] According to later data, the distribution of casualties was slightly different. For example, *Ha'aretz* of February 15, 1985, puts the number of fatalities for June 15, 1982 (9 days into the war) at 237 (36 percent).
Sources: Ha'aretz, June 18, 1982; August 23, 1982; June 12, 1983; February 15, 1985; June 7, 1985; September 25, 1986.

what precisely made an increasing number of Israelis willing to abandon their conformist conventions and take the "radical" measure of using the casualties as a political weapon against the state. In fact, the puzzle is even more perplexing because while many perceived the number of IDF casualties in Lebanon to have been excessively high,[3] objectively, it was not.

In fact, perhaps this casualty "paradox" is the best starting point of the discussion. Let us first note that, at least from a military point of view, Israeli fatalities in Lebanon were indeed relatively limited. For example, comparing the fatalities Israel suffered in Lebanon with those it had in other wars with the rate of civilian deaths in Israel, or with the number of casualties incurred by Israel's enemies in Lebanon (particularly in light of the territorial outcome), the IDF fatalities in Lebanon were moderate.[4] Table 12.1 reveals how many Israeli soldiers fell in each phase of the Lebanon war.

As Table 12.1 indicates, most Israeli casualties were incurred in the first year of the war. In this first year, most of the casualties were incurred in the first quarter ending with the election of Bashir Gemayel and the expulsion of the PLO from Beirut. Furthermore, in this first quarter, most of the casualties were incurred in the first two weeks of the dynamic two-front war of territorial conquest against the Syrians and the PLO. Unlike in other protracted wars, then, Israeli casualties were highest in the first stages of the war, and their number over time decreased as the war lingered on. Military reality and societal perception regarding casualties were simply at odds.

What accounts for this discrepancy? For one thing, it is clear that while the military estimate of whether any number of war fatalities is high or low

[3] See, for example, Schiff and Yaari, *Milhemet*, 38; Gabriel, *Operation Peace for Galilee*, 176.
[4] Israel suffered nearly 3,000 fatalities in the short 1973 Yom Kippur War. In two and a half years of the War of Attrition period, it suffered over 500 fatalities.

can be established according to relatively objective comparative standards, societal perceptions and estimates are subjectively influenced by factors other than sound comparisons and standards. Instead, societal estimates depend on indeterminate variables such as expectations and notions of the "worthiness" of sacrifice – that is, perceptions of how vital or justified are the causes or objectives for which the state risks its citizens' lives. These variables, however, are often themselves functions of yet other indeterminate variables such as the cultural context and the ingenuity of the pro- and anti-war forces in "selling" or denying the importance of the causes and achievements of war. Consequently, the discussion of the weight of casualties in the public debate would be pointless unless political and cultural aspects, which determine why and how casualties become significant, are also considered.

Expectations about casualties are also often formed on the basis of cues the leadership supplies. In the Lebanon war, these cues could hardly have been more misleading. Initially, Sharon and Eitan vaguely promised that the war would be over "before long."[5] On June 8, merely two days into the war, Begin solemnly declared in the Knesset that "if we push the [PLO artillery] line 40 kilometers away from our northern border – [our] work will have been done; all fighting would stop."[6] He also told the Knesset that Israel suffered twenty-five fatal casualties, seven MIAs, and ninty-six wounded, adding: "Maybe [in] other nations facing such an operation, people would say that these casualties are reasonable. We shall not say so. For us, these are very bitter casualties. Very painful [ones] . . ."[7]

Of course, Sharon's and Eitan's reference to a short timetable, the presumed limited territorial objectives of the "operation," and Begin's early confession that the first few IDF casualties were "very bitter [and] painful . . .", suggested to Israelis that the overall operation would be sparing as far as Israeli casualties were involved. Moreover, the best reference Israelis had of a limited operation in Lebanon, the 1978 Litani Operation, resulted in only sixteen IDF fatalities. Thus, initial casualties expectations were set at a low number, and any number of fatalities running over several tens was almost bound to be perceived as excessive.

The truth of the matter is that while the territorial and political objectives of the war were deliberately concealed, the timetable and casualty expectations presented to the public may have reflected an almost honest, if foolish, underestimation by the General Staff. In his memoirs, Eitan argued that the "128" fatalities the IDF suffered up to July 4, 1982, was greater than his own casualty expectations for the whole war.[8] Thus, disregarding the fact that the "128 fatalities" statement was grossly mistaken (as by that time the

---

[5] See *Ha'aretz*, June 7, 1982.
[6] Quoted in Schiff and Yaari, *Milhemet*, 187.
[7] Quoted in ibid., 185.
[8] Eitan, *Sippur*, 275.

number was about double), the mere confession is still a good indication of how erroneous was the casualty estimate of the military command. By the time Israel had achieved Sharon's and Eitan's first political objective of coercing the PLO to leave Beirut, the IDF had over two and a half times the "128" number of soldiers killed that Eitan had defined as unexpectedly high. By the end of the war, one may note, the number of IDF fatalities was more than five times that number.

Initial public acceptance and attitude toward the casualties were in all likelihood also influenced by official inconsistencies. On June 14, Eitan numbered the IDF fatalities at 170 and those wounded at 700. Only three days later, after relative calm on the front, General Moshe Nativ, the head of the IDF Manpower Division, disclosed much higher numbers – 214 dead and 1,214 wounded.[9] This discrepancy in casualty reports in the second week of fighting made the media, and for that matter, any Israeli, increasingly suspicious and far more attentive to the cost of war.

While these conjectures contributed to skepticism, suspicion, and a gloomy atmosphere, it is still clear that "worthiness" perceptions were built on more factors than the initial casualty expectations, or the disappointment with the short-and-frugal war notion that Sharon and Eitan had propagated, and that Begin ignorantly had corroborated. At least some of the deterioration in public readiness to placidly accept casualties was due to reasons over which neither the state nor society had much control – basic social conventions and the dynamic and habits of press coverage.

In this respect, the problem for the state started almost immediately with the release of the first IDF casualty list. Once this list was released, and it is doubtful whether the IDF could have waited much longer after Begin's June 8 disclosure about the twenty-five fatalities, subsequent lists of fatalities had to be released without delay, followed by the rather intimate media treatment of each casualty. The Israeli newspapers, in line with a long-established tradition, devoted to each dead soldier a chronicle that typically included a photograph and a column of some 15–60 lines (60–300 words) narrating the general circumstances of the soldier's death, his rank and role, his home town, and a few words by a family relative or acquaintance. These and the obituary notices often filled a whole page in the news section (in *Ha'aretz*, typically page 3, 4, or 5). Each time such casualty lists were published, the Israeli public received an additional depressing reminder of the accumulating cost of war that contributed to the growing doubts over its worthiness.

Still, the conjunction of circumstances – governmental blunders and cultural sensitivity to casualties – cannot fully account for the growth of the protest over the rising number of casualties. Had it not been for structural reasons – Israel's democratic institutions and its inconvenient instrumental dependence – it is likely that the Israeli public at large, including the educated

---

[9] Benziman, *Lo Otzer*, 254.

middle-class, would have come to terms with both the unexpected prolongation of the war and the number of casualties. However, the combination of vibrant democratic institutions, inconvenient instrumental dependence, and, particularly, a middle-class vested interest in aborting the war provided the necessary and sufficient condition for a revolt against the state.

The interest of the middle-class becomes obvious when one considers the joint impact of the social composition of the IDF spearhead units and officer corps, the combat doctrine that puts officers in the lead, and the nature of the Lebanese theater of operations. These three factors virtually guaranteed that the educated class would pay a high price in blood. Indeed, the distribution of IDF fatalities suggests that this was the case. For example, 145 out of the 664 Israeli dead in the Lebanon war were officers – some 22 percent. Furthermore, out of the first 386 fatalities – that is, in the first stage of the war – the percentage of officers killed was higher, some 26 percent, and that of sergeants some 36 percent.[10] In other wars, similar structural features that defined the composition of casualties were present, but Israel did not suffer similar repercussions. Why the educated liberal class did not remain docile during the Lebanon war is elaborated later. Suffice to note here that a critical portion of society had a substantial "expedient" reason to be deeply upset over the war.

It is hard to establish with any measure of certainty what impact each variable – misconstrued casualty estimates, inherent sensitivity of society, dynamic of casualty accumulation and publication, media coverage, and political ingenuity – had on the perceptions of the worthiness of the war and its cost. Moreover, it is clear that perceptions of worthiness influenced the attitude toward casualties, as much as the other way around. In other words, certain Israeli strata were motivated, willing, and able to use the casualties politically *during* the Lebanon war because they had reached the conclusion that the extended war did not have existential justification, and thus did not justify the cost. Indeed, the trust in the leadership and the readiness to sacrifice started to diminish as the far-reaching objectives of war were unravelled. "*They [the government]*" soldiers complained, "*broke the rules of the game . . .* they used the Israeli army not in order to defend our existence."[11] Such argumentation was particularly rampant among the spearhead reserve paratroopers and other elite infantry units that were generally considered to have constituted the "best and brightest." For example, five weeks into the war, a group of reserve soldiers of a Special Operations Unit, wrote to Begin:

It was always clear to me that if I went to war, it would be a just war fought over our life and existence as a people. Today, it is clear to me that I was deceived and

---

[10] Data are from the list published by *Ha'aretz*, on June 7, 1986, and from the IDF's spokesman office as quoted in Gabriel, *Operation Peace for Galilee*, 235. The ratio of officers to soldiers was around 6–8 percent in different fighting formations.

[11] Michael Jansen, *The Battle of Beirut* (Boston: South End Press, 1982), 125 (italics added).

TABLE 12.2 *Difference in Israeli Non-Justification of the Lebanon War (NJW)*

A. General justification of the war (PORI poll question: "Do you or don't you justify Israel's action in Lebanon?")

	Only 40 km (%)	All (%)	None (net NJW)[a]
July 1982	24	66	5
October 1982	37	45	9
March 1983	40	41	11

B. Justification of the war as a function of human cost and achievements (DAHAF poll question: "In the face of the cost and results of the war, was it right or wrong to fight the war?")

	Right (%)	Wrong (CI-NJW)[b]
July 1982	84.0	13
October 1982	67.0	29
March 1983	55.6	39

C. Comparison of NJW increase from July 1982 to March 1983

	Change (%)		Cumulative change (%)	
	Net-NJW	CI-NJW	Net-NJW	CI-NJW
July 1982	0	0	0	0
October 1982	4	16	4	16
March 1983	2	10	6	26

[a] NJW = Non-justification of the war.
[b] CI-NJW = Cost-integrated non-justification of the war.
Source: Barzilai, "Democracy in War," 349a; and PORI polls in Hann-Hastings and Hastings, *Index to International Public Opinion, 1982–1983,* 332–34.

called (to serve) in the first war in the history of Israel that was not a defensive war but rather a dangerous gamble over achieving political aims. All that while paying heavily with IDF casualties, and while harming innocent civilians.[12]

What, then, was the overall impact of the casualties? Public opinion data from the first year of war may allude to the answer. Table 12.2 provides a

---

[12] Quoted in *Ha'aretz*, July 9, 1982. In another letter quoted in the *Kibbutz* movement newsletter, a soldier wrote: "I am from a patriotic breed ... but I am not willing to risk my life when it is unnecessary. This time it is not necessary and my life is a very precious value. I am not afraid to die, but I do not want to die ... if you will insist on guarding your lives and if you will refuse to serve in Lebanon, we will get out of there and won't have to live in [the kind of] insane state, I have no wish to remain in." Quoted in Menachem Dorman (ed.), *Be'tzel Ha'milhama: Sichot Be'yad Tabenkin* [In the Shadow of War: Discussions in the Tabenkin Memorial] (Hakibbutz Hameuchad, 1983), 36.

measurement of this impact. Part C of Table 12.2 is based on a comparison of the differential and cumulative rates of increase in net non-justification of the war (Net-NJW in A) versus the cost-integrated non-justification of war (CI-NJW in B). Admittedly, such a comparison is limited because the wording of the questions in Table 12.2 differs somewhat, the questions were asked by two different polling agencies, and the question of part B of Table 12.2 introduces the human cost variable lumped together with the outcomes variable. I would still argue, however, that the comparison in part C of the table suggests that the casualties issue galvanized and radicalized the opposition.

The centrality of the casualties issue in the anti-war campaign should hardly surprise anyone, particularly when the structure and characteristics of Israel's instrumental dependence are considered. Still, it is noteworthy how quickly the casualty issue was evoked, and to what rhetorical extremes critics of the war were ready to go. On June 16, 1982, barely eleven days after the beginning of war, the Peace Now movement had already raised the issue of casualties in the newspapers, in an ad that read, "For what [are we] killed? For what [do we] kill?" Eight days later, its leaders sent Begin a cable demanding that he "not sacrifice [even] one more person."[13] Since that point, the issue of casualties never dissipated. Moreover, war casualties became an instrument to attack the war leadership personally. Begin, who was perceived as responsible for the war, and "vulnerable" because he was not suspected of being indifferent to human suffering, was effectively attacked by parents of soldiers who died while conquering the PLO Beaufort stronghold and by a group of protesters that earned the name the Digital Picket. Both groups attacked Begin in a style not common in Israel until then (with the possible exception of sporadic attacks on the government after the 1973 Yom-Kippur War debacle). The Digital Picket was formed in May 1983, almost a year into the war, and it earned its name for placing a large sign displaying the up-to-date number of IDF fatalities right across from Begin's state residence. It performed this simple yet ingenious act in spite of the advice of experienced Peace Now activists who thought it would fade quickly for lack of public support.[14] The bereaved families used their pain as a weapon. For example, a month after the outbreak of the war, one of them wrote in an open letter to "those [in the government] who raised their hand in favor of the war" that "[they] declared, cynically and shamelessly, the 'Peace for Galilee' operation," wishing them that his "bottomless sorrow ... [may] pursue [them] in [their] sleep and [their] waking ... [and be] like the mark of Cain on [their] forehead for ever."[15] Sharon, who was perceived as the chief perpetrator of the war, was attacked even more

---

[13] *Ha'aretz*, June 25, 1982.
[14] Personal interview with *Peace Now* activist Tzali Reshef, October 23, 1991, Jerusalem.
[15] *Ha'aretz*, July 5, letter to the editor.

severely than Begin for being responsible for what a growing number of Israelis considered unnecessary casualties. When he attended a memorial for
paratroopers (where his military roots were) at the end of September 1982,
families greeted him with the spontaneous chant: "Sharon murderer, Sharon
monster."[16]

All in all, the IDF casualties had two major effects on the war. They
provided a powerful argument against the war and thus helped opponents
of the Lebanese gamble to garner a critical mass, and as I explain shortly,
they influenced military decision-making in ways that proved detrimental.

The Debate about the Morality of the Military Conduct in Lebanon

As I have already noted, the desire to reduce IDF casualties spurred the war
architects to rely on higher levels of less discriminating violence. Moreover,
there could be little doubt that a significant portion of the Israeli educated
class was almost as much at odds with the outlook of the two military
architects of war, and particularly Sharon, over ends-means relations in war,
as it was over the justification of sacrifice.

There is also little doubt that during the Lebanon war, Sharon lived up
to his reputation as being unreserved about the use of violence. His July 11,
1982, meeting with senior IDF officers over the fate of southern Beirut is a
good example of both his brutal approach and the fact that he sensed, as
his choice of words suggested, that the former was not universally espoused.
Thus, Sharon called for brutal measures, but at the same time couched his
call in terminology that was intended to make the latter palatable. On the one
hand, he suggested "finishing off the southern part [of Beirut]," "obliterating
whatever can be obliterated," and offered a way do so by "waves of airplanes
[that] would come and bomb relentlessly at their own leisure." On the other
hand, he defined the refugee camps as "terrorists' camps" and explained to
the officers that "we don't touch the city, only the terrorists" (all that, while
he admitted that thousands of civilians lived within the designated area of
the targets).[17]

The strategy of firepower saturation was executed several times on a small
scale and a few times on a larger scale. As should have been expected, it did
not eliminate (though it may have lessened) the focus on the casualties issue.
Rather, it created additional friction within the IDF. The essence of this
process was well reflected in a letter that a group of ninety reserve soldiers
and officers wrote Begin. In this letter, they asked Begin not to mobilize
them again for further service in Lebanon, and to call the army back home

[16] Edelist and Maiberg, *Malon*, 122 (Moshe Dayan was received with similar chants in the
aftermath of the Yom Kippur War).
[17] Quoted in Schiff and Yaari, *Milhemet*, 259–62; and Naor, *Memshala*, 128 note 44. See also
Feldman and Rechnitz-Kijner, *Deception*, 57.

because "[they] have killed too many and too many of [them] were killed in Lebanon, [they] have conquered, blasted, and ruined too much."[18] Thus, once the sense of necessity had disappeared, reserve soldiers not only objected to the risk the war posed to them and to the possible price they could pay, but they also rejected the legitimacy of using violence. Evidently, expedient and moral considerations replaced the expected preoccupation with victory.

In fact, the utilitarian approach to using firepower, and in particular what seemed as Sharon's unscrupulous attitude, was not all that well received even within the senior command of the IDF. It came to the point that even the CGS, General Eitan – whose own attitudes concerning the enemy rights, limits on the use of force, and the "purity of weapons" were at best considered ambiguous[19] – had reservations about Sharon's battle philosophy. What motivated General Eitan to part with Sharon will remain a subject for speculation. Nevertheless, it seems safe to suggest that he objected to Sharon's ideas mostly because he was attentive to the opinions of the members of the IDF command, who opposed Sharon's approach, and because he sensed that an overly "dirty" war would hurt the IDF. Such a war could damage the subtle, informal component of authority that made senior IDF officers effective commanders, and it could tarnish the army's image within Israel, putting it under undesired political pressure.[20]

Indeed, Eitan was correctly concerned, as soldiers began to feel that "there was a purity of light weapons, not a purity of artillery ... [that they, themselves] made the moral calculation [but their] superiors did not."[21] No wonder, then, that Eitan found it necessary, as early as the second week of the war, to deny accusations regarding the magnitude of civil casualties in Lebanon, reassuring the Knesset Committee on Security and Foreign Affairs that the IDF took special precautions in order to avoid harming civilians.[22]

Naturally, the clash of the values of expedient frugality as far as IDF casualties were concerned, and altruistic humanity as far as civil Lebanese casualties were concerned, was also expected to develop in the Israeli Air Force (IAF), which was assigned much of the destruction. In general, brutal air-war strategy was not quite to the liking of the IAF. In fact, the IAF commander, General David Ivri, was particularly appalled at Sharon's

---

[18] *Ha'aretz*, July 9, 1982.

[19] When Eitan was asked by *Ba'mahane* (the IDF journal) whether a soldier should be harmed as a consequence of his adherence to the principle of the "purity of arms," he responded: "I would say that the life of our soldier is more valuable than the life of the enemy, whether this enemy carries a weapon or does not carry a weapon." Quoted in *Ha'aretz*, July 9, 1982.

[20] In general, Eitan proved to be a "democratic" CGS. He did not discourage opinions that differed from his own, and he was ready to accept policy recommendations of generals irrespective of their ideology. Interviews with Brigadier General Furman and Major General Or.

[21] Quoted in Jansen, *The Battle*, 22, from the *Jerusalem Post* issue of July 9, 1982.

[22] *Ha'aretz*, June 16, 1982.

unscrupulous instruction.[23] Luckily, the Air Force could receive orders only from the CGS, General Eitan, and the latter decided to moderate Sharon's directions. Furthermore, once the orders were given, the Air Force could soften the instruction further by carefully planning its bombing runs, and as a last measure, its pilots could insist on being assigned well-defined targets. Some of them actually did so. They did not drop their bombs unless they were convinced that they had located their military targets.[24] Obviously, no matter how rigorously the IAF planned and executed its bombing runs, collateral damage in such areas as Beirut was all but certain.

The ground forces of the IDF encountered similar problems. They tried to control the level of violence by being selective in targeting, modifying battle procedures and techniques, and by giving Lebanese civilians suitable warning time to leave combat zones, and encouragement to do so.[25] Yet, much like the case of the Israeli Air Force, there were no perfect solutions for the dilemma of suffering casualties versus inflicting brutality. Rather, all the solutions seemed to have led to the exacerbation of the clash of values. Thus, the important thing to note is not that the IDF went out of its way to preserve humane values, but rather that the IDF was trapped between two conflicting objectives that did not have a perfect solution – a desire to achieve its assignments with minimal losses and a desire not to antagonize its soldiers and Israeli civilians by resorting to indiscriminate violence.

In any event, the first indications of trouble within the army over conflicting values and moral issues occurred between mid-June and early July 1982. These early signs of ferment were mostly confined to reserve units, and as such seemed extinguishable. These units could be demobilized or not recalled, and their complaints could be conveniently dismissed as isolated and politically motivated. Indeed, the war leadership treated the first wave of ferment precisely in this way. Sharon and Eitan reduced the use of reserve forces to the minimum necessary, and Begin haughtily dismissed the ferment as a political ploy. For example, he responded to the early July 1982 letter of the soldiers of the Special Operations Forces (SOF) reserve unit, writing: "The defense minister does not need your [vote of] confidence … I presume that you did not vote for him and for [our] party … as soldiers you owe him unreserved obedience."[26]

---

[23] Personal interview with retired Lieutenant General Rafael Eitan, October 21, 1991, Tel Aviv. See also Edelist and Maiberg, *Malon*, 17.

[24] Interview with an IAF combat pilot; news conference with "etatist" reserve pilots, *Ha'aretz*, July 2, 1982; and Naor, *Memshala*, 130.

[25] See Gabriel, *Operation Peace for Galilee*, 171–76. The battle in *Ein-Al-Hilweh* refugee camp is a good example. See Schiff and Yaari, *Milhemet*, 168–82. In Beirut, the IDF assigned each building a number, and commanders tried to reduce civilian casualties by sparing buildings that met the criteria they had developed for civilian occupancy. Interview with Brigadier General Yaron.

[26] Quoted in *Ha'aretz*, July 9, 1982.

However, while such a strategy may have calmed Begin for a while, it did little to resolve the problem the state faced. On July 26, 1982, seven weeks into the war, the newspapers published a story about a "regiment commander of distinction" who asked to be released from his command because of his opposition to the assault on Beirut.[27] Colonel Eli Geva's full argument became known only later, but there could be little doubt that the mere knowledge of the unwillingness of a senior officer to remain in command during a war, because of ethical considerations, was bound to influence deeply the consciousness of Israelis. Indeed, if Geva's intentions were, as he argued, to "turn on a red light" and "break the silence conspiracy," then the gravity with which the military and civil authorities considered his act suggests that he was successful in achieving just that.[28] At first, he was given an opportunity to take his case to Eitan, then to Sharon, and finally to Begin, though he provided more than enough reasons for a discharge. In that respect, more than anything, the tolerance of higher circles reflected the hope that the Geva time bomb could be defused. However, once it was clear that Geva would not change his mind, the political strategy to deal with him changed abruptly from one of damage aversion to one of damage limitation. As General Eitan wanted to preempt Geva, deter other potential "defectors," and regain the initiative, he swiftly dismissed Geva, and the army followed through with a campaign to discredit him.[29]

The trouble for the defense establishment was that Geva was not merely an isolated case, but rather a symptom of the developing atmosphere within the IDF command structure. Admittedly his reaction was radical for a professional soldier, but other officers essentially shared his criticism. In fact, Geva revealed that prior to his "revolt," he had received a note from an IDF general that stated that "(Sharon) does not understand that what should worry him more than the problem of terrorists is the danger that he would 'lose' the army after Beirut, and [the danger of] what is happening among Jews [i.e., Israelis] . . ."[30]

---

[27] See also Feldman and Rechnitz-Kijner, *Deception*, 57–58.

[28] See Schiff and Yaari, *Milhemet*, 264; and Geva's interview in *Ha'aretz*, September 26, 1982.

[29] Eitan tried to discredit Geva's motives in his memoirs, which in essence represent the IDF position of the time of the incident. Geva was blamed for abandoning his soldiers, being a captive of the media and an opportunist who promoted himself, and a commander who lost his good sense of judgment following failures in the early phase of war (see Eitan, *Sippur*, 222–25, 267–68, 282–84). Discrediting Geva, however, was not easy. Geva, a son of a former IDF general, was one of the officers promoted most quickly, and had an impeccable record that included distinguished conduct as a commander of a tank company during the 1973 Yom Kippur War.

[30] See Geva's interview in *Yediot Aharonot*, September 26, 1982. Quoted also in Schiff and Yaari, *Milhemet*, 266 (see also pp. 264–66). The defense establishment leaders also heard sharp criticism concerning the social consequences of their policy in Lebanon from the commander of their most esteemed Special Operations Forces (SOF) unit.

To make matters worse for the state and the defense establishment, the media decided to follow the pressures from the reserve and line army and open up themselves for dissent rather than close ranks with the government and defense establishment. On August 10, 1982, when the Geva affair was still fresh, *Ha'aretz* published an article by an anonymous IDF officer, described as "a senior officer commanding a combat unit in Lebanon," that essentially justified Colonel Geva's moral choice. Following a short inquiry, Eitan identified the officer, a major named Gershon Ha'Cohen commanding an armored battalion, and forced him out of the Army.[31] But punitive measures were of very limited value. Eitan could probe, intimidate, preempt, retaliate, and discredit as much as he wanted, but he could not tightly control, let alone eliminate, the spreading uproar within the army. Similarly, Eitan could not control the reaction within civil society to the signs of strain within the army.[32] He could slow down the pace of deterioration somewhat, but he could not stop or reverse it. Moreover, his actions could easily become counterproductive. Certainly, officers who shared Geva's reservations, but were nevertheless unwilling to take similar actions (because they were not ready to risk their careers and promotions, "betray" the institution they served, break the military rules of conduct, or face peer pressure), could always find safe press channels to convey in anonymity their dissatisfaction with the war concept or particular military moves. Eitan Habber of *Yediot* described this phenomenon vividly:

We did not bring [to the public] during the war period but a minute part of what soldiers, commanders, senior commanders, [and] generals told us ... We expressed accurately what the ministers, who were afraid to talk at Cabinet meetings ... thought, and [what] *generals, who kept their mouths shut in the command group and during briefings, [and what] brigade commanders, who literally begged us to raise this havoc, [thought]* ...[33]

Finally, while the accumulation of moral dissatisfaction over combat methods and outcomes within the fighting forces was not sufficient to bring society to vigorously oppose the war, one moral outrage – the massacre of Palestinians by Phalangist forces in the refugee camps of Sabra and Shatilla – served as the spark that kindled the fire that eventually consumed Sharon's

---

[31] See *Ha'aretz*, June 7, 1985, the Weekend Supplement. Earlier, *Ha'aretz* published critical postcards that Major Ha'cohen had sent his wife.
[32] Eitan himself pointed out that other senior commanders shared Geva's views. See Eitan, *Sippur*, 223. Geva was openly supported by a few public figures, including former CGS, Haim Laskov, and former Liberal party Minister Moshe Kol. Another salient case of "defection," albeit with less dramatic effect, involved reserve Lieutenant Colonel Dov Yirmia, the IDF rehabilitation officer for the refugees in southern Lebanon. See Jansen, *The Battle*, 28, 30.
[33] Quoted in *The Journalists' Yearbook*, 1983, 16 (italics added). Habber also argued: "... what kind of a war [was it]? [It was] such [a war] that Cabinet Ministers and IDF generals – IDF generals! – [were] whispering in our ears that it [was] superfluous ..." See also Yaakov Erez in ibid., 15, 16.

political oxygen and undercut Israel's domestic political capacity to run the war effectively further.

On September 20, 1982, the newspapers exploded with headlines such as "War Crime in Beirut," accompanied by gruesome pictures of slain children, women, and men.[34] Then, journalists pointed their fingers at the architects of the war, Eitan and Sharon, suggesting that they be removed from their positions if they did not leave voluntarily. In fact, the newspapers had more in store for the government and the state. They warned the government, in a style that sounded almost as a threat, of the dire consequences of any effort to avoid investigating the massacre by evoking the mantle of the supremacy of the state. *Ha'aretz*, for example told the government that "The alternative (to nominating an inquiry committee) is a permanent rift in the nation: the half, that according to the polls supports Begin, is not strong enough to face alone the tests we are about to encounter in the near future."[35]

Meanwhile, the army was also thrown into a storm. Suddenly, it was revealed that Geva was not alone in the military, at least in that sense that he had had enough. Senior officers of the IDF had developed frustrations at an early stage of the war. However, until the massacre in the refugee camps, they restricted themselves to anonymous leaks at most, as their collective interest was not at stake and Sharon was far too strong for them to be confronted directly. The massacre changed everything. Officers immediately grasped its political significance. They were professional soldiers, but they were also thoroughly assimilated in Israeli society, knowing well, and often sharing the opinions of, other elites in Israel. In short, they instantly realized that the massacre threatened to tarnish their and the army's image.[36] Moreover, the two words "inquiry commission" resonated strongly in their ears. Every officer in the senior command remembered that in the aftermath of the 1973 Yom Kippur War, Moshe Dayan, the Defense Minister, came clean out of the final report of the Agranat Inquiry Commission, whereas the army and its officers were thoroughly tainted. Finally, officers suspected that Sharon was trying to use the army as a deflector against the imminent attacks after the massacre, and at the same time understood that he was rapidly losing power and becoming a liability within his own party. Thus, a few turned courageous. Brigadier General Amram Mitzna resigned from his post as the commander of the IDF Command and Staff College (only to reverse his decision later), arguing that he had lost confidence in Sharon as Defense Minister. Colonel Yoram Yair, the commander of the prestigious IDF line-paratroop brigade, went to Sharon and demanded that Sharon assume responsibility

---

[34] *Ha'aretz*, September 20, 1982.

[35] *Ha'aretz*, September 22, 1982.

[36] Interview with Brigadier General Furman. According to Yaakov Erez of *Maariv*, General Drori, OC Northern Command, responded gloomily immediately after the massacre, saying: "Damn, nobody will remember this war but for the worse." Interview with Yaakov Erez.

for the massacre. Colonel Yair also asked to meet with Prime Minister Begin. Eitan, sensing the signs of ferment, gathered the IDF's senior officers for a meeting. For all practical purposes, the gathering turned out to be a vote of no confidence in Sharon. Following this meeting, Eitan convened the same forum for a second time in order to meet with Sharon. In this second meeting, most officers were far less aggressive. However, their basic message remained unchanged. The press, of course, promptly published reports on the two meetings, the reason why they were convened, and the general arguments of the officers.

It is intriguing that none of the developments I have reviewed thus far – the accumulation of IDF casualties, the departure from the 40 km plan, the fear of an imminent assault on Beirut, or the disclosure of the Geva affair and the dissent within the army – had as a significant impact on the fate of the war as did the massacre in the refugee camps. At least conventional wisdom suggests that the impact of the massacre should not have been greater than that of any other milestone of the war, since it did not involve the tangible interests of any level in Israel and did not place Israeli soldiers in a direct moral dilemma. Unlike the case with IDF casualties, no individual interests were threatened, and although the massacre was appalling, it did not automatically constitute a sufficient cause for domestic upheaval because Israelis did not perpetrate it, nor did they believe that their leaders premeditated it. Moreover, the horrible Phalangists' conduct against innocent civilians did not constitute a radical departure from the internal Lebanese code of conduct.[37]

In fact, the massacre was not the first, nor the most despicable occasion on which Israel bore responsibility for innocent casualties in war. In October 1956 during the Sinai campaign, Israeli forces executed forty-seven Israeli Arabs who had returned home after curfew.[38] During the War of Attrition, Israel blasted the Suez Canal cities with artillery fire, causing a major refugee problem in Egypt. At one point, a deep IAF air raid went wrong and an Israeli jet bombed an Egyptian school, killing tens of children. In the spring of 1973, the IAF shot down a Libyan airliner, full of passengers, that had strayed into the airspace of the occupied Sinai peninsula and had failed to respond to the instructions of intercepting IAF jets.[39]

Admittedly, the number of casualties in these cases was far smaller than in the refugee camps. However, Israel bore direct responsibility for them, and still Israeli society remained, by and large, indifferent. In the Sabra and Shatilla massacre, Israeli forces were not the direct perpetrators,[40] but rather

---

[37] The August 1976 Christian massacre of Palestinians in Tel-Al-Zaatar and the 1978 nasty summer campaign of the Syrians against the Christians attest to the kind of atrocities civilians in Lebanon had already experienced.

[38] See Elam, *Memalei*, 53–70.

[39] See Merom, "Israel's National Security and the Myth of Exceptionalism," 427–28.

[40] It is important to note that three months before the massacre, *Ha'aretz* warned against precisely such an eventuality, suggesting the exclusion of Christians from fighting in any IDF

they carelessly let a revenge-seeking proxy mislead them, get out of control, and plunge into an orgy of killing of innocent civilian Palestinians.

Indeed, there is no reason to assume that the massacre itself would have induced similar reaction in Israel under different circumstances.[41] Thus, while it is obvious that the moral outrage of Israelis was genuine, one must go beyond the analysis of sheer expedient and direct moral calculations in order to explain why the massacre had such a decisive impact within Israel.

The Debate about the Identity of the State

In the search for an answer to the puzzle of why the massacre led to such public uproar and political consequences, one has to consider the relations between the war in general and the massacre in particular and the image of the state that opponents of the war had. My understanding is that the criticism of the war on both moral and utilitarian grounds, and the powerful social response to the massacre in the refugee camps, boil down to a protest against the identity the Israeli state seemed to have assumed as a result of the war.

For two reasons, a short review of the ideological creed of the Peace Now movement is a good starting point for understanding my argument. First, Peace Now was the largest anti-war protest movement and thus a good representative of much of the resentment to the war. Second, by the time of the war, Peace Now had already established a clear and distinct ideological trail that exposed the relations between questions of moral values and state identity.

What one must first note, though, is that three critical formative events influenced the creed of the Peace Now movement, which was created in 1978: the 1973 Yom-Kippur War, the 1977 rise of the Likud party to power, and the visit of President Sadat to Jerusalem in the same year.[42] The rude awakening brought about by the 1973 Yom Kippur War, and the blood tax paid by the generation of the founding fathers of Peace Now, convinced the latter that the prevailing political philosophy of the post-Six Day War was bankrupt. Members of the 1973 war-ravaged "generation" simply lost confidence in the wisdom of power politics and the exclusive judgment the political and military elite enjoyed in matters of national security. Consequently, they believed that Israel should try to compromise with its neighbors and promote international cooperation rather than think only in terms of power politics, competition, and military coercion. From the perspective of Peace Now, the

---

controlled territory. Second, the Commanding Officer (CO), Northern Command, General Drori, and the Beirut theater of operations commander, Brigadier General Yaron, repeatedly warned the Phalangist forces not to engage in a massacre. See Schiff and Yaari, *Milhemet*, 323, 327–28, 333, 337.

[41] This conclusion is based also on my impressions from interviews with anti-war activists and journalists.

[42] See Hermann, "From the Peace Covenant to Peace Now," 331–36.

1977 ascent of the Israeli political Right to power could not have come at a worse moment. The ideology and past rhetoric of the Likud leadership suggested that Israel was about to regress into a less-tolerant form of government, one that would be less restrained in the use of force, precisely when Israel faced new opportunities to break through the vicious cycle of the Middle East conflict. Thus, much as the 1977 Sadat visit to Jerusalem and the 1979 Camp David accords seemed to vindicate the new thinking of Peace Now activists, the 1982 invasion of Lebanon seemed to confirm their worst fears.

Tzali Reshef, one of the prominent activists of Peace Now, conceded without any misgivings that the movement protested against what it considered the establishment's behavioral aberrations rather than against the establishment itself. In an interview in *Ha'aretz* some nine months into the Lebanon war, he explained: "[Peace Now] constitute[s] today more of a world-view than an organization, we are a public of hundreds of thousands that *share a fundamental world-view* ... we try *to bring the state back to its old course and into the family of [civilized] nations.*"[43]

The components of this "world view" of Peace Now were transparent by the time of the Lebanon war. The movement was already opposing, on record, fait accompli policies and the use of what it considered excessive force. Thus, it attacked the settlement policy and the use of brutal force to put down upheavals in the occupied territories, and the (pre-war) massive bombing of Beirut, for being morally wrong, for undercutting the chances of achieving peace, and for consuming resources vital for more proper domestic purposes.[44]

In the light of such an agenda, it is little wonder that the war concept and leadership, and Sharon in particular, were on a collision course with the Peace Now constituency. After all, the war and its objectives were expressions of power politics, and no other leader personified everything that the Peace Now constituency stood against.

On the morning of September 17, 1982, the first day of the massacre, but before the event became known, President Itzhak Navon was quoted in *Yediot* warning the nation gloomily that "[he] hope[ed] that we will not overstate our power so that we will not end up with a disaster." Thus, when on September 20 the massacre hit upon the Israelis at home, the conflagration was all but inevitable. In the increasingly tense atmosphere in Israel, it focused the public debate on the issue that underscored much of the protest. Increasing numbers of Israelis had already rejected the basic tenets of the etatist approach – the idea of fighting and being sacrificed for a "national interest" that was unilaterally defined by the state, the idea that the state could

43 *Ha'aretz*, March 11, 1983 (italics added).
44 See *Peace Now* positions in Mordechai Bar-On, *Shalom Achshav* [*Peace Now*] (Tel Aviv: Ha'kibbutz Ha'meuhad, 1985), 133, 134–35.

manipulate at will information, and the idea that it could suppress moral considerations in times of a controversial war. Now, the growing frustration with the moral and political road Israel had taken under Begin and Sharon reached the flashpoint. The massacre, and the government maneuvers to evade taking the minimum measure of establishing an inquiry commission, simply linked all the contested issues into one unified concern about the nature of the state that gave birth to such aberrations. Indeed, there could be little doubt that a question of identity was at issue. The editor of the left-leaning *Davar* daily, Hanna Zemer wrote:

The Prime Minister ... should have gone to the Presidential Residence to tender his resignation and thus free Israel and the Jewish people from the curse of this government which has turned our image into that of a monster ... we will not remain silent ... *there will come a day when we all send back our Israeli identity cards, because this is not the way we want to be identified.*[45]

The reasoning and recommendations of the Inquiry Commission concerning the events of the massacre in the refugee camps also revolved around the relations between moral values, state power, democratic order, and identity. In the conclusion of their report, the members of the Commission espoused the view that the state must abide by moral values, not the least because that was the only guarantee that these values would be preserved, and the state would not change its identity. The espoused members explained:

The end never justifies the means, and basic ethical and human values must be maintained in the use of arms ... The main purpose of the inquiry was to bring to light all the important facts relating to the perpetration of the atrocities; it therefore has importance from the perspective of Israel's moral fortitude and *its functioning as a democratic state* that scrupulously maintains the fundamental principles of the civilized world.[46]

[45] Quoted in Jansen, *The Battle*, 135–36 (italics added). *Davar* was essentially a Labor newspaper and as such opposed to the Likud government. Still, Zemer's style was quite unprecedented.
[46] *The Commission of Inquiry into the Events at the Refugee Camps in Beirut*, final report (Jerosalem: MFA, 1983), 107 (italics added).

13

The Israeli Struggle to Contain the Growth of the Normative Gap and the Rise of the "Democratic Agenda"

In launching the Lebanon war, the Israeli state enjoyed three convenient domestic conditions. First, public support was virtually guaranteed for a campaign against the PLO since most Israelis regarded the Palestinian organization as a vicious enemy that deserved to be fought, and if possible, eliminated. Second, the war leadership had ample time to prepare the marketing of the war. Third, the solutions the Israeli leadership devised for the international reactions it anticipated were also highly compatible with the problems the government was likely to face inside Israel.

In the final analysis, however, the structural advantages the government enjoyed also had a serious downside, as they worked as blinding agents, confining the internal political debate almost exclusively to international considerations. Thus, for example, as was indicated in the pre-war deliberations, considerations and anxieties of ministers, and even members of the liberal press, power politics took central stage.[1] Indeed, with few exceptions, neither officials and ministers nor even columnists in the liberal press were particularly concerned with potential domestic objections to the war or with its social consequences, until the first wave of reservist protest.[2]

It was not that the possibility that the war could unleash some opposition at home was utterly absent from the mind of all decision-makers. Indeed, the gap between the declared "40 kilometers" objective of the war, on the one hand, and the military actions and the unfolding real goals, on the other, seem to indicate that at least Sharon sensed a potential domestic problem. Still, political awareness of potential domestic problems in the wake of the war should not be overstated, and Sharon's deceptive policy should be seen as designed, first and foremost, to mislead his fellow ministers. As far as Sharon

[1] See Feldman and Rechnitz-Kijner, *Deception*, 28–41.
[2] Three notable exceptions were General Sagui, Chief of Military Intelligence; General Or, CO Central Command; and *Ha'aretz* journalist Uzi Benziman. See Schiff and Yaari, *Milhemet*, 121–22; and *Ha'aretz*, June 10, 1982.

was concerned, domestic complications were obstacles to be overcome rather than problems defining the limits of political feasibility. In any event, the concerns the war plans invoked within the government were almost entirely confined to the international level. Ministers were skeptical about Israel's ability to bring about the complete destruction of the PLO. They were skeptical about the feasibility of imposing a Christian regime in Lebanon. They feared that the conflict would spin out of control into a major confrontation with the Syrians. They feared a premature cease fire that would be imposed by the superpowers. Finally, they feared the consequences of the war for the infant and fragile Israeli-Egyptian peace (as did some press columnists).[3]

The second important point to note is that at least in the short run, the international and domestic needs of the Israeli leadership were essentially compatible, and thus made the government's planning a bit easier. This compatibility was well reflected in the choice to wait for a pretext in order to start the war and then launch it as an escalatory chain reaction. Thus, while the army was long prepared for the war, the Israeli government waited for some event – which turned out to be the assassination attempt against Israel's ambassador to the United Kingdom – that could be exploited in order to retaliate and thereby lure the PLO into an artillery duel. Such a duel was bound to damage northern towns and settlements in Israel and thereby provide the *casus belli* from an international point of view. However, it also emphasized a real or presumed defensive motivation for domestic consumption. For similar reasons, the government propagated the idea that the territorial objectives of the war were limited, concealed most of the war's objectives, and tried to run the war on a tight schedule. These could preempt international initiatives designed to deprive Israel of its projected gains, but at the same time they could also reduce potential adverse social reactions to the scope of the war within Israel.

In this respect, it is not surprising that the Israeli leadership portrayed the Palestinian deployment in Lebanon as constituting an existential threat, and depicted the PLO, perhaps more sincerely, as an organization of savages. The Israeli leadership argued that "our stay in Lebanon serves our struggle over the land of Israel," and tried to further sustain the existential argumentation by dramatizing the quantity of military spoils Israel had captured from the PLO.[4] Begin, quite ingeniously, gave a final twist to the marketing of the war when at the last moment he ordered a change to the name of the war to "Operation Peace for Galilee," a name that contained an idea that hardly anyone in Israel could oppose.

In addition to these built-in and designed advantages, the government could also count on a significant level of independent and arranged factional

[3] See *Ha'aretz*, June 6 and 9, 1982; and senior columnist Matti Golan in *Ha'aretz*, June 22, 1982.
[4] Quoted in *Ha'aretz*, July 9, 1982 (from Eitan's interview with *Ba'mahane*).

support. Thus, while the anti-war movement was quick to seize the agenda, it was by no means alone in the public arena. A month into the war, in early July 1982, a group of reserve pilots and other officers called upon Peace Now to cancel its planned demonstration because the former "felt, [that] while the fighting is on, this public debate impairs [their] capacity to function with full efficiency."[5] Workers of the large, state-controlled industries – namely, Israel Aircraft Industries (IAI) and Israel Military Industries (IMI) – pledged organized support for the war and the state. In mid-July, the government orchestrated a pro-government demonstration equal in size to the massive Peace Now demonstration of July 3. Toward the end of July, a few professors and reserve officers, who formed the "Peace and Security" association (which at that time was deeply inferior to Peace Now in terms of size and organization), lent the government vehement support against the anti-war protest. Reserve IAF Colonel Eliezer Cohen, for example, accused the journalists of Israeli TV of brainwashing the people, and added bombastically that "we [have] reach[ed] in this matter, the red line of self annihilation."[6]

Another reliable base of support for the government was in the religious Zionist community, which was largely conservative, community oriented, and supportive of the state and power politics.[7] The members and elite of this group, and in particular its ideological core, supported the Likud government and its policies in general, and the Lebanon war was no exception. Israel's Chief Rabbinate, under Rabbi Shlomo Goren (a retired Chief Rabbi of the IDF) argued that the war was not simply a "just war" but also a *mitzvah* war – that is, a war of religious prescription. At one point, the military rabbinate distributed, maps of Lebanon with the Biblical names of villages. And the zealots of the Gush Emunim movement, who settled in the occupied territories, were more than happy to find Biblical references suggesting that current Lebanese territories had belonged in the past to the ancient kingdom of Israel. No less indicative of this support of the government was the reaction of the *Bnei-Akiva* youth movement of the National Religious Zionist movement to the Sabra and Shatilla massacre. The youth movement proclaimed its "shock" over the carnage, but hastened to add that it was fully confident in the "purity of arms" of the IDF.[8] While ostensibly a repudiation, this cautious formulation expressed more than anything the belief that Israel bore no responsibility and that thus there was no need to investigate the events that had led to the massacre.

Finally, at least until the end of August, and in spite of unexpected difficulties, the government and the state could count on support arising from the

[5] *Ha'aretz*, July 2, 1982.
[6] *Ha'aretz*, July 26, 1982.
[7] See data in Arian et al., *National Security*, 71–73.
[8] *Ha'aretz*, September 24, 1982.

reality they had created, and collect the political dividend from the achievements of the IDF – the territorial gains in Lebanon, the Syrian military defeat and (temporary) political setback there, and the dramatic expulsion of the PLO from Beirut.

The Domestic Reaction of the Government and the State

The fact that Israel started the Lebanon war from favorable structural and conjectural conditions did not prove in the long run of great significance. In the final analysis, the basic domestic structure of Israel still led the war into an impasse. When the government was struggling to square its demands of legitimacy and social readiness to sacrifice with the human and moral costs of the protracted war, it was really seeking an illusive balance that might very well have been unattainable in the first place.

Unfortunately for the government, the efforts to forge an optimal set of policies for the potential and actual domestic consequences of the war relied on a false assumption that the level of public knowledge, perceptions, and reaction to what would happen in the war were controllable. Now, the roots of the government's failure were not in a shortage of the means of control. In fact, the Israeli state possessed enough censorship powers not only to impose stringent control over information and opinion, but also to regulate culture. Rather, the roots of failure were in Israel's instrumental dependence and democratic tradition that prevented the effective use of these existing powers.[9] There simply were very few coercive measures that the state could employ in order to control the free flow of information and ideas without stimulating the growth of the normative difference to unmanageable proportions. That does not mean that the government gave up on the effort to control information and perceptions, or that the fundamental attitude of state officials was modified. At least the defense establishment, under the leadership of Sharon and Eitan, tried to do its best to prevent, and then to control, the information flow. In short, during Sharon's and Eitan's tenure, the army partially assumed a political responsibility – to help keep the normative difference under control.

Obviously the media were the prime target of the efforts to prevent the normative gap from growing. Because of the nature of the war, the nature of relations between the press and the leadership, and the nature of the personalities involved, the domestic scene was bound to be conflict and confrontation

---

[9] Israeli governments had at their disposal the *Emergency Regulations*, which were residues of the British Mandate. However, both the scope and ineffectiveness of these regulations were demonstrated during the Lebanon war. In October 1982, using Mandatory law, the Israeli civil censor banned Hanoch Levin's play, "The Patriot," for being "deeply offensive to the basic values of the nation, the state and Judaism." Yet, theaters continued to perform the parody without ever being indicted. See *Ha'aretz*, October 27, 1982.

ridden. The Lebanon war began when the state-press coordination mechanism, the Committee of Newspaper Editors – that was always put to work
in times of military crises – was all but ignored. Furthermore, military reporters and commentators who for years had enjoyed direct access to the
IDF command were suddenly shunned. Indeed, this became immediately
clear when the war started, as they were not welcome in the operation's
command post.[10] Thereafter, they were not provided with the help the army
used to extend to Israeli journalists in times of war, but rather encountered
stricter control over their work (in particular TV correspondents) that originated in the office of the IDF spokesman.

Meanwhile, the defense establishment took other measures to influence
public opinion. Three weeks into the war, retired General Yesha'ayahu
Gavish, who served as the military commentator of the (only) Israeli, state
controlled, TV channel, was replaced by reserve General Aharon Yariv.[11]
Apparently, Sharon grew dissatisfied with Gavish's commentary, and thus
Yariv was brought in. The latter could offer the credibility of a politically
non-affiliated expert without being suspected of compromising state and government interests. Essentially, Yariv could be trusted to act in a predictably
loyal manner after long years of loyal support for the state. In early July
1982, in a move designed to reduce the domestic outrage over the destruction in southern Lebanon and the suffering of the refugees there, Sharon
was reported to be considering the nomination of Aryeh Eliav – a celebrated
left-wing activist and humanitarian – to head a rehabilitation effort there.
And on July 13, 1982, Sharon decided to cancel the recording of a designated
TV interview program in a rehabilitation center for wounded IDF soldiers,
although Eitan and his Chief Medical Officer authorized the broadcast. Apparently the etatist wish to control information also trickled down to lower
levels. For example, there were reports that troops in Lebanon did not regularly receive the more progressive morning press, and that occasionally the
"Op-Ed" pages of the dailies were missing.[12]

Although it is true that the army was drawn into the political game largely
because of Sharon and Eitan, it would be inappropriate to omit the fact
that the army and its officers were conceptually predisposed to side with
the government, and had considerable institutional and personal interest in
convincing the public that the war was justified and successful. Obviously
a failed war threatened to taint the army's image and put an end to the
promotion and careers of some of the senior officers. Thus the decision to
exceed the traditional obligation of obeying the government by committing

[10] See *The Journalists' Yearbook, 1983*, 8–10; and Levi Itzhak Ha'yerushalmi in *The Journalists'
Yearbook, 1985*, 105.
[11] See *Ha'aretz*, June 27 and 28, 1982.
[12] Benziman, *Lo Otzer*, 260; and a letter from a career officer to his wife, quoted in *Ha'aretz*,
June 28, 1982.

the army to an auxiliary political role was not simply imposed on the army, but rather accepted by it.

All in all, the limited political role the army assumed proved a rather bad bargain. There was no way to keep information from reaching the public and soldiers, and because some of the information suggested that democratic procedures were violated, the military was placed in the uneasy position of a co-conspirator in the eyes of civilians and in particular reserve soldiers. Some officers did not hesitate to turn to the media in order to influence decisions that they thought were wrong, costly, or career-threatening, while others, and sometimes even the same officers, were unhappy with the media because certain disclosures also threatened their interests. The tension between the army and the media was building up steadily, and, as expected, it was more intense among the higher circles of the military.

The magnitude of the friction between the press and senior IDF officers was exposed most vividly in an interview the editors of *Yediot Aharonot* conducted with members of the General Staff shortly before the massacre in the refugee camps, and which they – prominent media figures – found unusually disturbing.[13] In the interview, most officers earnestly insisted that in principle they supported the freedom of the press. Nonetheless, excluding Jackie Even and Ori Or, the generals blamed the media for the ferment within the forces, implicitly suggesting that the media should automatically support the state and the army. Reserve General Meir Zorea argued that "there [were] reports that undermined the fighters' morale" and that "during the fighting there [were] certain things better left unsaid, even if they [were] true." General Yohanan Gur added that "the press also bear etatist responsibility and must distinguish between what, and what not to [write and publish], when, and when not to [write and publish]." Finally, General Moshe Bar-Kochba, formulating somewhat radically the officers' view that the media possess too much power but display too little responsibility, argued:

Two things make a superb fighter: professional expertise and mental strength, namely the willingness to fight and to sacrifice the most precious of all – life. The willingness to fight ... is a matter of influence, explanation, and an understanding of, as well as an identification with the purpose. *The media are an important component determining the extent of identification ... This is more important than any particular weapon system. [During the war in Lebanon] our media did not pass the national test of encouraging the army in a war* [which was] more just than any other [in the past]. (Ibid.) (italics added)

General Eitan's opinion in this interview is particularly interesting. He was not only a staunch etatist by conviction, but he was also likely to stand at the eye of the storm in case the war turned sour. Thus, verging on the brink of outright preaching, Eitan abandoned his laconic style and moralized toward the editors of *Yediot* for the media's failure to report what he considered

[13] *Yediot Aharonot*, September 17, 1982.

to have been "the achievements of war" and for the "poison [Israel] had never faced before" – i.e., the media's allegedly exaggerated reporting of the violent consequences of the IDF's war.[14] In particular, Eitan was irritated by a question that indirectly negated the necessity of the war – the question whether the traditional roles of Israel and its enemies, those of David versus Goliath, were not reversed in the Lebanon war? Frantically blasting what was implied in this question, Eitan responded:

Tomorrow, it will probably appear in the newspapers that we are 'Goliath' and those poor Arabs are 'David.' Yet the truth is that it is the other way round…they are Goliath. Saudi Arabia, Jordan, Iraq, Syria, the Palestinians, Libya, Algeria and all these states. So, be careful with this metaphor, since foreigners, Americans, would say: 'Oh, finally the Jews also say that they are Goliath, and now they can manage (on their own).' One has to be very cautious (in dealing) with that…(ibid., note 13)

After retirement, yet still during the last stage of the Lebanon war, Eitan again articulated his basic views regarding the "suitable" role and "etatist responsibility" that a free press must assume in a democracy during times of war. Actually, his views may have only hardened during the Lebanon war. In any event, in his 1985 memoirs he insisted:

It is unnatural that the media feed the Israeli citizen and the world and supply them with ammunition against Israel. *At best, the media in Israel acted as if they were neutral observers in the battlefield, free from involvement and national responsibility.* It is not a question of military censorship, that forbids or permits, according to its own consideration [what to publish and what not to publish], but a question of *reporters and editors, whose national responsibility – and not [their] wish to demonstrate freedom of speech in Israel to the world, or to charm some stranger – should guide them.*[15]

Having noted the army's drift into the political struggle, and its role in the state struggle in the marketplace of ideas, it is important to emphasize that the actions and messages of its officers by no means constituted the most brazen or vicious attacks against the media. On the contrary, compared with the rhetoric of some circles in government, the military criticism of the press was presented with finesse. Right-wing factions within and outside the ruling coalition attacked the press and the anti-war movement with remarkable ferocity. Apparently, such attacks reflected a measure of authentic, if sadly misguided, conviction that opposition to state policy in times of war constituted no less than treason. However, the attacks reflected also vested political interests. A war perceived as a failure would have brought the Likud closer to losing power, and those attacking the press knew well that electoral disaster

---

[14] Ibid., note 13. Eitan also attacked the media for the "negative use of the freedom of the press," and particularly *Ha'aretz*, for publishing Major Ha'Cohen's justification of Colonel Geva's moral choice. See *Yediot Aharonot*, September 15, 1982.

[15] Eitan, *Sippur*, 226 (italics added). While these words reflect his post-war resentment toward the media, his earlier attitude was fundamentally similar.

would harm their personal and collective interests. They either stood to lose their power positions, or as in the case of the settlers in the occupied territories, they stood to lose significant benefits, including a measure of legitimacy and vast material support for their cause from public sources.

Therefore, right-wing figures suggested that Israel needed more "nationalistic" culture, that the media should learn how to glamorously cover a war as did the British in Falkland,[16] and that the TV coverage should be restricted according to Article 47 of the broadcasting legislation.[17] At the same time, they described Israeli TV as a "greenhouse for defeatists," and blasted the reporters for broadcasting "forbidden things" because they harbored distorted left ideology.[18] For obvious reasons, Sharon took a leading role in these attacks, and as time passed, his assaults became increasingly malicious, and only more so after he was forced out of the Defense Ministry. In May 1983, he lamented that in previous wars, mothers did not complain about the death of their dear ones, the number of casualties was not counted every day, and the radio and TV did not report about burial ceremonies ten times a day.[19] In June, he argued that the ability of the anti-war constituency to turn the cost of war into a political weapon was a revelation of Israel's weakness. Then he added his own version of a backstabbing theory – "the media and the opposition," he argued, "joined together in an intentional campaign of demoralization . . . [that had] reached such a point that . . . there is no military that can face such intentional demoralization."[20]

Ultimately, then, the political reaction of the Likud-run state was the same reaction many other regimes with strong etatist agenda and mild-authoritarian tendencies had adopted in response to their inability to narrow the normative gap between the state and segments of the educated middle-class. The government strengthened the state's alliance with the lower classes. This started with vicious hints and often outright accusations against the press, Peace Now, and others for "back-stabbing" the nation, and it ended with Sharon's periodical attacks on the Kahan committee for the "colossal" damage it inflicted on "the Jewish people, the Israeli state, and [him] personally."[21]

The Secondary Expansion of the Normative Gap

In the light of the position of state officials, army command, and right-wing circles toward the media, it seems clear why the anti-war agenda gained a

---

[16] Likud Member of Knesset (MK) Micha Raisser. See *Ha'aretz*, Weekend Supplement, July 1, 1983.
[17] See *Ha'aretz*, July 8, 1982.
[18] Likud MK Roni Millo. See *Be'eretz Israel*, 148 (1984), 14–15.
[19] *Yediot Aharonot*, May 24, 1983.
[20] Interview with *Yediot Aharonot*, Weekend Supplement, June 17, 1983.
[21] See, for example, *Ha'aretz*, February 15, 1983; and *Yediot Aharonot*, June 10, 1983.

third dimension that linked the war to a threat to the democratic order in Israel. Yet, in order to gain a more comprehensive understanding of the development of the third, domestic order, dimension, one needs to discuss in some detail the decision-making and actual military policy of the government, and Sharon in particular.

Let us start by reiterating that Sharon was not utterly oblivious to the idea that the scope of his Lebanese ambitions had the potential of offending many in Israel. In fact, it is this insight that was probably most responsible for developments that raised the threat the war posed to the democratic order. Sharon's wish to minimize the potential opposition to his policy, inside and outside the government, spurred him to act in ways that seemed to shake democratic conventions.

It all started with Sharon's understanding that the government was not likely to support his ambitious plans, and that without this support these plans were doomed. Thus his first objective became to receive political legitimacy for the war – that is, a clear authorization from the government. He proved capable of that as indeed he managed to maneuver the government into backing a war concept its members had repeatedly rejected almost until the first shots. The key to Sharon's success was two-fold. First, he cultivated his alliance with the undisputed leader of the Likud, Begin, using the latter like a Trojan horse in order to penetrate the government and defeat its objections to his plans. Indeed, once Begin agreed to go along with Sharon's war plans, the beginning of hostilities remained only a matter of timing, as the government was unlikely to oppose Begin as fiercely as it was to oppose Sharon.

Once Sharon achieved his first objective, he earnestly went about obtaining his second objective: extending the legitimacy his war was given and eliminating obstacles to a deep invasion. Sharon's major strategy to achieve this second objective was to "sell" the extended war to the government in a piecemeal fashion. Initially, the government was told that the "operation" was limited. Then Sharon asked the government to authorize pieces of his overall war plan one at a time. In this way, the ministers, who refused to authorize a deep invasion prior to the war, became increasingly committed to the latter with every move they authorized or did not veto. Their political fortunes increasingly became tied to the outcome of the war effort. Had they decided to "defect" and abort the war, they would have been blamed either for treachery that led to a failure and/or for dereliction of duty that facilitated the war in the first place. In short, the piecemeal strategy turned the ministers into political hostages of both the war and Sharon. They essentially stumbled into a gambler's dilemma, raising their bid in each round, fearing that if they quit they would lose everything while hoping that persistence would eventually get them the jackpot.

However, when Sharon so skillfully embroiled the government in a war it did not want, he was also trading off the short for the long term. Thus, while he received immediate government support, he was also setting the stage for a later social backlash. Much as was the case with the use of IDF firepower, Sharon managed to achieve his operational objective, but at the same time saw the seeds of the war's failure and his own political demise. All of this because in conducting the war piecemeal, he over-stretched the boundaries of democracy, even if he formally remained within them.

Indeed, Sharon's actions left an uneasy feeling of subversion against democracy. At first, this feeling was particularly prevalent among the soldiers who had to bear the consequences of his political strategy. This uneasiness began to spread when Israel seemed to have conspired to break the June 11, 1982, cease fire with the Syrians. Then, a little more than a week later, came the "crawling" stage – essentially an effort to establish firm control over the key Beirut-Damascus road by daily nipping pieces from Syrian-controlled territory in the region of Bahmadun-Alei. The crawling tactics, much like the violation of the June 11 cease fire, was accompanied by discrepancies between the official line of the IDF spokesman's office and the events in the theater of operation. Furthermore, because of the conniving nature of the crawling operation, the infantry forces that executed the daily crawl operated without full firepower support, and as a result their battle casualties seemed to have exceeded those absolutely necessary. Consequently, the soldiers became disgruntled, feeling that they were considered expendable, and suspecting that the crawling was unauthorized by the government. When Major General Moshe Levi, the deputy CGS, visited the reserve paratroop units that bore the brunt of the crawling battles, he found himself engaged, as he disclosed to the press, in "complex and difficult talks."[22] What was underlying General Levi's carefully chosen words was no less than open ferment among the troops, and their explicit warnings that the consensus over the war was fading quickly. Indeed, to make sure that General Levi did not miss the point, soldiers took the trouble to mention that they were not disciples of the Left. Sharon himself never visited these troops. Not because he did not intend to, but rather because he was advised by Brigadier General Menahem Einan, the commanding officer in charge of the theater of operation, that his visit would not be a wise move.

By the time the IDF closed in on Beirut, the growing confidence gap between the IDF and its soldiers could no longer be ignored. Officers and soldiers, particularly reservists, became increasingly irritated by "the mendacity" – a term that described their feelings that they were "mercenaries of those scoundrels in the government."[23] The representatives of the IDF spokesman unit felt the consequences of this credibility gap. The frustration,

[22] *Ha'aretz*, July 9, 1982. On the meeting of General Levi with the seething forces, see Schiff and Yaari, *Milhemet*, 246–48; and Benziman, *Lo Otzer*, 254.
[23] Personal interview with Dr. Yossi Ben-Artzi, a middle-rank reserve officer during the Lebanon war and *Peace Now* activist, October 17, 1991, Haifa.

rage, and animosity of the soldiers was turned against them with little subtlety.[24]

While the soldiers could only sense the tension between military actions and democratic processes, the upper echelon of the IDF had a good vantage point to actually observe both this tension and its consequences. First, senior officers received hints from Sharon that the government need not be privy to the precise nature of their discussions with him.[25] Second, while Sharon conferred with the General Staff during the war, some of the officers suspected that their meetings constituted no more than rehearsals for Sharon's selling bids in the government.[26] All the more so, as Sharon's impatience with opinions other than his, especially opposing ones, was well known. Finally, IDF commanders experienced moments of direct clash between the military decisions and governmental policy, and realized that they themselves were likely to pay a substantial price for cooperating with the undermining of democracy. This realization became particularly pronounced in mid- to late-June 1982, when Sharon was planning for the assault on Beirut. The account of Yaakov Erez from *Maariv* needs no further comment:

We visited the headquarters of the [IDF] division encircling Beirut, and found ourselves in the Order Group where the attack on West Beirut was planned, when ... we heard on the radio the Prime Minister announcing [in] the United States that there was no intention of ordering the IDF into West Beirut ... eyebrows were raised in amazement ... A battalion commander arrived and asked the brigade commander: 'What should I tell my soldiers?' And the brigade commander told him: 'I just asked the division commander the same question and I have no answer.' I saw generals waving their hands, unable to respond ... [27]

Indeed, Sharon's policy put the upper echelon of the IDF in an increasingly trying position. At first, senior officers saw no wrongdoing in manipulating the government and the people, or at least they were willing to tolerate such practices. As time passed, however, they were forced to reconsider their position and change their attitude, as they had to face, increasingly on a daily basis, their soldiers' rage over what was considered as deception. In short, senior officers soon discovered that there was nothing gratifying in having to maneuver between grudging subordinates and a deceptive superior, or in having to come to the forces with apprehension, excuses, and in low profile.

These pressures, and in particular the frustration spilling over from lower ranks to senior officers, did not produce a critical backlash, even though they were constantly growing, until the IDF was getting ready to take over urban Beirut. Only Beirut brought things to an extreme, and thus in the final analysis it became Sharon's nemesis. The point is that the success of the whole

[24] Personal Interviews with Yaakov Erez and Eitan Habber, in October 1991, Tel-Aviv.
[25] See, for example, in Schiff and Yaari, *Milhemet,* 261.
[26] Telephone interview with Brigadier General Furman, October 24, 1991.
[27] *The Journalists' Yearbook, 1983,* 13.

war, and Sharon's own fortunes, depended on what would happen there. And what would happen in Beirut, depended on how much democratic rules and procedures would be bent.

Clearly, a failure to force the PLO out of Beirut would mean a certain political disaster for Sharon and the government, for there was no way they could justify the war and the IDF casualties without reaching this objective. However, in order to force the PLO out of Beirut, Sharon needed either to convince the PLO that he would use the IDF if necessary, or to actually use the IDF. While the IDF could do the job, the opposition to attacking Beirut was almost unanimous in Israel, mainly because people feared that it would involve heavy casualties. Thus, Sharon found himself in dire straits. His preference was obviously to present the PLO with a credible threat that would not have to be executed. The more credible the threat, the greater were the chances that he would not have to carry it out, and that in turn implied no Israeli casualties. But credibility could be gained only at the price of making all the necessary preparations in order to attack Beirut, including convincing Israeli soldiers that they were about to assault the city. There was simply no margin for ambiguity that could lead the PLO to question Israel's resolve. In case the Palestinians did not surrender, Sharon would have to use the IDF. Yet, this exactly, or the anticipated consequences of such an act, was what fuelled the opposition to attacking Beirut in the first place.

In the hopes of circumventing this dilemma, Sharon tried to gear up the IDF for swaggering purposes, though because of international and domestic political considerations he was also ready for an all out assault. The action began with the IDF's nibbling at pieces of Beirut and using its artillery and air power in order to convince the PLO that it had little choice but to evacuate Beirut. In doing so, however, Sharon exceeded the authorization he was given by the government. In early August, Begin himself admitted that he "knew about all the (IDF) actions (in Lebanon) . . . sometimes before they were carried out, and *sometimes after.*"[28] Government sources happily leaked Begin's feelings, and the press duly reported the tensions within the government. Sharon was walking a tightrope. His frustrations produced the kind of conduct and rhetoric he would have probably liked to avoid. Apparently caught off guard, he boasted in an interview to Oriana Fallaci: "Had I been convinced that we had to enter Beirut, nobody in the world would have prevented that. *Democracy or no democracy, I would have gone in even if my government did not want it.* I mean, I would have convinced them."[29]

Obviously, these words were a gift to Sharon's opponents and to those who opposed the war. *Yediot*, in one of its wide-circulation weekend editions,

---

[28] Schiff and Yaari, *Milhemet*, 277 (italics added).
[29] Quoted from *Yediot Aharonot*, September 3, 1982 (italics added).

was more than happy to provide this fresh and incontrovertible evidence of Sharon's authoritarian proclivities. The interview became its major news item, receiving a fat headline and full coverage in the Weekend Supplement. Events in Beirut were thus bringing the domestic impact of the war to new heights.

Thus, as the military operation progressed, as the decision-making concerning events in the battlefield became independent of the political system, and as more governmental sources criticized the legitimacy of Sharon's actions, he was increasingly perceived as the link between the war and the threat to the democratic order. Hence the struggle to render a verdict on Sharon's role in the Sabra and Shatilla massacre and hence the struggle – once the recommendations were formulated – to force the government to execute them fully and get rid of Sharon. Indeed, once the government seemed hesitant to implement the recommendations, and as it was clear that Sharon would try to rescind them, a new wave of protest erupted.

At this stage, the most significant event was the demonstration Peace Now organized on February 10, 1983, that called for the immediate implementation of the committee's recommendations – the removal of Sharon from office. The demonstration drew a violent response from a Jerusalem mob, composed of a public that had been incited for months by right-wing politicians. The police, whose lower ranks leaned to the nationalist right, were largely indifferent to the abuse of the demonstrators by the mob. Toward the end of the demonstration, a hand grenade was thrown at the demonstrators, and as a result, Emil Grintzweig, a Peace Now activist, was killed and several others were wounded. If anybody needed conclusive evidence that the war brought a real threat to the Israeli democratic order, the murder of Grintzweig supplied it. The day after the murder, Zeev Schiff, the senior military commentator of *Ha'aretz*, wrote on the front page what many opponents of the war felt: "Sharon's struggle over his survival [in the government] and his political future, [was] parallel to the struggle over democracy ... whose outcomes [would] significantly decide the moral image [of Israel] ... "[30]

In the final analysis, then, the period between the massacre in Lebanon and the murder in Israel, was the most intense in terms of the democratic component of the anti-war agenda. However, reverberations of the struggle over the character of the political order in Israel continued past these months. Moreover, the damage to the trust of citizens in the political order in Israel was spilling over to state-society relations, as indeed was indicated by ominous signs. Bit by bit, acts of civil defiance and the rejection of state intervention that were opposite to pre-war conventions in Israel, multiplied. I have already noted the erosion in the traditional commitment of Israelis to serve in the military reserve. Israelis increasingly dodged call-ups, and a few of them were even ready to openly refuse to serve in Lebanon.

[30] *Ha'aretz*, February 11, 1983.

Moreover, these ominous signs did not stop in this particular meeting of the civil and military spheres. A few families of IDF fallen soldiers, who traditionally embraced the military involvement, excluded the state from the funerals of their loved ones by refusing a burial with full military honors. Similarly, a few other families insisted that the inscription on the grave should bear no reference to "Operation Peace for Galilee."

14

Political Relevance and Its Consequences in Israel

The Israeli forces that opposed the Lebanon war succeeded in halting Sharon and bringing the war to a grinding halt. Both were no small achievements, particularly considering that they were gained in the marketplace of ideas and as they involved a society that was conditioned to support almost any tough security stand. This, however, does not indicate that the Israeli society as a whole, or even a majority in Israel, opposed the war. Rather, all indicators suggest that the anti-war coalition remained a minority with limited social and political reach. The anti-war coalition stripped the government of the mantle that an active security policy had provided, and shook the overwhelming popular support it had initially enjoyed. But the former did not turn around the opinions of most Israelis, concerning the war, its conduct, or the leadership that brought the failed war upon them. The Kahan Commission of Inquiry more or less captured this social reality when its members wrote that they did not "deceive [themselves] that the result of this inquiry will convince or satisfy those who have prejudices or selective consciences, [for whom] this inquiry was not intended . . . "[1] Indeed, during the first year of the war, which was the most intense period in terms of the war and state-society strife, the majority of Israelis did not oppose the government war policy, nor, as Table 14.1 suggests, did their confidence in Begin, the government, and even Sharon change all that radically.

Satisfaction with Begin remained unchanged, even after the massacre in the refugee camps and the ensuing turmoil within Israel. Begin's popularity fluctuated somewhat from October 1982 until July 1983, but it remained high. In fact, it never even came close to his record low approval rating of 26 percent during his first term in office.[2] Sharon's popularity fluctuated somewhat differently. He lost nine points of "satisfaction" and gained "six points of dissatisfaction" in October, following the massacre. However,

[1] *The Commission of Inquiry*, 107.
[2] See Eliyahu Salpeter in *Ha'aretz*, October 29, 1982.

TABLE 14.1 *Satisfaction with Begin's and Sharon's performances*

Poll question: "Are you satisfied or dissatisfied with Menahem Begin/Ariel Sharon as Prime Minister/Defense Minister?"

	"Satisfied" and "more-or-less satisfied" combined (%)		Dissatisfied (%)	
	Begin	Sharon	Begin	Sharon
September 1981	65	62	31	15
March 1982	66	73	26	19
April	–	63	–	28
June	69	–	28	–
July	–	73	–	23
September	–	70	–	27
October	69	61	27	33
December	67	53	30	43
January 1983	69	61	27	34
March	72	out of office	25	–
April	74	–	21	
June	63	–	33	
July	66	–	30	

Source: PORI polls in Hann-Hastings and Hastings, *Index to International Public Opinion, 1981–1982*, 365–66; *1982–1983*, 376–77; *1983–1984*, 383.

his overall satisfaction rate (fully and partially satisfied combined) was almost double his dissatisfaction rate. In December 1982, his public standing plummeted, and this was echoed in his frustrating efforts to find a lawyer who would take his case before the final deliberations of the Kahan commission.[3] But even at this point, in a particularly bad month for Sharon,[4] the overall number of those satisfied with Sharon, was still greater by a margin of ten points than the number of those overall dissatisfied with him (53 percent to 43 percent). By January 1983, the trend was all but reversed, and Sharon's approval rate returned to "normal." In fact, Sharon was greeted with exceptional warmth when he attended, immediately after the Kahan commission recommended that he be dismissed, a reception of the Israeli Bar Association (IBA), and Dr. Amnon Goldenberg, the President of the IBA, all but repudiated the recommendations of the commission.[5]

Political support for the government reveals a similar reality. Of course, performance assessments of the government are not formed on the basis

[3] Sharon, *Warrior*, 515.
[4] In December, the Inquiry Commission released an interim report warning Sharon (and others) that he might eventually be held accountable for the events in the refugee camps. This was after Israeli security forces had the worst day for fatalities in Lebanon in mid-November, when an explosion demolished a headquarters in Tyre, causing seventy-five Israeli deaths.
[5] *Ha'aretz*, February 13, 1983.

of a single issue, but as the war was the most salient topic in Israel (and the economy was in bad condition), they are correlated rather strongly to the level of popular support for the war policy. Table 14.2 shows that the public perception of the government's performance did not drop below the pre-war 40 percent approval rating ("very good" and "good" evaluations combined), at least not until May-June 1983. In addition, it reveals that an absolute majority of Israelis wanted the government to stay in power, at least until May 1983, and that until July 1983, the Israelis, by margins of two and three to one, thought that the Likud government was the best among all alternatives. In short, at least for the ten to eleven months following the invasion, support for the government remained relatively high – in spite of the casualties (some 70–75 percent of the war total), the massacre in Sabra and Shatilla, the prolongation of war, the conclusions of the Commission of Inquiry, and the vocal protest against both the war and the government.[6]

Of course, this discussion ignores the fact that although Begin and the government remained favored by Israelis at large, they nevertheless suffered a substantial decline in public opinion polls in the period April-June 1983. This decline, however, had much more to do with the growing economic hardship Israelis faced, and Begin's conspicuous functional deterioration, than with the deadlocked war. Indeed, this is indicated by the comparative and individual rate of change in popular satisfaction with the performance of the Treasury Minister, Yoram Aridor, Prime Minister Begin, and the government (see Table 14.3). Over time, Aridor was by far the greatest loser, while Begin fared a bit better than the government (Table 14.3, section B). However, between April and June 1983, as Begin's disfunctionalism became apparent, satisfaction with his performance declined most rapidly (Table 14.3, section A). In between, the government scored politically when in May it seemed to have achieved one of Israel's major war goals – an agreement with Lebanon (though the paper was less than a peace treaty and soon turned out to be utterly worthless). Indeed, in June, a *Dahaf* public opinion poll showed that an Israeli majority of 51 percent (versus 37 percent) thought that the war was a success.[7] Let us note also that Sharon's successor, Moshe Arens – the person probably most responsible for Israel's pointless stay in Lebanon in 1984 (against the army's best advice) and hardly less hawkish than Sharon – enjoyed the kind of high approval rating usually reserved for Defense Ministers in Israel.

Support for the government, beyond the summer of 1983, was eroding, and public justification of the extended war was diminishing. However, in spite of the economic deterioration and the electoral setback for the Likud,

[6] PORI's had the advantage in the scope of questions, follow-up record, and a consistent minimum sample size of 1,200. However, results of other polls indicate the same trend. See, for example, Yaniv, *Dilemmas*, 246–50.

[7] *Ha'aretz*, June 8, 1983.

TABLE 14.2 *Confidence in the Israeli government, April 1981–June 1982*

A. Performance rating
Poll question: "What is your opinion on how the government deals with Israel's problems today?"

	"Very good" and "good" combined (%)	"Not good" (%)	"Not so good" (%)
April 1982	40	18	36
July	56	10	29
October	44	15	39
December	40	17	33
January 1983	47	15	39
March	43	15	39
April	43	13	38
June[a]	32	23	40

B. Relative confidence
Poll question: "Would you like to see the government serving to the end of its term, or should the government be replaced before the end of its term?"

	Continue serving (%)	Be replaced (%)	Depends (wavering) (%)
April 1982	55	34	5
June	61	29	3
October	57	31	3
December	60	32	4
January 1983	64	26	2
March	68	25	3
April	66	23	3

C. Relative confidence in an alternative government
Poll question: "Do you or don't you believe that a Labor government could have coped better with the problems facing the current government?"

	No (%)	Yes (%)	Maybe (%)
April 1982	49	27	9
July	60	20	7
October	55	23	9
December	51	27	12
January 1983	56	21	9
March	57	24	7
April	57	21	7
June	50	28	8

[a] In June, the question was phrased slightly differently.

Source: PORI polls in Hann-Hastings and Hastings, *Index to International Public Opinion*, 1982–1983, 142, 176–77; 1983–1984, 155.

TABLE 14.3 *Approval rating decline, October 1982–June 1983: Aridor, Begin, and the government*

	Satisfied[a]			Dissatisfied		
	Aridor (%)	Begin (%)	Government (%)	Aridor (%)	Begin (%)	Government (%)[a]
October 1982	57	69	44	31	27	15
December	51	67	40	41	30	17
January 1983	56	69	47	35	27	15
March	55	72	43	36	25	15
April	47	74	43	45	21	13
June	39	63	32	54	33	23

A. Incremental Change (January 1983 as a baseline)

	Positive evaluation			Negative evaluation		
	Aridor	Begin	Government[a]	Aridor	Begin	Government[b]
March 1983	−1	+3	−4	+1	−2	0
April	−8	+2	0	+9	−4	−2
June	−8	−11	−11	+9	+12	+10

B. Cumulative Change (January 1983 as a baseline)

	Positive evaluation			Negative evaluation		
	Aridor	Begin	Government[a]	Aridor	Begin	Government[b]
March 1983	−1	+3	−4	+1	−2	0
April	−9	+5	−4	+10	−6	−2
June	−17	−6	−15	+19	+6	+8

[a] "Satisfied" + "more or less satisfied," and in the case of the government, "very good" + "good" (as the poll question concerning the government measured its performance rating rather than satisfaction with it directly).
[b] "Not good at all" (excluding "not so good").
Source: PORI polls in Hann-Hastings and Hastings, *Index to International Public Opinion, 1982–1983*, 176, 377; *1983–1984*, 155.

the pursuit of the excessive objectives of the war continued to enjoy wide public support. In fact, the state retained much of the leeway it had as far as security decisions were concerned even when its decisions meant an obvious delay of Israel's withdrawal from Lebanon (and at least in the short run, additional costs).[8]

---

[8] For example, in March 1983, Israelis were divided evenly (41 percent versus 40 percent) over the justification of extended versus limited (40 km) war. Nevertheless, 67 percent (versus 26 percent) thought that Israel should not withdraw prior to reaching a normalization

Public opinion data, then, reveals the facts. The majority in Israel preferred to remain uncritically committed to the state, even as it became clear that the government had miscalculated miserably in leading Israel into Lebanon. The struggle to save Israel from itself was one of a resourceful, creative, highly mobilized, and effective coalition of societal forces, but nevertheless it comprised a minority. The social identity and the composition of this minority coalition is rather clear, as is indicated by the signatures on the anti-war manifestos, the newspaper articles on the protest, and the various interviews activists of different protest groups granted the media. Essentially the war was checked by members of the educated urban middle-class, members of Kibbutzim, intellectuals, and key journalists who were often military correspondents. However, what is surprising about this coalition, and thus needs further reference, is that mainstream journalists, particularly from among the military correspondents, assumed a very significant role in the struggle against the war, although they had traditionally acted in conformist ways.

The Conformist Soft-Left and the Press

At the time of the Lebanon war, the Israeli media, and particularly the written press, was well positioned to influence public opinion. Israelis were compulsive news consumers, the daily newspapers had extremely high circulation rates, and the public generally trusted the press coverage of security matters.[9] Under such conditions, it is tempting to assume that the media was an actual creator of public opinion. The role of the Israeli media as a major engine of protest, however, has a special significance for a different reason than the media's potential or actual power. What is surprising in the Israeli case is the historical precedent – the fact that the Israeli media decided at all to use its power, and turn against the war, and that it did so quite early.

Traditionally, Israeli journalists put national considerations above their professional ones. Certainly until the mid-1970s, the inherent rivalry between the free press and the state, at least as far as national security matters

---

agreement with Lebanon. In September 1984, 26 percent of Israelis thought that Israel should leave Lebanon unconditionally, but another 20 percent conditioned such a withdrawal on a Syrian quid pro quo, 8 percent rejected any withdrawal, and 29 percent were for a limited withdrawal. In January 1985, 35 percent supported unconditional withdrawal, but 45 percent supported withdrawal "only if suitable security measures are reached." See Elizabeth Hann Hastings and Philip K. Hastings (eds.), *Index to International Public Opinion 1982–1983* (Westport, CT: Greenwood Press, 1984), Vol. of 1982–1983, 232–34, Vol. of 1984–1985, 226. See also other poll results in Yaniv, *Dilemmas*, 196.

[9] *Yediot Aharonot's* circulation in the early 1980s was around 200,000 per day, and more than double that number for the Weekend Edition (correspondence with Eitan Habber). Edelist and Maiberg argue that almost 50 percent of Israelis read *Yediot Aharonot*, while some 30 percent read *Maariv*. See Malon, 44, 58. The Arian et al. study shows that more than two-thirds of the Israeli public believed the media were "objective, responsible and, credible in covering defense matters." See *National Security*, 42.

were concerned, was almost nonexistent in Israel. Hirsh Goodman, the military correspondent of the *Jerusalem Post*, suggested that the historical "enlightenment" of the Israeli press occurred in two stages. Up to the 1973 Yom Kippur War, military reporters were restrained by the *formal "pact"* they had signed with the IDF that elaborated their privileged access to information in return for certain obligations, and by their *self-imposed* censorship, which came out of their strong identity as Israelis. According to Goodman, the Yom Kippur War sharply reduced their sense of self-censorship, and the Lebanon war partially released them from the chains of their relations with the Israeli military establishment.[10] This analysis, however, even if correct in a very general sense, must be accepted with reservations. The auto-emancipation of the press after the 1973 war was not all that impressive or robust. Only a few, if any, critical reports on national security matters (such as those of present-day journalists Amir Oren or Reuven Pedhatzur) can be found in the leading newspapers in the period after the public discourse of the 1973 war subsided. If anything, it was not so much the Israeli identity of reporters that changed, but rather the concept of country. The "country," became identified more closely with society, whereas before it was perceived as the state.[11] This transformation was reflected during the Lebanon war, most notably in the issues the reporters chose to cover, the contents of their articles, and their conviction that they should use their power in order to expose the events of the war and thereby prevent needless casualties.[12]

Yet the "transformation" of journalists' order of loyalties could not have affected their role that much had it not been for conjectural causes – namely, the presence of Sharon. It is the latter who was largely responsible for the reporters' readiness to match their newly gained cognition with action. His actions, authoritarian style, visceral hatred of journalists, and readiness to let hatred dictate his relations with the press were the factors that convinced journalists that he must be stopped. In fact, some journalists tried to prevent Sharon from becoming Defense Minister in the first place, and once he was appointed, targeted him in the belief that in doing so they were serving society. Moreover, the military correspondents of the major dailies closed ranks because they expected Sharon to try to divide and rule them, and thus he was also responsible for their newfound esprit de corps.[13] Naturally, these foundations of distrust and the intense efforts to keep the press out of the Lebanon war only intensified the friction between the press and the state. Military correspondents knew that the tight information regime the

[10] Hirsh Goodman, *Migvan* 77 (1983), 28–29. See also Eitan Habber in *The Journalists' Yearbook, 1983*, 14 and 15.
[11] Interviews with Yaakov Erez and Eitan Habber. See Chapter 13, note 24.
[12] Ibid., see also Habber in *The Journalists' Yearbook 1983*, 10.
[13] Indeed, Sharon started to isolate reporters from the defense establishment immediately upon assuming the position of Defense Minister (August 1981). See Schiff and Yaari, *Milhemet*, 35.

IDF imposed had little to do with operational considerations. They also suspected that there was a deliberate effort to conceal, from the public and the government, information about the real dimensions and precise aims of the war.[14] Thus Sharon, and to a lesser extent the military, were responsible for the activation of the cognitive transformation journalists went through in the post-1973 period. Yaakov Erez of *Maariv* confirmed this argument when he admitted that "we owe thanks to the current defense minister [Sharon]. In the way he treated us, he created our independent thinking."[15]

In any event, the result of the rift between the defense establishment and the media became obvious almost immediately. Within days of the invasion, the journalists – freed from their past commitment and invigorated by a new sense of mission – noted that "there had not been yet a military campaign in Israel in which the press, and thus the public, were kept out of the information circles for so long."[16] In the second week of the war, Zeev Schiff noted that the objectives of the war were changing rapidly. Within a month, most of the military correspondents refused to serve as informal state agents, assuming instead the role of loyal emissaries of a defenseless society.[17] Accordingly, they brought home doubts of soldiers about the official narrative of the war, the legitimacy of the unraveling goals, and the morality of war.[18] At the same time, the journalists also briefed ministers about the IDF's moves in Lebanon and the mood among the troops.[19] In early August 1982, the press discussed the dispute within the government over the siege of Beirut. In mid-September, the National Convention of the Journalists of Israel was assembled for an emergency meeting in order to discuss the continuous violations of the freedom of the press. The convention protested Sharon's discriminatory behavior toward military reporters, and threatened to boycott meetings with state officials.[20] Then came the massacre in the refugee camps, which provided journalists with a unique opportunity to focus their attacks on Sharon and Eitan. As already noted, the press made the best out of this opportunity. Once the final report of the Commission of Inquiry was published, the press joined forces with other groups in order to assure the full implementation of the Commission's recommendations. Soon after the first anniversary of the war, on July 8, 1983, *Ha'aretz* printed a sharp criticism of the war by Shlomo Argov, the Israeli ambassador who was injured gravely by a Palestinian terrorist and who thereby provided Begin with

---

[14] See *The Journalists' Yearbook, 1983*, 8–9, 10, 15.

[15] Quoted in ibid., 16.

[16] Yoel Marcus in *Ha'aretz*, June 10, 1982.

[17] See Schiff in *The Journalists' Yearbook 1983*, 16–17.

[18] See Schiff's first reporting on soldiers' confusion over the changing objectives and the widening credibility gap, which grew out of the inconsistency of military actions and official declarations, in *Ha'aretz*, June 29 and July 9, 1982.

[19] See Schiff, *The Journalists' Yearbook, 1985*, 82.

[20] See *Ha'aretz*, September 17, 1982.

the pretext to initiate the events that led to the war. During the rest of the war, the press made sure that the war would be accurately perceived as a failed security policy.

Journalist Hirsh Goodman summarized the press's role in the Lebanon war as follows: "Never did the Israeli press have so much influence as in this war in Lebanon ... Never was our press so critical, so alert, and so relentless in its search for the truth ... Never did it bear so much responsibility."[21] His analysis is essentially accurate, but it should not lead the reader to erroneous conclusions. The flow of influence in press-society relations was not simply from the former to the latter, nor did the press initiate the criticism of the war. In fact, once the war started, the first reaction of most journalists, including that of the military reporters – who knew about the coming war, did not like it, and opposed it prior to the breakout of hostilities – was to rally round the flag.[22] This reaction was partially the result of the fact that the journalists, much like most Israelis, had a strong national instinct, but it was also the result of their fear that they would convince nobody of the vices of the war, and thus that their criticism would be ill-received.[23]

In hindsight, it is clear that the media, which revealed to the public sinister aspects of the war and other data that the war leadership wanted to keep nebulous, played a pivotal role in the destruction of the image given to the war by its architects. Yet the criticism, as noted in Chapter 12 and here, did not originate in the press. Rather it originated in the reserve units, whose soldiers pressured the media to tell the real story of the war to the public, and take sides. Once that was done, field reporters discovered that their critique was supported by many inside and outside of the army. Reserve and conscript forces, as well as career officers of all ranks, urged reporters to be their proxies and abort particular war plans such as the assault on Beirut, or help terminate the war altogether. Only then did the media gain the stamina necessary to forge its own anti-war line and confront the state.

Understanding the peculiar development of media criticism in Israel explains much of the success of the anti-war coalition to bend the hand of the state. However, one cannot ignore the role of other groups that often operated through the media. Indeed, looking back on the development of protest and influence in Israel during the Lebanon war, it becomes clear that factions of the loose anti-war coalition proved very astute. Many instances demonstrate that these groups operated effective social networks, exploited skillfully their societal positions at the crossroads of life in Israel, undermined

---

[21] Quoted in *Migvan* 77, 29.
[22] I learned about the advance knowledge of military reporters from interviews with Habber and Erez. See also an interview with Ze'ev Schiff in *Koteret Rashit*, December 1, 1982, 24.
[23] Interview with Habber. The printing workers of *Yediot Aharonot* orchestrated bulletins calling for Habber's resignation. Habber also received several telephone threats.

the image the state tried to build for the war, and pushed factions within the government to take action and check Sharon.

The talent of the anti-war factions was revealed particularly in making strategic decisions such as when and how to mobilize against the war, which issues to exploit, how to best use resources, and how to operate effectively outside the established political arena. These were revealed in what Avshalom Vilan from the Kibbutz movement described as the constant "search for the political benefit," what Peace Now activist Tzali Reshef described as a strategy of going for the numbers even at the cost of thematic concessions, and what Dr. Janet Aviad from the same movement described as a consideration of "when society would accept or reject us."[24] The talent of the anti-war faction, primarily of Peace Now, but also of other groups and even individuals, was revealed in the good sense of compromise and timing, in a good choice of methods of protest, and in the emphasis on the largest possible number of demonstrators.

Thus, Peace Now did not demonstrate during the first month of the war because its members were divided over the question of whether to demonstrate at all, and/or so early. Much of the opposition to an early demonstration came from members who were on active reserve duty in Lebanon, and therefore suffered from a genuine dissonance.[25] They did not like the war as civilians, but as soldiers on duty they felt uncomfortable with the idea of demonstrating while the fighting was still on. Some of them also feared that they could not muster an impressive number of protesters, and thus risk condemning their protest to irrelevance. In any event, a demonstration while the movement was split would have been a disaster. The calculations of the Peace Now leadership changed as a result of the June 26, 1982, demonstration of the left-wing Committee Against the War in Lebanon. This demonstration numbered some 10,000 protesters, including a significant number of Peace Now members. For certain central activists of Peace Now, this was perceived as a potential make or break point. They feared that once the streets were "left" to the extreme Left, the protest would be quickly tagged as radical. Such a development promised to strip Peace Now of its more leftist supporters, as well as deter its moderate middle-class followers – those that could muster the big numbers. In short, such a process would have endangered not only the Peace Now movement, but also the whole campaign against the war. Thus, Peace Now activists decided to demonstrate on July 3, 1982, in order to keep the lead and avoid a potential catastrophic alienation of moderate supporters. This was indeed the first massive demonstration against the war.

[24] Telephone interview with *Peace Now* activists Avshalom Vilan, October 16, 1991; personal interviews with Tzali Reshef and Dr. Janet Aviad, October 23, 1991, Jerusalem.

[25] Interviews with Dr. Aviad (see note 24) and Dr. Ben-Artzi, October 17, 1991.

Peace Now calculated well again in deciding to launch a large demonstration following the Sabra and Shatilla massacre. This time, the movement faced a different problem. In order to muster unprecedented numbers, it had to join forces with Labor and the other left-wing parties. However, the good reputation and the appeal of Peace Now were largely based on its insistence on remaining unaffiliated, and thus a joint demonstration threatened to tarnish the clean image of Peace Now. That in turn could have cost the movement the respect and support of a sizeable segment of Israeli society. Eventually, the argument for greater numbers in the demonstration carried the day, and Peace Now orchestrated, with the Labor and other political players, a demonstration of truly exceptional dimensions (see Table 15.3). This one decision contributed immeasurably to the appointment of the inquiry commission, and thus paid off handsomely.

Perhaps equally as important, different protest groups proved very wise in choosing issues and targets. Thus, Peace Now and other groups focused on Sharon not simply because he embodied whatever they were against, but also because they knew that he was "driving people nuts," and thus likely to serve as a vehicle for mobilization and an elevated level of cohesion.[26] The use of the war fatalities against Begin, which was discussed in Chapter 12, also indicates a keen political sense. It was intended to apply emotional pressure and hit the soft underbelly of the Prime Minister, and indeed it turned out to be a very potent weapon.

Still, with all their ingenuity, the members of the anti-war coalition needed more than the power to reveal, organize, and protest, if they were to destroy Sharon and his bid in Lebanon while the public at large was not committed to their cause. They needed the government unity to wither away. As noted, the tolerance of members of the government for Sharon and his policies decreased, partly as a result of indirect pressure from the soldiers, the press, and the protest. However, Sharon and the war were losing momentum inside the government for other reasons as well. Some ministers – notably Itzhak Berman, Mordechai Zippori, Simcha Ehrlich, and to a lesser degree Yosef Burg and Zvulun Hammer from the NRP – were concerned about Sharon's plans relatively early on, since they had indications that they were being deceived.[27] Already in June, ministers had learned about the IDF violations of the cease fire with the Syrians from soldiers' and officers' telephone calls to their homes.[28] Some ministers, such as Zippori and Burg, were in particular privy to such information because their children served in Lebanon. Indeed,

---

[26] Interview with Dr. Aviad.
[27] See Schiff and Yaari, *Milhemet*, 17–18. Itzhak Berman, a Liberal minister in the government, was the first to stand up to Sharon. Zippori deeply opposed the intimate relationship with the Christian Maronite leadership and the concept of the extended war. Armed with a good grasp of military matters, he did his best to expose Sharon's grand plans to his often embarrassingly sluggish fellow ministers.
[28] See Naor, *Memshala*, 115.

Zippori made his best to enlighten Begin about the real atmosphere among soldiers,[29] and Burg, whose son served as an officer in the "brigade that was not mobilized," was the one who informed Begin about the re-mobilization of that brigade on the eve of the planned assault on Beirut. In fact, Sharon was so irritated and frustrated by this kind of networking that in August, when Burg called for a termination of certain military moves because they contradicted the government's decisions, Sharon furiously snapped at him: "It is impossible to conduct a war by the report of family members."[30] In addition, ministers also received information directly from journalists. The ministers often felt they were in the dark, and were hungry for information, and the military correspondents were more than happy to keep them informed about the real developments in the battlefield and the opinions among the troops. While briefing ministers, the military correspondents also shared their own views with them. The events following the massacre in the refugee camps are a good example of this communication line. Journalists knew about the massacre even before the defense and political authorities in Israel did. Journalist Schiff informed Zippori of the stories he had heard about a "slaughter" in the refugee camps during the first day of the massacre, and the latter requested Shamir, the Foreign Minister, to check the veracity of these stories.[31] When journalist Ron Ben-Yishai was told of the savagery of the Phalangist forces in the refugee camps, he called Sharon in order to alert him to circulating stories about a massacre, and then checked their accuracy. After having discovered that the stories were well founded and that the IDF had been too slow to respond, Ben-Yishai wrote a sharp letter to Begin urging him to get rid of Sharon.[32]

In the final analysis, however, the disintegration of the united front of the government probably owed more to a general sense of frustration among ministers and to expedient calculations than to a pristine process of gradual enlightenment or to a few courageous ministers. On the one hand, the ministers felt that they were shut out of the decision-making process and that consequently they were drifting into political oblivion. On the other hand, they felt that Sharon had became an unbearable political liability. They concluded that in order to stay afloat, Sharon had to be dumped.[33] Thus, Sharon brought his own demise upon himself. He was the one most responsible for the unlikely alliance between members of the educated middle-class, the media, army officers, and government ministers, whose interests

[29] Personal interview with Zippori, October 26, 1991. See also Zippori's interview in *Koteret Rashit*, February 23, 1983, 15.
[30] Quoted in Naor, *Memshala*, 137.
[31] Schiff and Yaari, *Milhemet*, 334.
[32] Schiff and Yaari, *Milhemet*, 346.
[33] On February 10, 1983, sixteen ministers decided, against the single vote of Sharon, to fully accept the recommendations of the investigation committee.

converged over a single issue: Sharon's political survival. It was this coalition that finally brought down Sharon, and thereby also sealed the fate of the war.

The Consequences of Political Relevance

While it is clear that as the Lebanon war progressed, and particularly as of the Battle of Beirut, the anti-war movement gained numerically and established a firm control over the agenda, it is also clear that the immediate objectives of the protest were only partially achieved. Sharon was deposed, the Israeli military posture became defensive, and the political demands of Israel shrank, but Israel did not pull back immediately, nor did the majority of Israelis change their outlook on the necessity of the war, nor did they demand an end to the war. Nevertheless, the protest against the war was at the root of the Israeli change of heart. It influenced operational decisions, undermined the consensus the state struggled to retain, helped break the spirit of Begin, helped push Sharon out of decision-making, and ultimately forced the political and military establishments to search for ways to retreat from Lebanon while all of the ambitious political objectives of the government, save the destruction of the PLO, were abandoned.

It is easier to make the case for the overall, cumulative impact of the protest against the war than to demonstrate in particular cases how societal forces forced the hand of the state. Nonetheless, the capacity of the anti-war agenda and its propagators to compel the state, or its organs, to take undesired decisions can still be demonstrated in at least three instances: the case of the "brigade that was not mobilized," the case of the creation of the Kahan Inquiry Commission, and the case of TV reporter Dan Smamma. As the case of the brigade that was not mobilized was already dealt with in Chapter 11, let us only briefly recall it, restate its significance, and move on to discuss the two other cases in greater detail.

As I have already noted, Sharon and Eitan hesitated to mobilize the reserve paratroop brigade because they had serious concerns about the brigade's possible "negative influence...on other units."[34] These worries, in turn, ended up convincing the war leadership to have "one less block of houses" as Sharon put it.[35] But this inclination to escalate the level of brutality – due to the concern about the political effect of mobilization, exposure to the war, and possible casualties – only further undercut the legitimacy of the war. In short, it is clear, as was corroborated by Sharon, that the schism over the war was deep enough to affect the conduct and prospects of the war.

The second case concerns the forcing of the government to nominate an inquiry commission and fully execute its recommendations. As already noted,

---

[34] Naor, *Memshala*, 133–34 (originally from *Al Hamishmar*, June 17, 1983).
[35] Ibid.

the disclosure of the Sabra and Shatilla massacre led the press and members of Israel's different social elites to demand adamantly the appointment of a judicial inquiry commission. Even Israel's President, Itzhak Navon, who traditionally kept out of controversies (because of the nature of his position), immediately called for such a decision. Begin, however, who was deterred by the potential political consequences of complying with the mounting pressures, decided to reject all the demands. His judgment proved to be a serious miscalculation as it only aggravated the level of frustration and contributed to an avalanche of protests. On September 24, former Supreme Court Justice H. Cohen, not particularly choosy about his words, wrote Begin a public letter asking whether Israel was "ruled by blinds or whether the government [was] deaf," adding ominously: "If the government wants this state to go on functioning as a lawful state, it is its simple and immediate duty to establish an inquiry commission."[36] Then a series of political and institutional "defections" followed. Minister Berman resigned, explaining that he was disenchanted with the war and the way it was being conducted, and in particular with the refusal of the government to nominate a judicial commission of inquiry. He went one step further than Justice Cohen, suggesting that should Begin fail to nominate a commission of inquiry, "there will be a crisis of confidence between a large portion of the population and the state and its different organs."[37] Berman's foreboding assessment was echoed by the press, and his resignation was followed by the resignation of professor M. Milson, the head of the civil administration in the (occupied) territories, who was about the only academic of some reputation willing to serve in such a position under the right-wing Likud government. Next came indications of a breakdown within the army – the demands of Brigadier General Amram Mitzna and Colonel Yoram Yair to dismiss Sharon. Meanwhile, some of the most important Israeli professional associations joined the call for the establishment of a judicial inquiry commission, and Peace Now joined forces with the Zionist left-of-Likud opposition parties in order to orchestrate the "400,000 demonstration." Finally, the ministers of the NRP, one of the junior partners in the government coalition, sided with the demand for an investigation. These events – that led to the creation of the Inquiry Commission and the removal of Sharon from the Defense Ministry – are summarized in Table 14.4.

The case of the Inquiry Commission is significant for two reasons. First, there is little doubt that the appointment of the inquiry commission was largely the result of societal pressure. After all, Begin flatly rejected the call for such an inquiry commission and had to reverse his decision after a short and intense period of societal pressure exerted on the government. Indeed, both Sharon and Eitan concluded that the committee was born out of public

---

[36] Quoted in *Ha'aretz*, September 22, 1982.
[37] *Ha'aretz*, September 24, 1982.

TABLE 14.4 *Sabra and Shatilla: A timetable of societal coercion*

September 16–18, 1982	• Phalangist massacre
September 20–28, 1982	• "War Crime in Beirut" headlines; shock pictures; President Navon calls for investigation
	• **The government reject the calls**
	• Former Justice Cohen calls for an investigation
	• Professor Milsón resigns in protest
	• Minister Berman resigns and gives a newspaper interview
	• Publication of Brigadier General Mitzna's and paratroop Colonel Yair's calls for Sharon's resignation
	• The Israeli Bar; 27 former ambassadors; 100 poets and literary editors; 200 Weitzman Institute scientists; Kibbutz movement; former right-wing minister of Justice Tamir; and the Writers' Association call for an inquiry commission
	• The "400,000" demonstration
September 29, 1982	**The government yields, and accepts the demand for an inquiry commission**
December 2, 1982	Warnings issued by the Commission to Begin, Sharon, Shamir, Eitan, and other senior IDF officers
Febuary 8, 1983	Final conclusions of the Commission of Inquiry
Febuary 10, 1983	• Peace Now demonstrates for full implementation of the Commission's recommendations. Protester Emil Grintzweig is murdered in a hand grenade attack
	• The government votes 16 to 1 to accept the recommendations of the Commission of Inquiry

pressure.[38] Second, the act of compelling the government was also a clear act of bending the arm of the state. The idea that one's own judicial system can investigate the responsibility of the government for the immoral consequences of *international conduct* is simply one that most states flatly reject.

The third case concerns the relations between the defense establishment and the media. As I have already explained, the media pampered both the army and the state (as far as security matters were concerned) until the Lebanon war. The change in the media attitude during the war was mostly limited to attacks on the political elite, its war policy, and the new identity brought upon the Israeli state and its citizens. The army as an institution

---

[38] In Sharon's words, the committee was created because "the public atmosphere . . . was impossible to ignore." See *Warrior*, 509. Similarly, Eitan concluded that the "400,000" demonstration influenced Begin, and particularly his ministers, to appoint the committee. See *Sippur*, 305.

was criticized only mildly, and mostly as a means of attacking the political system. But the army and its officers could not have come out of a controversial war totally clean, and in any case, officers easily identify with the state. Furthermore, the tension between the defense establishment and the media was already growing before the war as a result of the fundamental attitude of Sharon and Eitan toward the latter. When it turned out that the war was neither short nor frugal in terms of casualties, the media went their own way, and the friction between them and the state increased. The depth of the frustration in the army command became evident in the September 1982 interview that the General Staff granted *Yediot Aharonot* (quoted in Chapter 13). But things reached critical proportions only toward the end of December when the army decided to confront Dan Smamma, a correspondent of the Israeli state-run TV.

Smamma was singled out for a report in which a group of soldiers was filmed singing a macabre verse to a popular children's tune to the effect, "For Sharon we'll fight the war, don't expect us home no more."[39] This version of the song had already appeared in writing in Amnon Abramovitch's column in *Maariv*. But *Maariv* was privately owned, and there was practically nothing the state could do against any newspaper without unleashing a struggle it could not possibly win. Besides, audio-visual messages are considered to have a much greater impact than the written word. Furthermore, early in the war, the military censor had already rejected one of Smamma's reports that covered a debate, within a paratroop unit, over the possibility of attacking Beirut. With the prolongation of the war, the IDF authorities decided that an officer from its spokesman unit would escort every TV reporter. Smamma was instructed not to ask soldiers about their feelings and to interview officers only. Moreover, the interviews required advance notice and could not take place before the spokesman's representative had briefed the officer chosen to be interviewed.[40] Smamma did not break with the actual words of the IDF instructions, but apparently had a hard time complying with their spirit. So, on December 28, 1982, the IDF decided to get tough and freeze relations with Smamma for what was considered his "excesses."

By the time of the Smamma incident, journalists and editors had already expressed their collective displeasure with the state's, and particularly Sharon's, disrespect for the freedom of the press. In such an atmosphere, the military's decision regarding Smamma brought already strained relations between the press and the state to the flash point. The (only) Israeli TV network immediately retaliated by canceling a planned interview with Major General David Ivri, the Chief of the IAF. All the media, and particularly the cadre of military reporters, hastened to declare their total support for Smamma. Even

[39] The Hebrew verse literally translates into: "We shall fight for Sharon, and we will return in a coffin."

[40] *Ha'aretz*, September 7, 1983.

the politically cautious director of the Israeli TV, Tuvia Sa'ar, and the director of the Israeli Broadcasting Authority, Yossef Lapid – who was appointed by the Likud – stood collegially by Smamma.

It took the IDF about a week to lift the ban on Smamma under the guise of some face-saving formula. However, not until the departure of Sharon and Eitan from office, in February and April 1983, were the damaged relations between state and press restored. Sharon's and Eitan's successors, Moshe Arens and Lieutenant General Moshe Levi, were much more cautious in their relations with the press. They abandoned the confrontational approach of their predecessors, relaxed the rules governing the coverage of the military, and rebuilt, to a degree, the damaged relations of the military establishment with the media.

As far as the general impact of the protest against the war is concerned, it is clear that the latter reduced Israel's freedom of action, contributed to the erosion of the solidarity within the government, and forced on Begin, Sharon, and their government decisions they dearly wished to avoid. In his memoirs, Sharon argued that by "mid-July [1982] domestic and international pressure was building-up ominously …"[41] His factotum in the Unit for National Security, Major General Tamir, added that "as of July [1982] it was difficult to convince the government of Israel to grant authorization for the operation of the IDF inside urban Beirut, because [of] the Israeli people…"[42] For whatever reason, Sharon also suggested that Begin was disturbed enough by the political consequences of the stalemate in Beirut that he (Begin) found it necessary to tell the government that "if we continue to remain at the gates of Beirut as we are doing now, we may bring disaster on ourselves…we are at a turning point that may lead to a national crisis. Our people will not tolerate weeks and months of unnecessary mobilization of the army, with extended service, where we are being shot at and our boys are being hurt."[43] Whether or not Begin actually said these words, the fact remains that this is precisely what happened to Israel in Lebanon. Indeed, in early July, Begin, sensing the political hazards of the rising tide of protest, added the small Telem party to the coalition. And in September, shortly before the massacre, Begin tried, but failed, to reschedule the general elections to an earlier date. As for Sharon, he believed that his fellow ministers had deserted him when "the war was no longer popular" and when "media attacks had become savage and demonstrations were rocking the streets."[44]

The particular timetable of the course of influence of society on the state and the outcomes of war is short and fairly clear. Within a month, a strong anti-war coalition base was formed. Between late June and September 1982,

[41] Sharon, *Warrior*, 425.
[42] Tamir, *A Soldier*, 167.
[43] Sharon, *Warrior*, 486.
[44] Ibid., 486, 487.

the anti-war coalition broadened its popular base and consolidated its power. Anti-war sentiments spread into the consciousness of the educated class at large, and the protesters succeeded in turning the cost of war, its moral consequences, and the legitimacy of the decision-making process before and during the war into major items on the national agenda.[45] The public at large did not flock to the side of the anti-war protesters, but the agenda was increasingly under their control, and spontaneous support for the war was about to decline sharply.

On September 14, 1982, Bashir Gemayel was assassinated. As a result, Israel lost control over events in Lebanon. Three and a half months into the war, the anti-war coalition, exploiting the carelessness of the military and Sharon, was able to seize the moment and turn the massacre in Sabra and Shatilla into a political means of destroying Sharon's political power. Begin's "gentiles killed gentiles" and "blood libel" strategy was promptly exposed for what it was – a frantic effort to deflect criticism, avoid assuming responsibility for the massacre, and capitalize politically from the general paranoia of Israelis. Meanwhile, a series of "defections," dissension, and leaks of government officials and state agents further crippled the autonomy of the state. Within eight months of the inception of the war, Sharon was forced out of the Defense Ministry. Within a year, Israel's political elite all but admitted that it had reached the end of its political capacity to pursue its ambitious war objectives in Lebanon. In mid-May 1983, Shamir, the Foreign Affairs Minister, declared that "the IDF has completed a glorious operation in Beirut, [and therefore] it should be returned home."[46] At the end of May, Israel's State Comptroller, Itzhak Tunik, joined in the criticism, saying that "in our worst dreams we did not imagine that whatever took place would ever happen. We could have been 11 months past the operation."[47] Immediately following, Begin confessed in the Knesset that "this [was] a difficult moment for the nation."[48] In mid-June, about a year into the war, Sharon revealed in an interview that he had failed to assess correctly the level of national consensus and the staying power of the government.[49] When July drew to a close, Begin's depression was common knowledge, and rumors had it that he was about to resign. In August, Begin announced his resignation and retired from political life into seclusion. In November 1983, Shamir, the new Prime Minister, decided to visit the IDF forces in Lebanon. He was criticized so fiercely by reserve soldiers during this visit that upon returning he immediately declared that Israel was getting closer to winding up the

---

[45] Tzali Reshef explained that Peace Now sought to influence public opinion so that the latter would in turn influence decision-making. See *Ha'aretz*, September 23, 1983.
[46] *Ha'aretz*, May 12, 1983.
[47] Quoted in *Ha'aretz*, May 31, 1983.
[48] *Ha'aretz*, June 1, 1983.
[49] *Yediot Aharonot*, June 17, 1983.

"operation" in Lebanon and that in the future it would reduce the number of soldiers deployed in Lebanon.[50]

Israel decided to finally withdraw only some two and a half years after the beginning of the Lebanon war, and it took another six months to get the IDF soldiers to the pre-June 1982 line of deployment (inside Lebanon). However, at a very early stage, following the uproar in Israel and the forcing of Sharon out of the Defense Ministry, any sensible observer – Lebanese, Shiite, Syrian, or other – could not have failed to observe that Israel was exhausting itself and that its ejection from Lebanon was only a matter of time. The events and declarations of the summer of 1983 only consolidated such conclusions. Consequently – if only out of considerations of Israeli domestic politics – the incentive of Israel's enemies in Lebanon to harass its forces became irresistible. These developments and the uproar within Israel convinced the army, well in advance of the government, that it would be better to get out of Lebanon. As soon as the war became stagnant and support within Israel had dwindled, senior IDF officers encouraged reserve officers under their command to speak up against remaining in Lebanon, while instructing the soldiers to regard their military objective in Lebanon as "staying alive."[51] In summary, after the first year of war, Israel's stay in Lebanon constituted no more than a protracted rear-guard battle, an effort – motivated by expedient partisan considerations and misguided national security perceptions – to salvage as much as possible from the ruins of a crumbling adventure.[52]

[50] See *Ha'aretz*, November 9 and 10, 1983.
[51] Interview with Dr. Ben-Artzi.
[52] See Yaniv, *Dilemmas*, 148–284, particularly pp. 199–206, 262–68.

PART IV

15

Conclusion

Small wars kept recurring in world history because powerful states were often tempted to exploit their military superiority in order to subjugate and oppress others. Conquerors and oppressors were usually well equipped to win such wars. Given overwhelming military superiority, one needed little more than mediocre military talent in order to crush insurgent populations. The cohesion of insurgent communities and the acumen of their military leaders made counterinsurgency campaigns more costly and slower, but rarely did either change the end results of the confrontation.

Still, it would be misleading to argue that power asymmetry was alone responsible for the victory of the strong party in small wars. Underlying successful counterinsurgency campaigns was also a cultural capacity to exploit the military advantage to its limits and pay the necessary price – that is, the readiness to resort to extremes of personal brutality, and occasionally tolerate significant losses. As I noted in Chapter 3, social and political developments in Western states in the nineteenth century eroded this cultural capacity. Thus, while technological and organizational innovations increased the relative military power of democratizing states, social developments reduced their oppression potential. To put things in a broad theoretical perspective, the (realist) iron rule of power has eventually broken down in the context of democratic small wars. After 1945, democracies discovered that military superiority and battlefield advantage have become fruitless, if not counterproductive, in protracted counterinsurgency campaigns.

I have explained this modern power paradox in terms of a struggle between two forces on three realms over three issues. The two forces are the "state," on the one hand, and part of the educated middle-class, which is a proxy of "society," on the other. The three realms are the instrumental, the political, and the normative. Their manifestations are in the state's military dependence on society, the voice of "society" in politics, and the different value priorities of each. The three contested issues concern sacrifice, combat behavior, and the (domestic) powers of the state.

Specifically, I have submitted that the mobilization of the sons of the educated class (that is, the expansion of *instrumental dependence*) and their exposure to battlefield risks and brutality is the first major critical development in democratic small wars. I have further argued that this group (which is *politically relevant*) acts as a destructive catalyst in the process that dooms the efforts to win these wars, and that this process (of expanding the *normative difference*) is extremely potent because of the synergetic interaction between its moral and expedient dimensions. Moral considerations form the basis for rejecting domestic and war policies that are marketed in terms of national necessity. In a sense, moral concerns often jumpstart the opposition to war. Yet altruistic considerations alone cannot decide the fate of the war. Without expedient reasoning, the opposition to the war is almost certain to remain marginal, no matter how vocal it is. Expedient criticism is, then, indispensable because it is responsible for the formation of a critical anti-war mass. At the same time, expediency alone cannot decide the fate of the war because its origins can be controlled by the state. All it takes is a readiness to contain the cost of the war by resorting to elevated levels of battlefield brutality. However, excessive brutality lends credit to and strengthens the moral criticism of the war. Thus, neither moral nor expedient reasons alone doom the efforts of democracies to win small wars. Morality is too limited a motivator, whereas expediency is too containable. Together, however, they become invincible. The state cannot prevent moral criticism without decreasing the effectiveness and efficiency of its war effort (in terms of casualties) – something that increases the expedient motivation to oppose the war. And the state cannot control expedient criticism without resorting to brutal behavior – something that necessarily increases moral criticism. Hence the essence of the *illusiveness of the balance of tolerance.*

The second critical development in small wars consists of a shift in the center of gravity from the foreign battlefield to the domestic marketplace of ideas. It is often the fear of this shift, as much as the shift itself, that leads into another critical development in the war. State leaders and officials try to regain control over the public discourse by elevating the levels of deceit and repression. As both involve the abuse of fundamental principles that underlie democratic life, the state inadvertently supplies the anti-war opposition with a winning card: The war can be depicted as threatening democracy itself, a matter of enormous moral as well as expedient significance that expands the normative difference beyond repair.

This book, then, argues that what fails democracies in small wars is the interaction of sensitivity to casualties, repugnance to brutal military behavior, and commitment to democratic life. Presented differently, it claims that the failure of democracies in small wars consists of the inability to resolve three related dilemmas. The first dilemma is how to reconcile the humanitarian values of a portion of the educated class with the brutal requirements of counterinsurgency warfare. The second dilemma is how to find a domestically acceptable trade-off between brutality and sacrifice. The third dilemma is how to preserve support for the war without undermining the democratic order.

It is possible at this point to further refine the logic of the core argument. While democracies are ill-adapted to win small wars, one should take into consideration the fact that small wars and democracies are, like other social phenomena, variables of some elasticity. Mostly before 1945, proto and limited democracies have encountered limited societal opposition while fighting small wars, and by and large they overcame domestic dissent and won abroad. Thus, we can conclude that the strength of the relations between democracy and small wars is dependent on the scope of the liberal and democratic "content" of the incumbent, as much as it depends on the intensity of the small war. This content decides whether, and to what extent, a state can compromise its democratic identity in order to effectively fight small wars, and thus also how likely it is to be successful. States that are less liberal and less democratic can be expected to encounter fewer and lesser domestic obstacles than their more democratic peers when they fight brutally small wars. Moreover, their a priori capacity to compromise their more limited democratic order without creating a secondary expansion of the normative gap further increases their chances of enduring and prevailing in small wars. The ways the Turks fight the Kurds and the Russians fight the Chechens – as much as the capacity of both incumbents to pursue their wars and prevail – should not come as a surprise, nor as incompatible with the argument made in this book.

Domestic and International Causality Reconsidered

The fact that democracies have failed in small wars primarily because of their domestic structure does not mean that international and other causes do not contribute to such failures. Indeed, unlike in the Algeria and Lebanon conflicts, international conditions may play a significant role in the conduct and outcome of small wars, and their role may in fact be on the rise. I have raised the issue of non-domestic influence on small wars twice in my discussion of multiple-level analysis in the Introduction. First, I did so with reference to the work of Putnam, and second, in the discussion of the attainability of the winning balance-of-tolerance and its position on the tolerance plane. In the next section, I discuss again multiple-level analysis, albeit from an empirical perspective, focusing on the case of the Vietnam War.

The United States and the Failure in Vietnam
While the American military was plagued by many problems that impaired its performance in the Vietnam War, still it handled rather well the offensives of the North Vietnamese Army (NVA) and of its insurgency arm in the South,

the "Vietcong" (VC).[1] In military encounters, and particularly in major pitched battles such as Khesanh and during the 1968 Tet offensive, American forces displayed great tenacity, exploited their superior firepower well, and took a terrible toll on their enemies.[2] At the same time, the American three-pronged strategy – punitive bombing of the North, relentless air attacks on the supply lines of the Ho Chi Minh trail, and counterinsurgency operations in the South – failed to produce a "breaking point" and the overwhelming political results the Americans had expected.[3] In spite of enormous pain, the motivation of the NVA and the VC was not destroyed, their operational capacity survived, and their ultimate goals remained unchanged. In the end, in spite of significant battlefield successes,[4] all the Americans achieved was to buy their South Vietnamese allies a few more years of political independence.

This outcome can be partially explained by international-level variables. For example, there is no denying that Soviet and Chinese material support played a crucial role in the war.[5] It is doubtful whether, without this support, the NVA and the VC could have survived the American onslaught, let alone fight on a broad front in the South. Similarly, it is clear that systemic factors imposed critical limits on American military planning and action. For example, an American invasion of North Vietnam may have increased the chances of getting the war in the South under control. However, in large measure, such an invasion was not initiated because the Americans feared it would lead to a superpower conflagration.[6] It is equally clear that the outcome of

---

[1] On the Vietnam War, see George McT. Kahin, *Intervention: How America Became Involved in Vietnam* (NY: Anchor Books, 1987); Michael Maclear, *The Ten Thousand Day War: Vietnam 1945–1975* (NY: Methuen, 1981); Gabriel Kolko, *Anatomy of War* (NY: Pantheon Books, 1985); John Bowman (ed.), *The World Almanac of the Vietnam War* (NY: Pharos Books, 1986); Guenter Lewy, *America in Vietnam*; Stanley Karnow, *Vietnam: A History* (NY: Penguin, 1984); Marvin E. Gettleman, Jane Franklin, Marilyn Young, and Bruce H. Franklin, *Vietnam and America: A Documented History* (NY: Grove Press, 1985); Blaufarb, *The Counterinsurgency Era*, particularly pp. 205–78; Pape, "Coercive Air Power in the Vietnam War," 103–46; Krepinevich, *The Army and Vietnam*; and George C. Herring, *America's Longest War* (Philadelphia: Temple University Press, 1986).

[2] Total Communist fatalities are conservatively assessed at 500,000–600,000. They represent 2.5–3 percent of the pre-war population affiliated with the North (including those in the South). See Mueller, "The Search for the 'Breaking Point' in Vietnam: The Statistics of a Deadly Quarrel," *International Studies Quarterly*, 24:4 (1980), 503–08, and note 9. Bui Tin described the Communist losses during Tet as "staggering," adding that: "[Our] forces in the South were nearly wiped out by all the fighting in 1968 ... We [also] suffered badly in 1969 and 1970." See interview with Stephen Young, *Wall Street Journal*, August 3, 1995, 8.

[3] See Mueller, "The Search for the 'Breaking Point'," 497–519.

[4] See Blaufarb, *The Counterinsurgency Era*, 269–71, 276–78; and Rosen, "Vietnam and the American Theory of Limited War," 102–03.

[5] See Herring, *America's Longest War*, 148–49; and Ellis, *From the Barrel of a Gun*, 228.

[6] See Mark Clodfelter, *The Limits of Air Power* (NY: The Free Press, 1989), 42–43; Herring, *America's Longest War*, 132, 140, 178; Krepinevich, *The Army and Vietnam*, 261; Gettleman

the Vietnam War was partially decided by motivational and organizational factors, particularly the exceptional devotion and readiness to sacrifice of the NVA and the VC and their leaders tight control of the population in the North and supporters in the South.[7] In terms of the formula I submitted in the Introduction to this book, international and motivational factors pushed to extremes the level of the balance-of-tolerance that American society had to reach if the United States was to win the Vietnam War (see Chapter 1, Figure 1.1).

This being the case, it is nevertheless clear that a discussion that omits the role of American domestic politics cannot come close to explaining the outcome of the Vietnam War, or the capacity of the Communists to achieve their objectives in spite of great military inferiority. Perhaps the easiest way to start supporting this contention is to recall the three material conditions that had to be met before the war could be concluded as the North wished.[8] First, the massive American ground presence in Vietnam had to be terminated. Second, the American use of air power had to be stopped. Third, American support of the South Vietnamese Army (SVA) had to be critically curtailed.

The first important point is that none of these conditions was achieved in the battlefield, and not for lack of trying by the NVA and the VC. In fact, they tried hard to achieve their objectives in the battlefield, but in the end had to alter both their strategy and timetable. Thus, after years of rejecting negotiations, they decided to complement their military struggle with diplomacy, and what is more important – compromise – as indeed the 1973 Paris peace agreements prove.

The second important point is that these three conditions were eventually met because of developments inside the United States, and in particular, because of the pressure American society put on the state.[9] Evidence of the

et al., *Vietnam and America*, 287; Tom Wells, *The War Within* (Berkeley: University of California Press, 1994), 99; and Lyndon B. Johnson, *The Vantage Point* (NY: Holt, Rinehart and Winston, 1971), 153.

[7] See Mueller, "The Search for the 'Breaking Point,'" particularly pp. 511–15, and in a "Rejoinder," *International Studies Quarterly*, 24:4 (1980), 530.

[8] See Clodfelter, *The Limits of Air Power*, 170–71.

[9] For discussions of domestic aspects of the Vietnam War, see Milton J. Rosenberg, Sydney Verba, and Philip E. Converse, *Vietnam and the Silent Majority* (NY: Harper and Row, 1970); Lawrence M. Baskir and William A. Strauss, *Chance and Circumstance: The Draft, the War, and the Vietnam Generation* (NY: Knopf, 1978); Thomas Powers, *The War at Home: Vietnam and the American People 1964–1968* (Boston: G.H. Hall, 1984); Nancy Zaroulis and Gerald Sullivan, *Who Spoke Up? American Protest Against the War in Vietnam, 1963–1975* (NY: Doubleday, 1984); Bernard Edelman (ed.), *Dear America: Letters Home From Vietnam* (NY: Pocket Books, 1985); Melvin Small, *Johnson, Nixon, and the Doves* (New Brunswick, NJ: Rutgers University Press, 1988); Charles DeBenedetti, *An American Ordeal* (Syracuse, Syracuse University Press, 1990); Mitchell K. Hall, *Because of Their Faith: CALCAV and Religious Opposition to the Vietnam War* (NY: Columbia University Press, 1990); David L. Schalk, *War and the Ivory Tower: Algeria and Vietnam* (NY: Oxford University Press, 1991); David W. Levy, *The Debate over Vietnam* (Baltimore: The Johns Hopkins University Press, 1991); Melvin

pivotal role of domestic politics is readily seen in many aspects of the Vietnam War, of which the U.S. air strategy and the composition of the ground troops are particularly revealing.

The strategy of bombing the North, attacking the NVA/VC supply lines, and relying on saturation bombing in the South was grounded in military logic that emphasized technological superiority and the American way of making war – namely, the preference for subjecting the enemy to devastating firepower.[10] However, the extensive reliance on air power also reflected political calculations. Specifically, it reflected the wish to limit the scope and role of American ground troops in order to control the number of casualties, and check a possible deterioration of domestic support for the war.[11] Similarly, the biased nature of the draft, to the disadvantage of the less educated and the poor, reflected pre-Vietnam calculations of national efficacy.[12] However, the particular composition of the American army in Vietnam also reflected domestic politics. President Johnson relied on draftees and enlisters, sharply limited the call-up of reservists and National Guardsmen, and (unlike in Korea) used both sparingly because of his fears of the consequences of acting otherwise.[13] It is also obvious that Johnson found the draft convenient because a system of deferments and exemptions protected the sons of the educated middle-class.[14] For political reasons, then, Vietnam was indeed a working-class war for at least much of its duration.[15]

Perhaps the strongest indication of the role of American domestic politics in the outcome of the Vietnam War is that all of the domestic maneuvers to uphold support for the war were in vain. In spite of the efforts to tailor the strategy and force structure according to societal fundamentals, the Johnson

Small and William D. Hoover (eds.), *Give Peace a Chance: Exploring the Vietnam Antiwar Movement* (Syracuse: Syracuse University Press, 1992); Christian G. Appy, *Working-Class War: American Combat Soldiers and Vietnam* (Chapel Hill: University of North Carolina Press, 1993); Kenneth J. Heineman, *Campus Wars* (NY: New York University Press, 1993); Wells, *The War Within*; Richard R. Moser, *The New Winter Soldiers: GI and Veteran Dissent during the Vietnam Era* (New Brunswick, NJ: Rutgers University Press, 1996); James K. Davis, *Assault on the Left* (Westport, CT: Praeger, 1997); James E. Westheider, *Fighting on Two Fronts* (NY: New York University Press, 1997); Rhodri Jeffreys Jones, *Peace Now: American Society and the Ending of the Vietnam War* (New Haven: Yale University Press, 1999); and Maurice Isserman and Michael Kazin, *America Divided* (NY: Oxford University Press, 2000).

[10] See Herring's quote of General Depuy in *America's Longest War*, 151 (from Daniel Ellsberg's 1972 book, *Papers on the War*, NY: Simon & Schuster).
[11] See Clodfelter, *The Limits of Air Power*, 52; and Wells, *The War Within*, 153–54.
[12] See Appy, *Working-Class War*, 30–32; and Baskir and Strauss, *Chance and Circumstance*, 14–30.
[13] See Appy, *Working-Class War*, 36–37; Baskir and Strauss, *Chance and Circumstance*, 50–51, 52; and Herring, *America's Longest War*, 184. Note also that the largely volunteer-based marines fought a significant part of the ground war. See Appy, *Working-Class War*, 28.
[14] See Westheider, *Fighting on Two Fronts*, 23–24, 29–30, 35.
[15] Appy, *Working-Class War*, 6.

administration and then the Nixon administration found themselves at the mercy of the contradictions between the requirements of the Vietnamese battlefield and the values of American society. Consequently, both administrations went through the very same grinding process that French and Israeli governments experienced.

Opinion polls show that by early 1966, the Vietnam War had become a dominant public issue.[16] It also seems clear that by late 1967, "most administration officials probably agreed with McGeorge Bundy that the war's 'principal battleground' was 'in domestic opinion' "[17] – that is, that the war's center of gravity had shifted from Vietnam to the domestic marketplace of ideas. Those who missed this development were to be enlightened by the domestic reaction to the 1968 Tet offensive. Thus, when Nixon assumed power in 1969, American society was already in hopeless ferment, and the presidency had already turned to the slippery slope of unlawful and despotic conduct.[18] Unwilling to abandon Vietnam, but painfully aware of the magnitude of his domestic problem, Nixon searched for some magic solution. In an effort to reduce the power of the opposition to the war, he endorsed Defense Secretary Melvin Laird's advice and half-heartedly initiated a series of troop withdrawals.[19] He also made the draft more equal, proposed the Vietnamization program, and relied more heavily on bombing in order to compensate for the gradual elimination of ground forces. These measures resulted in some political gains. However, by that time, the events in Vietnam, the opposition at home, and all sorts of "defections" had already destroyed the credibility of the Vietnam policy. Equally damaging, the administration became ever more paranoid, repressive, and – as the 1969–1970 "secret" bombing and then invasion of Cambodia prove – deceptive. Thus, while Nixon's combined policies bought him time and some freedom of maneuver, they could not eliminate the anti-war sentiment and the protest potential, or change the ultimate outcome of the war.

The destructive force of the contradictions between the requirements of the Vietnamese battlefield and the values of American society can also be learned from the draft and deployment policies of Johnson and Nixon. Most notably, Johnson deepened while Nixon broadened (though also diluted) their military dependence on society, and both had to rely on short, twelve-month tours of duty because, as General Westmoreland argued, it was "politically impossible" to do anything else.[20] In doing so, they acted against their best domestic interests because they thereby promised extensive exposure of the American society to the war. Johnson drafted, enlisted,

---

[16] See Wells, *The War Within*, 70.
[17] Ibid., 220.
[18] See Herring, *America's Longest War*, 182–83.
[19] See Bowman, *The World Almanac of the Vietnam War*, 327–28.
[20] Quoted in Krepinevich, *The Army and Vietnam*, 206.

and flooded Vietnam with young soldiers, and thereby fuelled a debate over sacrifice and the equality of the draft. To his misfortune, the destructive effect of the biased draft was exacerbated because the war became entangled with the civil-rights struggle and social upheaval of the 1960s.[21] In 1969, the draft lottery that presumably overcame the discriminatory nature of the Selective Service System was introduced. However, the revised draft all but forced college and university students to become more critical and active against the war. The students, sensing that they were more likely to go to Vietnam, turned extremely hostile toward both Nixon and his policy, and consequently radicalized their struggle. This gradually eroded the resolve of the administration and strengthened the morale of the Vietnamese Communists, whose leaders closely followed developments on the American scene.[22] The war was eventually lost because it became unsustainable at home.

Of all domestic developments, the revision of the draft deserves further discussion because it highlights the critical place instrumental dependence and expedient motivations played in the process that destroyed the American war policy in Vietnam. Let us recall that initial protest against the war was largely confined to three groups: the black community (whose sons were overrepresented in ground combat units and among the casualties, at least until 1968), part of the clergy, and a limited number of academics and students in elite universities.[23] Let us also recall that the early struggle against the war was essentially moral, and that it should have attracted much attention, if not sympathy, because the protests included some mesmerizing acts such as self-immolation.[24] Yet, moral outrage – that was all but certain to grow because of the saturation bombing, body counting, the use of chemical agents, and revelations about troop behavior in places such as Cam Ne and My Lai – seems not to have decided, at least not alone, the fate of the war. It took the expedient interests of strong sectors of society to turn around the initial and rather comprehensive support for the war.[25] And, as American officials well understood, this happened increasingly as conscription widened, the number of troops in Vietnam grew, loopholes in the draft were closed, casualties

---

[21] See Isserman and Kazin, *America Divided*; and Davis, *Assault on the Left*, 39–42.

[22] Wells, *The War Within*, pp 4–5; and in Henry Kissinger, *The White House Years* (London, Weidenfeld and Nicolson and Michael Joseph, 1979), 444, 511, 513, 514, 515, 1013, 1019.

[23] Blacks, women, and Jews opposed the war in significantly larger proportions than other groups. On the service and casualties of the black community, see Westheider, *Fighting on Two Fronts*, 13–14; and Appy, *Working-Class War*, 19–20. On academic protest, see Heineman, *Campus Wars*.

[24] In early November 1965, Norman Morrison set himself on fire across from the Pentagon, and a few days later, Roger LaPorte did the same in front of the United Nations building.

[25] Initial support for the war included the public at large, the bureaucracy, both parties, major religious groups, and political dignitaries, including Truman, Eisenhower, Nixon, Goldwater, Adlai Stevenson, and most state governors.

accumulated, and above all, a growing number of university and college students felt at risk.[26]

Indeed, the critical weight of the expedient interest of the students can be gleaned from a combination of several observations. First, American ground forces were pulled out of Vietnam largely because of domestic, and in particular, campus pressure. Nixon favored only a limited withdrawal, and wished to leave a permanent contingent in Vietnam, whereas his National Security Adviser, Henry Kissinger, apparently opposed any troop withdrawal.[27] Yet, domestic protest eventually forced Nixon to pull the troops out of Vietnam down to the last soldier, not before it destroyed – as the hasty May-June 1970 withdrawal of forces from Cambodia suggests – some of his key war initiatives. Second, campus pressure did not develop because of doubts concerning the attainability of the war objectives, but rather because of doubts concerning the morality of the war, and even more so, concerning the necessity to risk one's life in it. Indeed, the protests grew hand in hand with the threat of service in Vietnam, while the opposition to the war lost steam once Nixon reduced the perceived threat to students by pulling troops out.[28] Third, having reduced the expedient motivation for protest, Nixon was able to maintain his "immoral" bombing autonomy, and even escalate the bombing right until the very end of the war.[29]

In summary, what the leadership of the black community and a limited intellectual constituency had long fought for, often on moral grounds, became attainable only after American instrumental dependence was deepened and the expedient interests of the educated middle-class were threatened. In the final analysis, then, there can be little doubt that domestic pressure, often of expedient origins, had a detrimental effect on the capacity of the United States to fight in Vietnam, let alone win the war. Moral considerations launched the opposition to the war, but they alone could not change the Vietnam policy. Successive waves of mobilization and the altering of the draft did that. These actions raised the perceived stake for students, and in turn the latter's protest spread and became wilder. America sank into turmoil, and eventually the only sensible solution was to pull the troops out. In the final analysis, domestic pressure, and expedient calculations in particular, dealt a death blow to the American war effort in Vietnam.

---

[26] See Heineman, *Campus Wars*, 183. Before 1969, 14.6–15.9 percent of the soldiers discharged from the armed forces had a full (or some level of) college education as opposed to 20.5–29.9 percent, after 1968. See Appy, *Working-Class War*, 26 table 2. At the same time, the overall number of American casualties declined sharply after 1968, and college-educated soldiers were, in any event, less likely to serve in combat assignments or to experience heavy combat.

[27] Wells, *The War Within*, 288, 345

[28] See Wells, *The War Within*, 403; and Kissinger, *The White House Years*, 1038.

[29] It must nevertheless be noted that Nixon's instructions to bomb the North toward the end of the war encountered moral opposition among B-52 aircrews. See Wells, *The War Within*, 561; and Clodfelter, *The Limits of Air Power*, 164.

The Role of Instrumental Dependence in a Comparative Perspective

Obviously there are many similarities among the French, Israeli, and American small wars I have discussed. In all of them, the war became a domestic battle over hearts and minds after the state had reached a high level of instrumental dependence and failed to strike an accepted balance between battlefield casualties and brutality. This failure was accompanied, or followed, by state efforts to circumvent social realities in undemocratic ways – be it by resorting to deception and/or repression – that proved counter-productive. In all cases, a few independently minded and foresighted people and a radical, yet marginal, predisposed minority were bound to oppose the war.[30] Opposition grew from a grassroots base, and state policy lost credibility because of high- and low-level "defections."[31] In all cases, a portion of the educated middle-class – with the help of the media, cultural icons, and other celebrities – made the difference by shifting the war's center of gravity from the battlefield to home. In all cases, representatives of mainstream institutions joined the opposition to the war only after a significant period of support for government policy. Finally, in all cases, the war destroyed the careers of leading politicians, toppled governments, and redefined politics. In France, the war consumed several governments, ended the career of a few prime ministers (including Guy Mollet), and destroyed the Fourth Republic. In the United States, Defense Secretary McNamara and other high officials left office, Johnson retired, and the Democrats lost the presidency. Then, Nixon drifted into a thinking and acting mode that forced him to resign before he would be impeached, and Congress gained in power. In Israel, Sharon was ousted, Begin became dysfunctional and retired, and a Unity government replaced the Likud-led coalition.

Yet there are also marked differences among the cases that I believe can be partially explained by reference to instrumental dependence. In particular, I submit that the different manifestations of variables – such as state tenacity, battle conduct, protest formation, and the state's reaction to developments at home – can to a large extent be explained by reference to the scope of mobilization and the social nature of the armies of France, the United Sates, and Israel.

Table 15.1 indicates the levels of the general and actual instrumental dependence of Israel, France, and the United States during the Lebanon, Algeria, and Vietnam wars, respectively. The crude difference in the numbers

---

[30] Concerning Vietnam, see Levy, *The Debate over Vietnam*, 46.

[31] The estimated number of draft violations during Vietnam stands at 570,000. See Baskir and Strauss, *Chance and Circumstance*, 11. For details concerning various incidents of high- and low-level "defections," soldiers' protest, and social pressure on decision-makers, see Moser, *The New Winter Soldiers*; Herring, *America's Longest War*, 176–78; Wells, *The War Within*, 4, 106–12, 198–99, 250–51, 365, 373–74, 417–18, 422, 428, 441, 526; and Isserman and Kazin, *America Divided*, 185, 268.

TABLE 15.1 *Levels of instrumental dependence of the democracies in Lebanon, Algeria, and Vietnam*

	Israel[a] (1983) (%)	France (1958) (%)	United States (1968) (%)
General[b]	16	2.45	2.0
Actual[c]	2	1.14	0.18

[a] Jewish and Druse population only (some 85 percent of total Israeli population).
[b] Standing and reserve armed forces as percentage of the population.
[c] Highest deployment in war as percentage of the population.
Sources: World Almanac and Book of Facts (1969) 592, 741–43; H. S. Steiberg, *The Statesman's Year-Book*, 1004–05; Heller et. al., *The Middle East Military Balance – 1983*, 92, 95.

is obvious, but one must not be tempted to draw conclusions before less tangible characteristics of the armies and the actual battlefield contingents of the three states are considered.

The French force in Algeria included a large number of conscripts and reservists who were integrated into society. It also included an officer corps that had spent much of its service in the colonies and the Indochina war (1945–1954) and spearhead units that were composed of foreign and professional troops. Almost by definition, unlike conscripts and reservists, these groups were rather detached from French society. The French also relied on a large number of proxy troops, mostly indigenous, who naturally were the least affiliated to French society. Finally, the French state regulated the risks to different groups through draft deferments that served mostly the sons of the educated middle-class, and through selective battle assignments that imposed a disproportional cost of war on the professional and foreign troops.

The American contingent in Vietnam was largely composed of young conscripts, among whom the educated middle-class was sharply underrepresented.[32] Of even greater significance, American reservists, by and large, remained at home. In fact, the Reserves and the National Guard served as sanctuaries for the sons of the educated middle-class. Finally, as in France, but to a greater extent, the United States relied on proxy troops, mostly South Vietnamese (but also from among Asian allies and the British Commonwealth).[33]

The Israeli army in Lebanon somewhat resembled the French contingent in Algeria. It included mostly conscripts and reservists. Much like in

[32] Men from lower income families and small-town America were apparently much more likely to die in Vietnam than those from urban middle-class backgrounds. See *Working-Class War*, 12, 14, 23.
[33] Only about one-third of the all-time highest number of allied troops in Vietnam were Americans. SVA fatalities were more than four times those of American forces. See Bowman, *The World Almanac of the Vietnam War*, 221; and Jeffrey J. Clarke, *Advice and Support: The Final Years, 1965–1973* (Washington: Center of Military History, United States Army, 1988), 275 table 14.

France, both groups were thoroughly blended into society. However, unlike in France, the professional segment of the IDF was also deeply integrated into society. Finally, unlike the other two cases, the Israeli educated class was heavily represented in the IDF command structure and among its combat units.

The numerical and substantive differences in instrumental dependence help explain the following aspects of variance: First, they seem to partially explain the tenacity and ferocity with which each state pursued its war goals. France, with significant, but well-regulated levels of actual instrumental dependence, and with the help of indigenous and other colonial troops, fought with full force and few inhibitions on personal brutality for eight years. The United States, with the lowest rate of actual instrumental dependence (among the cases), an extremely biased draft system, and a large contingent of South Vietnamese and other allies, also fought for eight years and with what seems to be equal brutality. Israel, on the other hand, with the highest levels of both general and actual instrumental dependence, the most "societal" army among the three cases, and hardly any indigenous or other allies to rely on, quickly adopted a passive and defensive posture in Lebanon. It was also more restrained as far as brutality was concerned, and much quicker to retreat. It pulled back the IDF toward the pre-war line of deployment, though this line was inside Lebanon, within three years of the war's start (Israel withdrew to the international border in May 2000).

Second, the differences in instrumental dependence seem to partially and indirectly explain the content of the anti-war debate in each case. As noted earlier, because the command and leading combat units of the French army were segregated from society, the government felt no need to restrain them, and they in turn acted in an extremely brutal manner. This in fact paid off in the battlefield. However, because conscripts and reservists observed but could not stomach the level of brutality, moral indignation became the leading theme of the French anti-war campaign. All of this, even though the relative rate of French casualties per-population and per-year of war should have made the expedient issue of sacrifice more dominant in the anti-war campaign (see the comparison in Table 15.2). In sum, the controlled nature

TABLE 15.2 *Fatalities in Algeria, Vietnam, and Lebanon*[a]
(fatalities per army deployment at peak, per year of war)

France	United States	Israel[b]
0.0065	0.013	0.0022
	0.0012 (post-PLO exile)	

[a] Including accidents and other causes.
[b] Jewish and Druse population only (about 85% of total Israeli population).

of the French dependence on society allowed for the brutalization of the war as well as a relatively lethargic domestic response to casualties. It also helped to make morality the leading theme of protest.

As noted, American instrumental dependence during the Vietnam War was the lowest and shallowest of the three cases, if only because of the sheer size of the American population, the availability of the baby boom generation, and the biased nature of the draft.[34] In the American case, morality and expediency seem to have played more equal roles in the protest against the war than in the other two cases. Although it seems that while the content of the American public debate was often moral, expedient interests – as suggested by fluctuations in protest following changes in the students' chances of being drafted[35] – dominated the motivation of the most important sector among the protestors.

In Israel, the state with the highest and deepest level of instrumental dependence among the three cases, the human cost of war quickly became the heart of protest. It is interesting that expediency dominated the struggle against the war although Israel fought a far less intense war than either France or the United States, as measured, for example, by mortality rate per army deployment at its peak (see Table 15.2). This is all the more significant because low mortality in the Israeli case reflected an a priori effort to minimize the risk to the forces because of the command's and the government's understanding that dependence on society limited their freedom of military maneuver. In fact, the Israeli authorities tried hard not only to limit casualties, but also (though with a few exceptions) to regulate the use of firepower in order not to over-antagonize the soldiers and the liberal constituency at home.

Third, the nature of instrumental dependence seems to partially explain the difference in the formation-pace, social scope, and intensity of the antiwar opposition. In particular, the Israeli case stands out in three respects. First, the early response of the Israeli society to the war was far more comprehensive than of either French or American societies (see Table 15.3). Second, theological opposition to the war was by and large absent in Israel, whereas many French and American clergy vehemently opposed the war from early on and in spite of the official position of their churches.[36] Third, while large-scale opposition to the war was quick to form in Israel, it was far less radical than in either France or the United States. It seems clear that the unusually

---

[34] Out of some 26,800,000 men of the draft-age "Vietnam generation," only about 2,150,000 served in Vietnam; 1,600,000 of these served in combat functions; and 15,400,000 were deferred, exempted, or disqualified. Data are from Baskir and Strauss, *Chance and Circumstance*, 5 figure 1. See also Isserman and Kazin, *America Divided*, 132.

[35] See the data in Baskir and Strauss, *Chance and Circumstance*, 5–6, 6–7; and Appy, *Working-Class War*, 36.

[36] On the response of American clergy, see Hall, *Because of Their Faith*; and Levy, *The Debate over Vietnam*, 95, 98–102.

TABLE 15.3 *Key Israeli demonstrations during the Lebanon War – a counterfactual calibrated comparison*

	Size of actual demonstrations in Israel		Size of comparable hypothetical demonstration in:	
Identity	Time (in weeks) from inception of war	Size	U.S.[a]	France[b]
Left	3	10,000	600,000	140,000
Peace Now	4	100,000	6,000,000	1,400,000
Soldiers against Silence	9	2,000	10,000	10,000
Center to Left (after Sabra and Shatila)	16	200,000[c]	12,000,000	2,800,000

[a] The largest U.S. demonstration during the war had 250,000–300,000 demonstrators. It took place almost five years after the beginning of the war. However, 2,000,000 people are said to have participated in the nationwide October 15, 1969, Moratorium Day protest activity.
[b] The largest French demonstration during the war had 500,000 demonstrators. It took place after seven years of war.
[c] The number of demonstrators was estimated by the press and the Left at 400,000. There is no way to substantiate this number, and it is clearly exaggerated. A conservative approach should reduce the original number by 50 percent.

wide scope, quick formation pace, but also mild nature of protest in Israel reflected the intimacy of state-society relations, intimacy that largely grew out of Israel's deep dependence on its society.

Finally, let us consider the variance in state reaction to the anti-war protest and the level of conflict between state and society. In both France and the United States, the state escalated the war in the face of opposing domestic demands. William Bundy perhaps best epitomized this disregard of society when he confessed in an interview with Tom Wells: "I stuck cotton in my ears when it came to domestic opinion."[37] Not surprisingly, such a conception of state autonomy and attitude toward society contributed to the extreme reaction of both the American and French societies. When strong groups felt incapable of making their voices heard, let alone be influential, their frustration led to resisting the draft, and in a few cases even to siding with the enemy and acting violently. Considering the fundamental attitude of the state and social response to the latter, it is no wonder that in both cases, the state turned increasingly deceptive and repressive in a hopeless effort to narrow, if not overcome, the growing gap between policy and the public.[38] In the Israeli case, the fundamental positions of state and society

[37] Wells, *The War Within*, 157.
[38] The repressive attitude of the American state became obvious during the 1968 Democratic Party convention in Chicago, and in the May 1970 killing of four students at Kent State University (a few days later two black students were also shot dead in Jackson State College, Mississippi). On CIA and FBI spying, FBI agitation, IRS prodding, police violence, other

and the subsequent developments were somewhat different. Israeli leaders and officials of state organs were more reluctant to go to the same extremes of ignoring and confronting society, and correspondingly the levels of state deception, and even more so repression, were lower than in the French and American cases. In this respect, the mellowing effect of instrumental dependence seems rather straightforward: Simply put, the Israeli authorities could not afford to risk the destruction of the fabric of their military power by brutally confronting their critics from among the educated middle-class. On the other hand, the tightly knit state-society relationship also limited the forms of protest in Israel. Thus, while the clash between the Israeli state and society was accompanied by high rhetoric, the actual level of anti-war activity – in terms of radicalism, confrontation, and violence – was nowhere close to that in France or the United States.

In sum, the argument made concerning the effect of instrumental dependence on inter-case variance seems compelling. Nevertheless, a caveat is necessary: The variance I noted cannot be attributed to any single cause, including instrumental dependence. Indeed, the content of criticism and the precise manifestations of social protest, and state reaction to both, were also functions of other factors and idiosyncrasies. For example, American domestic reaction to the Vietnam War was also influenced by the civil-rights struggle and general upheaval in the 1960s. Similarly, the shape of social protest and state action in Israel were partially influenced by the small size of the Israeli population and by the fact that the war was also part of Israel's conflict with the Arab world. Finally, the French readiness to fight and ignore society was partially the result of past humiliations (in World War II and Indochina) and of the feelings Algeria evoked as a part of constitutional France, whereas the opposition (particularly against torture) was partially the result of the experience of German occupation. In short, the nature of instrumental dependence accounts for important aspects of the variance among cases, but it often does so in conjunction with idiosyncratic factors.

Democracy and the Use of Force

Before concluding this book, I would like to raise two issues that concern the meaning of my arguments in the context of the relationship between democracy and military power. The first, largely theoretical issue, concerns the presumed benevolence of democracies. The second, more concrete issue, concerns the relations between the legacy of small wars and military intervention.

---

forms of harassment, as well as the criminal mind-set of the Nixon inner circle, see Davis, *Assault on the Left*, 229–37, 325, 345, 354, 448–49, 514.

How Peaceful are Democracies?

The question of whether democracies are war-averse, and if so why, is of great interest for many scholars. Current conventions suggest that democracies are peaceful in a limited way: They refrain only from fighting each other. This is the nucleus of the Democratic Peace Theory (DPT)[39] – which is supported by statistical analysis and reasoned by arguments that originate in Kant's discussion of perpetual peace.[40]

Essentially, the reasoning of the DPT concerns motivation and values on the one hand, and institutional structure on the other. Proponents of the theory maintain that because of expedient and/or moral calculations, free citizens refuse to sacrifice their blood and money and oppose violence against foreign communities, except in extreme circumstances. They also point out that such views become politically meaningful because of the institutional structure in democracies. Or, more specifically, they discuss how institutions such as representative government, political parties, and the media translate altruistic and egotistic regard for human life into benign international behavior.

Although the DPT claims to explain only the relations among democracies, it nevertheless sets the stage for a general discussion of democracy and war, if only because its proponents often present the origins of democratic benevolence without reference to the identity of international rivals.[41] In a nutshell, they claim that democracies are *inherently* cost-averse, morally constrained, and institutionally organized in ways that generally restrain the resort to force.[42] Paradoxically perhaps, critics of the DPT seem to accept such comprehensive arguments, something that can be learned from the realist contention that democracies are too slow to react to shifts in the balance of power and often too soft in their response to aggression.[43]

How does this work relate to the debate about the presumed benevolence of democracies? First, it is in agreement with some of the general arguments

---

[39] See, for example, Bruce Russett, *Controlling the Sword* (Cambridge: Harvard University Press, 1990), 124–32, and *Grasping the Democratic Peace* (Princeton: Princeton University Press, 1993), 24–42; and Michael W. Doyle, "Liberalism and World Politics," *American Political Science Review*, 80:4 (1986), 1151–69.

[40] See Michael W. Doyle. "Kant, Liberal Legacies, and Foreign Affairs," *Philosophy and Public Affairs*, 12:3 (1983), 205–35.

[41] Exceptions include John M. Owen, "How Liberalism Produces Democratic Peace," *International Security*, 19:2 (1994), 87–125; and Russett, *Grasping the Democratic Peace*, 35, 40.

[42] See also Mark W. Zacher and Richard A. Matthew, "Liberal International Theory: Common Threads Divergent Strands," in Charles W. Kegley (ed.), *Controversies in International Relations Theory* (NY: St. Martin's Press, 1995), 123.

[43] See Randal Schweller, "Domestic Structure and Preventive War: Are Democracies More Pacific?" *World Politics*, 44:2 (1992), 235–69, particularly pp. 238, 268. See also Gilpin, *War and Change*, 209. For a Realist analysis of U.S. foreign policy, see Hans J. Morgenthau, *American Foreign Policy* (London: Methuen, 1951), 222–42; and Martin Indyk, "Beyond the Balance-of-Power: America's Choice in the Middle East," *The National Interest*, 26 (1991/1992), 42.

of the DPT. Much like DPT advocates, I maintain that the key factors that shape the international outcomes of democratic policy are to be found on the domestic level of analysis. Similarly, I find that the prime motivations behind democratic social preferences concerning the use of force abroad include a mix of expedient and moral considerations. Consequently, much like DPT advocates, I consider international-level phenomena (outcomes of small wars) to be caused by a bottom-up (society-state) process, or explainable at the unit level. Above all, I support the idea that democracies ought to be considered as a breed of their own when certain forms of international conflict are concerned.

At the same time, the analysis and some of the findings of this work differ with the DPT. First, while I consider democracy to be special, and have a peculiar impact on world politics, I do so in a different context and on a different scale. Most obviously, while the DPT deals with peaceful relations among democracies, I deal with violent relations between democracies and non-democratic actors. Thus, my analysis is complementary to the DPT in the sense that it explains relations beyond the democratic "zone of peace" – that is, relations between democracies and actors in the zone of turmoil.[44]

A second, and perhaps more subtle, difference between the DPT and this work concerns the depiction of democracy. Implied in the DPT is the idea that the relations between state and society in democracy are harmonious almost to the point that the two, or at least their preferences, are practically indistinguishable. I promote a different view in which society and the state coexist but also experience extreme rifts. As I have demonstrated, the state can launch a war independently of society and society can rise against state policy, and consequently the two may clash. Moreover, whereas the DPT tends to present the democratic citizenry, in a crude and comprehensive manner, as peaceful, I find inherent peaceful proclivities only in a minority of the citizens. Empirically, this finding is expressed in the fact that democratic states got involved in small wars with a high level of popular and institutional support, and retained both for a considerable length of time. Moreover, as I have shown in the French and Israeli cases (and as seems to be true of the American case as well), the most important political players and institutions, including the media and opposition parties, were either mute or supportive of the state's war initiative. In all cases, this state of affairs eventually changed, but the change happened only after an extra-institutional minority had created a climate that all but forced dominant players to become critical of the war.

[44] See James M. Goldgeier and Michael McFaul, "A Tale of Two Worlds: Core and Periphery in the Post-Cold War Era," *International Organization*, 46:2 (1992), 467–91; and chapter 3 in Max Singer and Aaron Wildavsky, *The Real World Order: Zones of Peace/Zones of Order* (Chatham, UK: Chatham House, 1993).

Democracy and International Intervention

My study has found that democratic states harvest bitter fruits following their decision to involve society in small wars while ignoring its criticism. Logic tells us that this should somehow alter institutional choices and security policy in democracies if only because pain and failure are among the most fundamental mechanisms of learning. In fact, considering the level of pain and the magnitude of failure in small wars, the experience of democracies must be of a formative quality. Indeed, I believe that it is already plain that failed small wars have revised the thinking of democratic leaders, state institutions, and ordinary citizens about the relations between domestic politics and foreign policy.[45] These wars have also inspired international actors that contemplated challenges against democratic states.

That having been said, it would be wrong to conclude that failures in small wars invariably give the advantage to the challengers of democratic states or deter democracies from intervening in Third World conflicts. For at least two reasons, democracies have not given up military intervention, nor is such intervention bound to fail. First, in spite of the participatory nature of democratic politics, foreign and security policy remain in democracies largely within the domain of the state. And the institutional instincts of states are likely to continue preventing the subordinating of foreign policy to a broad veto power of society. Second, and of even greater significance, democratic leaders have retained the autonomy to use force by adapting the means, objectives, and strategy of intervention to domestic constraints. In particular, democracies rely on professional troops (means), avoid certain interventions and limit others that can regress into small wars (objectives), and act swiftly and with massive force once they decide to intervene (strategy).[46]

The reliance on professional troops was partially "imposed" on states such as France and the United States following failure in small wars.[47] For liberals, it was conceived out of a wish to limit state power. However, instead of leaving the state with a diminished military capacity, the professional army has proven to be a formidable means that preserves, with the help of advanced military technology, state autonomy. Indeed, French presidents have used the post-Algeria Foreign Legion in Africa, including the Ivory Coast and

---

45 See the lessons Robert Pfaltzgraff drew from Vietnam, in Scott Thompson and Donaldson D. Frizzell (eds.), *The Lessons of Vietnam* (London: Macdonald and Jane's, 1977), 276.
46 These principles are embedded in the Weinberger Doctrine, which President George Bush reiterated in 1993. They have also been endorsed by Colin Powell. See Thomas Halverson, "Disengagement by Stealth: The Emerging Gap between America's Rhetoric and the Reality of Future European Conflicts," in Lawrence Freedman (ed.), *Military Intervention in European Conflicts* (Oxford: Blackwell, 1994), 76, 83. See also Wells, *The War Within*, 580–82.
47 See in Donald Vought, "American Culture and American Arms: The Case of Vietnam," in Richard A. Hunt and Richard H Schultz Jr. (eds.), *Lessons from Unconventional War* (NY: Pergamon Press, 1982), 180.

Chad,[48] while American presidents have employed the post-Vietnam professional army in Lebanon, Grenada, Panama, Somalia, the Persian Gulf, and the Balkans.

The least one can say is that American liberals not only miscalculated badly, but that they were also late to realize that. It took some two decades before people such as syndicated columnist Mark Shields and former anti-Vietnam activist Daniel Ellsberg called for the resumption of the draft, once they had observed how President George Bush (senior) deployed American forces against Iraq (in the summer of 1990). It was only then that Shields, Ellsberg, and others seem to have understood that those most likely and best able to check the president's war powers would not do so unless they had a personal stake. To reverse the slogan of the American Revolution, they realized that there is "no representation without taxation." In fact, concerns regarding the president's autonomy to use force will probably only increase as technological innovations further reduce both the size of the force needed for intervention and the risks soldiers face in battle.

The wish to avoid places and situations that can lead states into small wars seems to govern the overall pattern of democratic intervention.[49] Fear of small wars, then, helps determine in which circumstances democracies will intervene and in which they will remain on the sidelines.[50] In general, democracies adopt cautious and modest intervention objectives and try to avoid, or at least limit, intervention in civil or ethnic conflicts because these contain the seeds of small wars.[51] The successive 1990s crises in the Balkans, and the early one in particular in Bosnia (1992–1993), are good examples. In Bosnia, the Americans consistently refused to commit ground troops because they feared that this could drag them into another Vietnam.[52] The Europeans, fearing the domestic consequences of intervention, were more hesitant, and

---

[48] See Dominique Moïsi, "Intervention in French Foreign Policy," in Hedley Bull (ed.), *Intervention in World Politics* (Oxford: Clarendon Press, 1984), 67–77.

[49] Geographical distance and isolation may also play some role because they allow better control over the flow of information from the battlefield to home, the resort to higher levels of brutality, and thereby control over casualties. See, for example, the article of Phillip Knightley, "Fighting Dirty," *The Guardian*, March 20, 2000.

[50] The lesson of Vietnam – being more selective in intervention – was integrated into the Nixon and Weinberger Doctrines. It can also be found in scholarly works such as Blaufarb, *The Counterinsurgency Era*, 310; Robert E. Osgood, *Limited War Revisited* (Boulder, CO: Westview, 1979), 50–51, 68; and Thompson and Frizzell, *The Lessons of Vietnam*, v.

[51] See, for example, James Gow, "Nervous Bunnies – The International Community and the Yugoslav War of Dissolution," in Freedman, *Military Intervention in European Conflicts*, 31; and Jane M. O. Sharp, "Appeasement, Intervention and the Future of Europe," in ibid., 49–50.

[52] In August 1992, Lawrence Eagleburger cooled down calls for military action in Bosnia by raising the specter of "another Lebanon or Vietnam." See Steven L. Burg and Paul S. Shoup, *The War in Bosnia-Herzegovina: Ethnic Conflict and International Intervention* (Armonk, NY: M.E. Sharpe, 1999), 210.

held back on even a limited use of air power (beyond patrolling) until well into 1994. When members of NATO eventually decided to commit a limited number of troops to Bosnia, it was only after they had secured cooperation from Slobodan Milosevic and had ensured that their forces' mission would be limited to peacekeeping.

The rules about where not to intervene are occasionally ignored. However, the pain that is almost certain to follow tends to revive these rules. This is exactly what happened in Somalia in 1992–1993.[53] The Americans and other U.N. members were drawn into intervening because the heart-breaking pictures of devastating hunger in civil-war-torn Somalia could not be reconciled with the liberal conscience or complacency. The U.N. contingent that was led by American forces was far superior to the militias of recalcitrant factions that wreaked havoc in the country. Nevertheless, when local warlord Mohammed Farrah Aidid decided to challenge the international contingent, the Western powers found themselves at the mercy of the illusive balance of tolerance. The warriors of Aidid shrewdly used civilians as a protective blanket, inflicted a few casualties on the international force, and in particular on American troops, and induced the latter to relax their stringent rules of engagement. In short, Aidid's forces managed to depict the intervention as too costly and too dirty.[54] Unable to eliminate Aidid without substantial bloodshed, and unwilling to shed much of their own blood, the United States and other members of the U.N. force gave up both the hunting of Aidid and the establishment of order in Mogadishu.

As I noted, retaining democratic freedom to intervene also includes the adaptation of strategy to domestic constraints. In particular, once democracies decide to intervene in situations that can degenerate into small wars, they try to act decisively and with overwhelming force but without resort to their ground troops. They fear that acting gradually or with ground forces would lead to the prolongation of war, the accumulation of casualties, and the loss of control over developments at home. Indeed, this is how the United States and other members of NATO have acted in several recent crises, two of which – the Iraqi invasion of Kuwait and the Serb ethnic war in Kosovo – are particularly revealing.

For the intervening states, both the Gulf War and the Kosovo crisis had the potential of becoming small wars. At the same time, the margins in both cases were wide enough to avoid such a development. As both cases also suggest, defiant Third World leaders may be slow to realize how painful experiences make democracies effective. Thus, while Saddam Hussein and Slobodan Milosevic seem to have had some crude understanding of the

---

[53] On low tolerance for casualties and its effects, see Barry M. Blechman and Tamara Cofman Wittes, "Defining Moment: The Threat and Use of Force in American Foreign Policy," *Political Science Quarterly*, 114:1 (1999), 27–28.

[54] See ibid., 26.

balance-of-tolerance problem democracies face, they were utterly unaware of the capacity of democratic leaders and institutions to learn from past failures in small wars.[55] Indeed, both confused the unwillingness of American society to sustain small wars with the illusion that they could somehow deter the administration from fighting at all, or effectively. They, and even more so their armies and people, paid dearly for their reckless reading of history. In both cases, the Americans and their allies waged a full-scale, high-technology low-casualty war that hardly involved society. In the Iraqi case, they also followed the air campaign with a massive, yet limited, ground offensive. In both cases, the armies of the defiant challengers were destroyed. At the same time, the coalitions stopped short of involving themselves in occupation duties, partly because their members feared that this would draw them into a small war.

Concluding Remarks

The twenty-first century started following dramatic changes in international relations. The spirit and technology of capitalism and democracy won over Communism, the rigid global bipolar structure disappeared, and the United States became the undisputed hegemon in a "new world order." This new order, however, proved somewhat chaotic. In the absence of superpower patronage and binding ideologies, states and communities fell prey to ethnic and other conflicts. Demands for self-determination and independence multiplied, and a few states collapsed, while others were created. The significance of these developments cannot be overstated. The post-Cold War international order frequently creates situations that invite military intervention, and these can often develop into small wars, if only because of the expected technological and power asymmetry between protagonists.[56] Moreover, the actors who are most likely to consider intervention and best able to project power are Western democracies. Above all, there are compelling reasons to believe that the formula and variables I have identified will influence the patterns of the relations between these democracies and the actors in the zone of turmoil.

In spite of the aforesaid, it is still clear that the application of the formula and the precise influence of its different variables will remain contingent on future developments that are outside the scope of this book. Thus, while we can safely surmise that democracies will continue to eschew small wars, we do not know to what degree they will preserve their autonomy or precisely when they will intervene with force. All we know is that the influence of future developments on the balance-of-tolerance is likely to be uneven and contradictory. For example, the development of military technology is likely

---

[55] See also Michael I. Handel, *Masters of War* (London: Frank Cass, 2001), 14.
[56] See ibid., xxii.

to increase the autonomy of the state because it reduces the need to rely on society, permits armies to fight "cleaner" wars, and condenses the time frame of operations. At the same time, however, other developments – for example, the strengthening of democratic societies, increased sensitivity to human life, and better media capacity to independently cover military actions in real time – act to decrease state autonomy.

It also seems clear that in addition to these developments, the future of military intervention will also be affected by "constructivist factors." Norms and institutions that states develop while interacting among themselves and with their societies are going to increasingly shape states' identities and the constraints on their behavior. In fact, democratic states, more than other states, are increasingly constrained not only by their own civil societies, but also by behavioral standards that they develop as a group, and that even spread, albeit more slowly, beyond them. As I have noted, "sisterly vigilance" played a minor role in the cases I studied. It did not significantly alter the choices or behavior of France, Israel, and the United States, nor did it decide the outcome of their small wars. The rise of human-rights agenda, humanitarian intervention, ad hoc international indictment of senior officers and leaders for war crimes, and the establishment of a permanent international tribunal for the prosecution of individuals for war crimes and other crimes against humanity seem to change this state of affairs. In theoretical terms, constructivist factors add a new dimension to the two-level game democratic states play and lose in small wars. Democratic leaders now face their external enemies, their own societies, and the scrutiny of a society of states. Indeed, the efforts of the leaders of weaker parties in asymmetrical conflicts to provoke their enemies into acting brutally, as seen in the Palestinian territories and elsewhere, are designed to attract the attention of international audiences more so than that of the society of their democratic enemies. In a similar vein, the American preference to rely on international rather than unilateral intervention is probably motivated by the need to keep in check sisterly vigilance as much as it is by rational burden-sharing calculation.

It is for future studies to sort out how various states, and democracies in particular, will adapt to these new constraints, and how they will exploit future opportunities. Yet, for the generation that is unlikely to witness world war, but is also disillusioned with visions of the "end of history," the study of asymmetrical conflict and peering into the soul and guts of democratic countries in war will remain a critical base for future research. Military conflict is here to stay, and the friction between realist state-impulses and the proclivities of liberal-democratic "society" promises to illuminate the nature of armed conflict in the age of democratic military supremacy in ways that are certain to draw further attention.

Postscript

It is impossible to resist the temptation of referring to the Israeli-Palestinian conflict, if only because its special characteristics provide a challenging opportunity for a review of the utility of this book's thesis and concepts. Yet such an enterprise is also intimidating. Venturing analysis and qualified predictions concerning an evolving conflict of this sort is akin to skydiving on a stormy night, when even the best of parachutes may fail. Indeed, while I act on temptation as I submit the following preliminary analysis, I do so with trepidation.

In order to understand the implication of my thesis for understanding recent developments in the Israeli-Palestinian conflict, it is essential to present first its contours and special features. The most fundamental point to note is that the conflict involves a powerful democracy, an occupied community, and sharp asymmetrical relations. Considering these, it is not surprising that whenever the conflict turns violent, it assumes characteristics of a "democratic small war." The Palestinians resort to insurgency and terror, and Israel responds with counterinsurgency. Indeed, the cycle of violence that started in September 2000 reminds one of events that occurred in other small wars, and particularly during the French war in Algeria.

At the same time, the Israeli-Palestinian conflict is also marked by significant idiosyncrasies. First, while the actual violence is bilateral, it cannot be totally separated from the much larger Israeli-Arab conflict, which is of a different nature than small wars. Second, the violence of the Al Aqsa Intifada erupted following a process of conflict resolution in which Israel had been committed in principle to abolishing the occupation and accepting Palestinian independence. Third, the boundaries of the small war are almost absent, or at least they are not clearly demarcated. The battlefield is not overseas, nor is it limited to the territory of the oppressed community. Rather, while Israeli forces are deployed and operate offensively in Palestinian territories, the Palestinians strike civilians with ferocious terror everywhere inside Israel. Finally, some of the insurgents' demands concern the core sovereignty and territory of the democratic oppressor (unlike the cases of India, Mandatory Palestine, Algeria, Kenya, Vietnam, Lebanon, and so on).

While these idiosyncrasies render the application of the logic of democratic small wars to the Israeli-Palestinian case more challenging and complex, they do not preclude it. Understood in this book's terms, the events unfolding since the failure of the July 2000 conference Camp David conference are as follows. Following the deadlocked, the Palestinians chose confrontation. After a short period of violent demonstrations and clashes with IDF forces, Palestinian militant organizations decided to step up guerrilla operations and terror attacks against Jewish civilians inside Israel and in the occupied territories. Soon enough, Tanzim forces that were affiliated with the Palestinian Authority (PA), joined the violence. The PA apparently financed and authorized, or at least encouraged, some of the terrorist actions. Until April 2002, Israeli forces had been relatively restrained, although their actions continually escalated. In the beginning, the IDF tried to contain violent Palestinian demonstrations with limited force that was nevertheless occasionally excessive and deadly. As the terror became wilder, the IDF tried to isolate Palestinian cities. Having failed to prevent further terror, it then initiated targeted assassinations against those suspected of masterminding these attacks. All in all, Israel was drifting from isolation to selective eradication that grew more brutal and involved much Palestinian suffering and occasional innocent casualties.

In April 2002, the conflict reached a boiling point. In response to a bloody wave of terror, which climaxed in a massacre of Jews attending a Passover Seder dinner in Netanya by a suicide bomber, the IDF invaded PA controlled territories in the West Bank. Acting on quality intelligence and with large forces that included a significant component of reservists, the IDF set out to destroy the Palestinian guerrilla and terror networks. The latter, one must add, had become much more deadly as a direct consequence of the Oslo accords, in which Israel relinquished its tight control over the Palestinian population and territory. The Palestinians exploited the Israeli partial withdrawal to secure external financial support, stockpile arms and explosives, and build the foundations for an insurgency and terror campaign. Yet their operational successes had roots in more than money, guns, and opportunity. They were results also of softer factors such as a growing sense of national pride, constant incitement, and the coming of age of a new emboldened generation of refugee-camp adolescents. Summed up in the terms of this book, material conditions and higher motivation permitted the Palestinians to define for Israel a higher winning balance-of-tolerance. If Israel was to defeat the Palestinians or keep them docile, its society had to display higher levels of tolerance for violence and casualties.

Because conflict often involves dialectic processes, the developments did not work strictly in favor of the Palestinians. Their terror, their proximity, and their demand for the "right of return" convinced Israelis that they faced an existential threat. Consequently, the tolerance of Israeli society to both violence and casualties increased, and the Israeli state and army enjoyed a potentially convenient level of battlefield autonomy. Indeed, by the time of the April 2002 escalation, the vast majority of Israelis supported the massive use of force.[1] Explained in Robert Putnam's two-level-game terms, the nature of the conflict and Palestinian choices convinced the Israelis to grant the state a priori ratification (legitimacy) for a rather extreme position. Taken together, the Palestinian capacity to push upward the winning balance-of-tolerance, and the elevated tolerance to brutality and casualties of Israeli society, had one inevitable outcome: The conflict became bloodier.

Yet the analysis of the violent interaction between the parties reveals only part of a complex picture. In particular, it leaves out much of what this book is all about: the domestic interaction between state and society. It is therefore worthwhile to turn to a discussion of the domestic developments within Israel, which are likely to define the future of the conflict no less than exogenous variables.

As I have argued throughout this book, the structure and level of instrumental dependence decide, in conjunction with other variables, whether and to what extent democratic states are drifting into a collision with their societies. In this respect, it is important to start the discussion with two observations. First, the escalation of the conflict left Israel no choice but to increase the level of its actual instrumental dependence. Second, for operational reasons and cost calculations (which I discussed in Chapter 2), Israel felt compelled to escalate the level of battlefield brutality.

As I have argued, Israel initially chose containment, then isolation, and once the latter failed, added selective eradication. Eventually, it invaded the Palestinian refugee camps, villages, and cities with great force. All along, Palestinian civilians were occasionally killed or maimed, the general population suffered, and property was destroyed. In particular, fields, orchards, and houses that provided cover for attacks against settlers and military patrols were levelled. At the same time, because the conflict escalated, more soldiers including reservists became involved, and their well-being became a prime consideration. Essentially, the military preference was to control the consequences of elevated levels of instrumental dependence by rotating units and assigning combat selectively. In principle, reservists were kept out of the most dangerous and potentially bloody battle zones. However, as often happens in war, intentions were occasionally defeated by harsh realities, and when this occurred, brutality ensued. The April 2002 battle over the Jenin refugee camp is a perfect illustration. Because the IDF command underestimated the potential for casualties – after Palestinian cities

[1] See the April 2002 *Peace Index* project of the Tami Steinmetz Center for Peace Research, Tel Aviv University, at www.tau.ac.il/peace/Peace_Index.

and refugee camps were conquered at a minimal cost – it employed reservists in the attack on the Islamic Jihad terror network in this camp. One
company fell prey to an ambush, and within a matter of hours thirteen
reservists lay dead. Almost instantly, the IDF changed its tactics. Instead
of the discriminating but risky house-to-house mop-up infantry operation,
bulldozers and attack helicopters were called in. The new casualty-thrifty
assault was over rather quickly, but at the cost of widespread ruin at the
scene of battle, a few innocent casualties, and an international demand for
investigation (following false Palestinian allegations that a massacre took
place).

   In the broader scheme of things, the Jenin battle was but one indication of how Israel faces the very same dilemma it and other democracies
encounter in small wars. The search for effectiveness and efficiency led to
brutality, but the latter generated a moral cost. Indeed, by late 2001-early
2002, it was clear that the media had abandoned its ultra-patriotic coverage
of events in favor of a less-decisive position that included revelations about
the ugly sides of occupation and the distress these caused among a few of
the soldiers.[2] Of even greater significance, "defections" assumed unprecedented proportions that were not encountered even during the controversial
Lebanon war. In mid-January, physician Yigal Shohat – a retired colonel,
former IAF fighter pilot, and crippled POW – called on pilots and soldiers
to refuse to bomb Palestinian cities, serve in the territories, and demolish
Palestinian houses.[3] Soon after, a group of reserve officers and soldiers signed
a petition in which they described the occupation as corrupting the entire
Israeli society, declaring that they would "not continue to fight beyond the
green line in order to rule, deport, destroy, block, exterminate, starve, and
abuse a whole nation."[4] Another milestone was reached in early February
2002. Ami Ayalon – a career Israeli Navy Seal officer who rose to the rank of
Commander of the Navy, and then Head of the General Security Services –
publicly explained that he was troubled by the fact that "there [were] . . . too
few refusals to obey [illegal] orders [in the territories]."[5] In late March,
a Beer-Sheba University lecturer and former IAF pilot (who had refused
to bomb Beirut in 1982) called upon fellow pilots to refuse to fly combat
missions that could result in civil casualties.[6] In mid-April, as events escalated, the number of reservists who had signed the petition increased to
over 450 (the number reached 520 in August 2002), and the number of
those concurrently serving jail terms for refusing to serve reached an Israeli

[2] See, for example, Schief in *Ha'aretz*, January 13, 2002; *Maariv*, Weekend Supplement,
February 8, 2002; and *Maariv*, March 18, 2002.
[3] Shohat, "Black Flag," *Ha'aretz*, January 17, 2002.
[4] *Yediot Aharonot*, 7 Days Supplement, January 25, 2002.
[5] Quoted in *Ha'aretz*, February 1, 2002.
[6] *Ha'aretz*, March 31, 2002.

all-time record of over 40.[7] In late April, for the first time since the start of the violence, four combat soldiers of the conscription army refused to guard Palestinian detainees *inside* Israel.[8] In January 2003, when the IDF initiated retribution in response to a Palestinian suicide bombing in Tel Aviv, a junior intelligence officer aborted a vengeful Israeli air strike against a less-than-clearly defined Palestinian terror target. The officer, positioned in a critical intelligence center, purposefully held onto information vital for the attack because he considered the latter morally flawed. The officer was promptly released from his unit. However, the commander had to discuss the issue with his fellow officers and soldiers because some of them supported the officer and had become disgruntled by the army's reaction and apparent breach of its own policy of "purity of arms" and its instruction to soldiers to disobey "manifestly illegal orders."[9]

As was the case elsewhere, members of prominent civil groups shared the moral criticism and supported the "defections" of soldiers. They included writers such as Sami Michael, journalists such as B. Michael, university professors (by January 2003, 324 of them had signed a petition supporting the service resisters), and a few celebrities such as singer and cultural icon Yaffa Yarkoni.[10] These were joined by a few non-radical political figures. Member of the Knesset Roman Bronfman, in particular, voiced unwavering support for the reservists who had signed the petition, describing them as embodying "the conscience of the country."[11]

Having presented these developments, which are exceptional in the Israeli context of small wars (but not elsewhere), it is important to consider two additional issues. First, at least as I write these lines, there are only faint signs of utilitarian intolerance of the cost of war (in terms of casualties and otherwise).[12] This lack of a clear and strong Lebanon-type expedient criticism is probably due to a few factors. Clearly, at least until May 2002, most Israelis have accepted the casualties because they considered the military actions as defensively motivated and existential. Moreover, these casualties have been

---

[7] The number of soldiers (mostly reservists) jailed for refusing to serve are as follows: 170 during the Lebanon war, 180 during the first *intifada*, 20 between the first and second *intifadas*, and about 50 up to April 2002 during the *Al Aqsa intifada*. See *Ha'aretz*, March 31, 2002; and www.seruv.org.il.

[8] *Yediot Aharonot*, April 26, 2002.

[9] *Maariv*, January 27, 2003 and January 28, 2003; and *Ha'aretz*, January 28, 2003.

[10] Yarkoni's criticism resonated particularly strongly because her half-century singing career revolved around Israel's glory and anguish in war. She – who was perceived as patriotic and perhaps as a "darling" of the establishment – suddenly lashed out at the army's treatment of Palestinians in no uncertain terms, drawing comparisons with the Holocaust.

[11] *Ha'aretz*, February 21, 2002.

[12] The motivation of IDF recruits declined marginally, and reservists overwhelmingly reported for duty for the April 2002 incursions. At the same time, parents of new recruits, apparently largely from among the educated middle-class, increasingly tried to convince their children to avoid service in combat units. See *Ha'aretz*, March 20, 2002.

accepted because they were perceived as part of an effort to prevent a much larger number of civilian casualties of terror. Finally, the IDF has managed to keep its losses low, particularly relative to the volume and nature of fighting. Let us note, however, that none of these factors is inherently stable. In fact, developments in early May 2002 may be telling. Following the deadly suicide bombing of May 7, the Israeli cabinet decided to mobilize reservists and invade Gaza. Shortly thereafter, however, the plan was cancelled (or at least postponed), apparently because of fears that invasion could involve a high number of casualties, and concerns that too many people might consider the latter unjustified.[13] In short, the absence of strong expedient criticism of the confrontation policy until May cannot be considered as inherently long-lived.

The second issue concerns the nature and composition of current dissent. It is clear that moral criticism remains a preoccupation of a minority, chiefly from among the educated middle-class. As such, it is consistent with the profile of protest and dissent in other democratic small wars, and for the following reasons it could also prove to be as powerful a catalyst.

First, dissent is robust, as indicated by the fact that while the savage Palestinian terror, and the provocative "right of return" demand, may have limited the scope of dissent, they were not powerful enough to prevent it. Second, extreme forms of dissent have become legitimate among moderate constituencies that enjoy great political relevance. Obviously, Israeli protest against the occupation is nothing new. However, while the radical (and marginal) Left dominated past protests, the 2002 "defections" and protest involved a rather mainstream constituency that had traditionally been obedient and docile. Indeed, those who had initiated the officers' petition describe themselves as "belonging to the center" and as having a "different profile from that of the [more radical] *Yesh Gvul* movement."[14] Most of them had probably supported the Oslo Accords, opposed the occupation, and despised the settlement policy long before the violence started in September 2000. However, they regularly undertook reserve duty in the territories and by and large remained publicly unengaged. In early 2002, they crossed a critical threshold. Having defined the occupation as illegitimate, they took the extremely radical step of refusing to serve, during what the public at large considered a time of national security emergency.

Third, the potential power of society in general, and that of moral criticism in particular, seem to resonate well among senior IDF commanders. Indeed, they seem to integrate the potential power of society and the marketplace of ideas into their operational planning.[15] In fact, such an attitude, as Amir

---

[13] See both Amir Oren and Amos Harel, in *Ha'aretz*, May 12, 2002.

[14] *Ha'aretz*, March 31, 2002.

[15] See Amir Oren's analysis, in particular concerning the views of Brigadier General Eival Giladi, who was in charge in 2002 of operational planning in the IDF, in *Ha'aretz*, March 23, 2002.

Oren noticed, is consistent with the views of the CGS, Moshe Yaalon, who had long been convinced that "the main [battle]field is on the TV screen and nowhere else."[16] Indeed, because the army is aware of the political relevance and potential power of the draft-resisting reservists, it has confronted them in a lenient and low-key manner. Summarized in theoretical terms, it seems that the army command believes that the center of gravity could easily shift to the marketplace of ideas at home, and that there the state would be at the mercy of "society."

What is the combined effect of these three tendencies? Would the sensitivity and adjustment of the military command to the power of society undercut the emerging normative gap and offset the critical change in the identity of the anti-occupation/war constituency? Or can Israel escape the process that doomed the military efforts of other modern democracies that fought protracted small wars?

The answer to this last question is, in my judgment, no, although the process may be less painful because of Israeli idiosyncrasies. In fact, there are already some indications that the process is moving ahead unhampered. For one, the government and state institutions seem to have taken the first despotic steps, which could lead into a critical secondary expansion of the normative gap. Most certainly, it is clear that politics and people have already been radicalized and polarized, and that state institutions have become less tolerant. The government certainly has become more critical of journalistic probing, and inclined to tighten its control over the electronic media, although much of the latter are already in the hands of its loyalists. The first signs of intimidation are also evident. For example, the notoriously militant Education Minister, Limor Livnat, has asked the Attorney General to consider indicting university professors for supporting students who refused to report to reserve duty in the territories. At the same time, the right- and left-wing opinion exchange has became more heated and radical; tolerance for dissenting opinions and left-wing presentations has decreased; and critics of government policy and of the occupation have been increasingly depicted as "traitors."[17] Radicalization and polarization may also be indicated by short-term fluctuations in public opinion. For example, support for soldiers' refusal to serve in the territories surged in early 2002, but so did support for soldiers' refusal to participate in hypothetical evacuation of Jewish settlements.[18]

In any event, if developments in the Israeli-Palestinian conflict conform to the process I have discussed throughout this book, then the outcome

[16] Ibid.
[17] A good example of intolerance is a petition submitted by forty-three academics at Beer-Sheba University against letting Dr. Yossi Beilin (a chief architect of Oslo) lecture on campus. See also MK Roman Bronfman's article against the abuse of free media in *Yediot Aharonot*, April 24, 2002.
[18] See the February 2002 peace index, at www.tau.ac.il/peace/Peace_Index.

is likely to fall in line with those of other small wars, irrespective of idiosyncrasies. Spelled out clearly, the Palestinians are all but certain to lose military encounters with Israel, but are nevertheless likely to realize most of their political goals. Specifically, they will have an independent Palestinian state, most Jewish settlements in the territories will be dismantled, and the settlers will be repatriated. At the same time, Palestinian goals that concern Israel's core sovereignty, particularly the demand for an Israeli recognition of the "right of return," will not be realized. There is more than one reason for this. One main reason, however, is that such goals are outside the context of a small war. Indeed, the outcome of the Israeli-Palestinian confrontation is expected to reflect the boundaries of the small-war element of the conflict.

What I describe may seem trivial today. After all, Israeli leaders have already affirmed the vision of a Palestinian state. Yet the truth of the matter is that this position itself is a result of an Israeli change of mind that occurred after years of limited conflict and only after Israeli leaders tried hard to secure a very different outcome. At least in this sense, it is worthwhile to recall the wars in Algeria and Lebanon. There, too, backed by force, democratic leaders tried every solution short of total withdrawal before they capitulated to domestic pressure that was related to military realities that surprisingly favored the occupier. At any rate, the issue in question does not concern the current trend, but rather Israel's capacity to reverse it as, for example, was suggested by the May 13, 2002, "no" vote to a Palestinian state in the Likud party convention.

If one accepts the logic and premises of this book, then the answer to this question is rather clear. On the one hand, the process of Israeli withdrawal from the occupied territories may stall at times, but it is unlikely to be reversed. On the other hand, for the process to move more quickly, the Palestinians would have to abandon the "right of return" demand (irrespective of whether they genuinely do so) and stop the terror, particularly inside Israel. That is not to deny that the use of force, including terror, has brought the Palestinians some gains. Brutal force has apparently convinced Israelis to get closer to the Palestinian position, and prefer separation to occupation, even if unilateral.[19] By 2002, however, terror had exceeded its utility. Israelis are more ready to support confrontation and tolerate higher costs of war than they would be if the Palestinians ceased questioning Israel's sovereignty and stopped terrorizing its civilians. Thus, the timetable of Palestinian independence, if not its actual prospects, is partially dependent on the Palestinian capacity to adhere to the "boundaries" of a small war.

Indeed, if the Palestinians confine themselves to confronting Israel without raising existential issues, then Israel will be likely to either settle the conflict or find itself at the mercy of its domestic structure. In the final analysis, its instrumental dependence, the political relevance of its educated middle-class,

---

[19] See the May 2002 peace index, at www.tau.ac.il/peace/Peace_Index.

and the free marketplace of ideas will impose the "choice" on Israeli policy. If Israel decides not to compromise, and bloody Palestinian terror inside Israel ceases, then the Center-Left would find it difficult to participate in a right-wing coalition, the limits on the intensity of opposition will be lifted, and a massive secondary expansion of the normative gap is almost certain to follow. A right-wing government could hardly avoid treating the Palestinians more harshly and the opposition at home more despotically. If that happens, however, utilitarian criticism and a call for the defense of democracy will be added to stronger and more widespread moral objections to oppression of Palestinians.

Israel might be further polarized, and both the Right and Left are likely to endorse more strongly the conclusion, which they have already formed, that democracy and the war over the occupied territories cannot coexist. The different preferences of each camp are more than likely to collide, possibly violently (particularly as far as the extreme Right is concerned). Eventually, however, the liberal educated middle-class is likely to prevail, and convince Israel to withdraw from the territories, even though that is unlikely to end the conflict between Israel and a future Palestinian state.

Bibliography

GENERAL

Abzug, Robert H., *Inside the Vicious Heart: Americans and the Liberation of Nazi Concentration Camps* (NY: Oxford University Press, 1985).

Addington, Larry H., *The Patterns of War since the Eighteenth Century* (Bloomington: Indiana University Press, 1984).

Aldridge, Brian, " 'Drive Them till They Drop and then Civilize Them': The United States Army and Indigenous Populations, 1866–1902," paper presented to the conference on *Low-Intensity Conflict: The New Face of Battle?* University of New Brunswick, Fredericton, Canada, September 27–28, 1991.

Almond, Gabriel A., *A Discipline Divided* (Newbury Park, CA: Sage Publications, 1990).

Amnesty International, *East Timor Violations of Human Rights* (London: Amnesty International Publications, 1985).

Andrews, Bruce, "Social Rules and the State as a Social Actor," *World Politics*, 27:4 (1975), 521–40.

Appy, Christian G., *Working-Class War: American Combat Soldiers and Vietnam* (Chapel Hill: University of North Carolina Press, 1993).

Arnstein, Walter L., *Britain Yesterday and Today* (Lexington: D.C. Heath, 1976).

Arreguín-Toft, Ivan, "How the Weak Win Wars: A Theory of Asymmetric Conflict," *International Security*, 26:1 (2001), 93–128.

Asprey, Robert B., *War in the Shadows: The Guerrilla in History* (NY: William Morrow, 1994).

Baldwin, David A., "Power Analysis and World Politics: New Trends Versus Old Tendencies," *World Politics*, 31:2 (1979), 161–94.

Bartov, Omer, *The Eastern Front, 1941–45, German Troops and the Barbarization of Warfare* (NY: St. Martin's Press, 1986).

Baskir, Lawrence M. and William A. Strauss, *Chance and Circumstance: The Draft, the War, and the Vietnam Generation* (NY: Knopf, 1978).

Beckett, Ian F.W., and Pimlott, John, *Armed Forces and Modern Counter-Insurgency* (NY: St. Martin's Press, 1985).

Bédarida François (trans. A.S. Forster), *A Social History of England 1871–1975* (London: Methuen, 1979).

Belfield, Eversley, *The Boer War* (Hamden, CT: Archon Books, 1975).

Bell, Boyer J., *The Myth of the Guerrilla* (NY: Knopf, 1971).

Benda, Julian (trans. Richard Aldington), *The Treason of the Intellectuals* (NY: Norton, 1969).

Benedix, Reinhard, *Nation Building and Citizenship* (NY: John Wiley & Sons, 1964).

Best, Geoffrey, *Humanity in Warfare* (London: Methuen, 1983).

____ *War and Society in Revolutionary Europe 1770–1870* (NY: St. Martin's Press, 1982).

____ "Restraints on War by Land Before 1945," in Howard, Michael (ed.), *Restraints on War* (Oxford: Oxford University Press, 1979), 17–37.

Betts, Richard, "Comment on Mueller," *International Studies Quarterly*, 24:4 (1980), 520–24.

Blaufarb, Douglas S., *The Counterinsurgency Era: U.S. Doctrine and Performance* (NY: The Free Press, 1977).

Blechman, Barry M., and Tamara Cofman Wittes, "Defining Moment: The Threat and Use of Force in American Foreign Policy," *Political Science Quarterly*, 114:1 (1999), 1–30.

Bley, Helmut (trans. Hugh Ridley), *South-West Africa under German Rule 1894–1914* (London: Heinemann, 1971).

Bowman, John (ed.), *The World Almanac of the Vietnam War* (NY: Pharos Books, 1986).

Bull, Hedley, "The Emergence of a Universal International Society," in Bull and Watson, *The Expansion of International Society*, 118–26.

Bull, Hedley, and Adam Watson, *The Expansion of International Society* (Oxford: Clarendon Press, 1984).

Burg, Steven L., and Paul S. Shoup, *The War in Bosnia-Herzegovina: Ethnic Conflict and International Intervention* (Armonk, NY: M.E. Sharpe, 1999).

Butler, David, "Electors and Elected," in Halsey, *British Social Trends*, 297–321.

Callwell, Charles E., *Small Wars: Their Principles and Practice* (University of Nebraska Press, 1996).

Campbell, Arthur, *Guerrillas* (London: Arthur Barker, 1967).

Carey, Peter, *East Timor: Third World Colonialism and the Struggle for National Identity* (London: RISCT, Conflict Studies 293/294, 1996).

Chaliand, Gerard (ed.), *Guerrilla Strategies* (Berkeley: University of California Press, 1982).

Chen, Yung-fa, *Making Revolution* (Berkeley: University of California Press, 1986).

Clarke, Jeffrey J., *Advice and Support: The Final Years, 1965–1973* (Washington: Center of Military History, United States Army, 1988).

Clausewitz, Carl von (eds. and trans. Michael Howard and Peter Paret), *On War* (Princeton: Princeton University Press, 1984).

Clodfelter, Mark, *The Limits of Air Power* (NY: The Free Press, 1989).

Cohen, Eliot A., "Constraints on America's Conduct of Small Wars," *International Security*, 9:2 (1984), 151–81.

Craig, Gordon A., and Alexander L. George, *Force and Statecraft* (NY: Oxford University Press, 1990).

Czempiel, Ernst-Otto, and James N. Rosenau, *Global Changes and Theoretical Challenges* (Lexington, MA: Lexington Books, 1989).

Danziger, Raphael, *Abd al-Qadir and the Algerians* (NY: Holmes & Meier, 1977).

Davey, Arthur, *The British Pro-Boers 1877–1902* (Cape Town: Tafelberg, 1978).

Davis, James K., *Assault on the Left* (Westport, CT: Praeger, 1997).

Davison, Basil, *The People's Cause* (London: Longman, 1981).

Day, Alan J. (ed.), *Border and Territorial Disputes* (Detroit: Gale Research Co., 1982).

DeBenedetti, Charles, *An American Ordeal* (Syracuse: Syracuse University Press, 1990).

Denzin, Norman K., and Yvonna S. Lincoln (eds.), *Handbook of Qualitative Research* (Thousand Oaks, CA: Sage, 1994).

Doyle, Michael W., *Empires* (Ithaca: Cornell University Press, 1986).

—— "Kant, Liberal Legacies, and Foreign Affairs," *Philosophy and Public Affairs*, 12:3 (1983), 205–35.

—— "Liberalism and World Politics," *American Political Science Review*, 80:4 (1986), 1151–69.

Drechsler, Horst, "South West Africa 1885–1907," in Stoecker, *German Imperialism in Africa*, 39–62.

Duffy, Christopher, *The Military Experience in the Age of Reason* (NY: Routledge & Kegan Paul, 1987).

Eckhardt, William, *Civilizations, Empires and Wars: A Quantitative History of War* (Jefferson, NC: McFarland & Company, 1992).

Edelman, Bernard (ed.), *Dear America: Letters Home from Vietnam* (NY: Pocket Books, 1985).

Ellis, John, *A Short History of Guerrilla Warfare* (NY: St. Martin's Press, 1976).

—— *From the Barrel of a Gun: A History of Guerrilla, Revolutionary, and Counter-Insurgency Warfare, from the Romans to the Present* (London: Greenhill, 1995).

—— *The Social History of the Machine Gun* (Baltimore: Johns Hopkins University Press, 1975).

Evans, Peter, Dietrich Rueschemeyer, and Theda Skocpol (eds.), *Bringing the State Back In* (Cambridge: Cambridge University Press, 1985).

Fall, Bernard B., *Streets Without Joy* (London: Pall Mall, 1964).

Farewell, Byron, *The Great Boer War* (London: Allen Lane, 1977).

Farouk-Sluglett, Marion and Peter Sluglett, *Iraq since 1958* (NY: KPI, 1987).

Ferguson, Yale H. and Richard W. Mansbach, *The State, Conceptual Chaos, and the Future of International Relations Theory* (Boulder: University of Denver Press, 1989).

—— "Between Celebration and Despair: Constructive Suggestions for Future International Theory," *International Studies Quarterly*, 35:4 (1991), 363–86.

Finer, Samuel E., "State- and Nation-Building in Europe: The Role of the Military," in Tilly, Charles (ed.), *The Formation of National States in Western Europe* (Princeton: Princeton University Press, 1975), 84–163.

Fischerkeller, Michael P., "David versus Goliath: Cultural Judgments in Asymmetric Wars," *Security Studies*, 7:4 (1998), 1–43.

Flora, Peter et al., *State, Economy, and Society in Western Europe, 1815–1975* (Chicago: St. James Press, 1983, 1987), Vols. 1 and 2.

Freedman, Lawrence (ed.), *Military Intervention in European Conflicts* (Oxford: Blackwell, 1994).

Friedman, Leon (ed.), *The Law of War: A Documentary History*, Vol. 2 (NY: Random House, 1972).

Fuller, John F.C., *The Conduct of War: 1789–1961* (Westport, CT: Greenwood Press, 1981).

Geneva Conventions of August 12, 1949, for the Protection of War Victims (Washington, DC: U.S. Government Printing Office, 1950).

George, Alexander L., "Case Studies and Theory Development: The Method of Structured, Focused Comparison," in Lauren, *Diplomacy*, 43–68.

Gettleman, Marvin E., Jane Franklin, Marilyn Young, and Bruce H. Franklin, *Vietnam and America: A Documented History* (NY: Grove Press, 1985).

Gilpin, Robert, *War and Change in World Politics* (Cambridge: Cambridge University Press, 1981).

Ginsberg, Benjamin, *The Captive Public* (NY: Basic Books, 1986).

_____ *The Consequences of Consent: Elections, Citizen Control, and Popular Acquiescence* (NY: Random House, 1982).

Goldgeier, James M., and Michael McFaul, "A Tale of Two Worlds: Core and Periphery in the Post-Cold War Era," *International Organization*, 46:2 (1992), 467–491.

Gong, Gerrit W., *The Standard of Civilization in International Society* (Oxford: Clarendon, 1984).

Góngora, Mario, (trans. Richard Southern), *Studies in the Colonial History of Spanish America* (NY: Cambridge University Press, 1975).

Gow, James, "Nervous Bunnies – The International Community and the Yugoslav War of Dissolution," in Freedman, *Military Intervention*, 14–33.

Gurr, Ted Robert, "War, Revolution, and the Growth of the Coercive State," *Comparative Political Studies*, 21:1 (1988), 45–65.

Hall, John A. (ed.), *States in History* (Oxford: Oxford University Press, 1986).

Hall, Mitchell K., *Because of Their Faith: CALCAV and Religious Opposition to the Vietnam War* (NY: Columbia University Press, 1990).

Halsey, A.H. (ed.), *British Social Trends since 1900* (London: Macmillan, 1988).

_____ "Higher Education," in Halsey, *British Social Trends*, 268–96.

_____ "Schools," in Halsey, *British Social Trends*, 227–67.

Halverson, Thomas, "Disengagement by Stealth: The Emerging Gap between America's Rhetoric and the Reality of Future European Conflicts," in Freedman, *Military Intervention*, 76–93.

Handel, Michael I., *Masters of War* (London: Frank Cass, 2001).

Hart, Jeffrey, "Three Approaches to the Measurement of Power in International Relations," *International Organization*, 30:2 (1976), 289–305.

Hartigan, Richard S., *The Forgotten Victim: A History of the Civilian* (Chicago: Precedent Publishing, 1982).

_____ *Lieber's Code and the Law of War* (Chicago: Precedent, 1983).

Hehn, Paul N., *The German Struggle against Yugoslav Guerrillas in World War II* (NY: East European Quarterly, distributed by Columbia University Press, 1979).

Heineman, Kenneth J., *Campus Wars* (NY: New York University Press, 1993).

Henderson, Bernard W., *The Life and Principate of the Emperor Hadrian* (Rome: "L'Erma" di Bretschneider, 1968).

Herring, George C., *America's Longest War* (Philadelphia: Temple University Press, 1986).

_____ "Franco-American Conflict in Indochina, 1950–1954," in Kaplan et al., *Dien Bien Phu*, 29–48.

Herring, George C., Garry R. Hess, and Richard Immerman, "Passage of Empire: The United States, France, and South Vietnam, 1954–1955," in Kaplan et al., *Dien Bien Phu*, 171–95.

Holsti, K. J., "The Concept of Power in the Study of International Relations," *Background* (*International Studies Quarterly*), 7:4 (1964), 179–94.

Howard, Michael (ed.), *Restraints on War* (Oxford: Oxford University Press, 1979).

—— *War and the Liberal Conscience* (Oxford: Oxford University Press, 1978).

—— *War in European History* (NY: Oxford University Press, 1976).

—— "The Forgotten Dimensions of Strategy," *Foreign Affairs*, 57:5 (1979), 975–86.

—— "The Military Factor in European Expansion," in Bull and Watson, *The Expansion of International Society*, 33–42.

Huberman A. Michael, and Matthew B. Miles, "Data Management and Analysis Methods," in Denzin and Lincoln, *Handbook of Qualitative Research*, 428–44.

Huntington, Samuel P., *Political Order in Changing Societies* (New Haven: Yale University Press, 1970).

—— "Political Development and Political Decay," in Bienen, Henry (ed.), *The Military and Modernization* (NY: Aldine/Atherton, 1971), 157–211.

Iliffe, John, "The Effect of the Maji Maji Rebellion of 1905–1906 on German Occupation Policy in East Africa," in Gifford, Prosser, and Wm. Roger Louis, *Britain and Germany in Africa* (New Haven: Yale University Press, 1967), 557–75.

Immerman, Richard H., "Perceptions by the United States of its Interests in Indochina," in Kaplan et al., *Dien Bien Phu*, 1–26.

Indyk, Martin, "Beyond the Balance-of-Power: America's Choice in the Middle East," *The National Interest*, 26 (1991/1992), 33–43.

Isserman, Maurice, and Michael Kazin, *America Divided* (NY: Oxford University Press, 2000).

Jeffreys Jones, Rhodri, *Peace Now: American Society and the Ending of the Vietnam War* (New Haven: Yale University Press, 1999).

Janowitz, Morris, "Military Institutions and Citizenship in Western Societies," *Armed Forces and Society*, 2:2 (1976), 185–204.

Jervis, Robert, "Bargaining and Bargaining Tactics," in Pennock, Roland J., and John W. Chapman (eds.), *Coercion* (NY: Atherton, 1972), 272–88.

Johnson, Lyndon B., *The Vantage Point* (NY: Holt, Rinehart and Winston, 1971).

Kahin, George McT., *Intervention: How America Became Involved in Vietnam* (NY: Anchor Books, 1987).

Kahler, Miles, "External Ambition and Economic Performance," *World Politics*, 40:4 (1988), 419–51.

Kaiser David, *Politics and War* (Cambridge: Harvard University Press, 1990).

Kaplan, Lawrence S., Denise Artaud, and Mark R. Rubin (eds.), *Dien Bien Phu and the Crisis of Franco-American Relations, 1954–1955* (Wilmington, DE: SR Books, 1990).

Karnow, Stanley, *Vietnam: A History* (NY: Penguin, 1984).

Katzenstein, Peter J., "International Relations Theory and the Analysis of Change," in Czempiel and Rosenau, *Global Changes and Theoretical Challenges*, 291–304.

Keller, Allan, *The Spanish-American War* (NY: Hawthorn Books, 1969).

Kennedy, Paul, *The Rise and Fall of Great Powers: Economic Change and Military Conflict from 1500 to 2000* (NY: Random House, 1987).

Khoury, Philip S., *Syria and the French Mandate: The Politics of Arab Nationalism 1920–1945* (Princeton, Princeton University Press, 1987).

Kiernan, Victor Gordon, *From Conquest to Collapse: European Empires from 1815 to 1960* (NY: Pantheon Books, 1982).

King, Gary, Robert O. Keohane, and Sidney Verba, *Designing Social Inquiry: Scientific Inference in Qualitative Research* (Princeton: Princeton University Press, 1994).

Kissinger, Henry, *The White House Years* (London, Weidenfeld and Nicolson and Michael Joseph, 1979).

Kolko, Gabriel, *Anatomy of War* (NY: Pantheon Books, 1985).

Krasner, Stephen D., *Defending the National Interest: Raw Materials Investments and U. S. Foreign Policy* (Princeton: Princeton University Press, 1978).

_____ "Approaches to the State: Alternative Conceptions and Historical Dynamics," *Comparative Politics*, 16:2 (1984), 223–46.

Krepinevich, Andrew F. Jr., *The Army and Vietnam* (Baltimore, MD: Johns Hopkins University Press, 1986).

Kugler, Jacek, and William Domke, "Comparing the Strength of Nations," *Comparative Political Studies*, 19:1 (1986), 39–70.

Laqueur, Walter, *Guerrilla* (Boston: Little, Brown and Co., 1976).

Lauren, Paul G. (ed.), *Diplomacy: New Approaches in History, Theory, and Policy* (NY: The Free Press, 1979).

_____ *The Evolution of Human Rights* (Philadelphia: University of Philadelphia Press, 1998).

_____ "Theories of Bargaining with Threat of Force: Deterrence and Coercive Diplomacy," in Lauren, *Diplomacy*, 183–211.

Levy, David W., *The Debate over Vietnam* (Baltimore: Johns Hopkins University Press, 1991).

Lewy, Guenter, *America in Vietnam* (NY: Oxford University Press, 1980).

Mack, Andrew, "Why Big Nations Lose Small Wars: The Politics of Asymmetric Conflict," *World Politics*, 27:2 (1975), 175–200.

MacKenzie John M. (ed.), *Popular Imperialism and the Military, 1850–1950* (Manchester: Manchester University Press, 1992).

_____ *Propaganda and Empire* (NY: Manchester University Press, 1984).

Maclear, Michael, *The Ten Thousand Day War: Vietnam 1945–1975* (NY: Methuen, 1981).

Manicas, Peter T., *War and Democracy* (Cambridge: Basil Blackwell, 1989).

Mann, Michael, *States, War and Capitalism* (NY: Basil Blackwell, 1988).

_____ "The Autonomous Power of the State: Its Origins, Mechanisms and Results," in Hall, *States in History*, 109–36.

Maoz, Zeev, "Power, Capabilities, and Paradoxical Conflict Outcomes," *World Politics*, 41:2 (1989), 239–66.

Marshal, Catherine, and Gretchen B. Robertson, *Designing Qualitative Research* (London: Sage Publications, 1989).

Materski, Wojciech (ed.), *Katyn: Documents of Genocide* (Warsaw: Institute of Political Studies, Polish Academy of Sciences, 1993).

McNeill, William, H., *The Pursuit of Power: Technology, Armed Forces, and Society Since A.D. 1000* (Chicago: University of Chicago Press, 1982).

Meinecke, Freidrich, *Machiavellism: The Doctrine of Raison d'Etat and its Place in Modern History* (London: Routledge and Kegan Paul, 1954).

Merom, Gil, "A Grand Design? Charles de Gaulle and the End of the Algerian War," *Armed Forces and Society*, 25:2 (1999), 267–88.

_____ "Israel's National Security and the Myth of Exceptionalism," *Political Science Quarterly*, 114:3 (1999), 409–34.

Migdal, Joel S., *Strong Societies and Weak States* (Princeton: Princeton University Press, 1988).

Mitchell B. R., *European Historical Statistics 1750–1970* (NY: Columbia University Press, 1975, 1978).

Moïsi, Dominique, "Intervention in French Foreign Policy," in Bull, Hedley (ed.), *Intervention in World Politics* (Oxford: Clarendon Press, 1984), 67–77.

Mooers, Colin, *The Making of Bourgeois Europe: Absolutism, Revolution, and the Rise of Capitalism in England, France and Germany* (NY: Verso, 1991).

Moravcsik, Andrew, "Taking Preferences Seriously: A Liberal Theory of International Politics," *International Organization*, 51:4 (1997), 513–53.

Moreman, T. R., "'Small Wars' and 'Imperial Policing': The British Army and the Theory and Practice of Colonial Warfare in the British Empire, 1919–1939," *Journal of Strategic Studies*, 19:4 (1996), 105–31

Morgenthau, Hans J., *Politics among Nations* (NY: Knopf, 1973).

_____ *American Foreign Policy* (London: Methuen, 1951).

Moser, Richard R., *The New Winter Soldiers: GI and Veteran Dissent during the Vietnam Era* (New Brunswick, NJ: Rutgers University Press, 1996).

Mosse, George L., *Fallen Soldiers* (NY: Oxford University Press, 1990).

_____ *The Nationalization of the Masses* (NY: Howard Fertig, 1975).

Mueller, John, *Retreat from Doomsday* (NY: Basic Books, 1989).

_____ "Rejoinder," *International Studies Quarterly*, 24:4 (1980), 530–31.

_____ "The Search for the 'Breaking Point' in Vietnam: The Statistics of a Deadly Quarrel," *International Studies Quarterly*, 24:4 (1980), 497–519.

Mullin, Chris, and Phuntsong Wangyal, *The Tibetans* (London: Minority Rights Group, Report No. 49, 1983).

Nordlinger, Eric A., *On the Autonomy of the Democratic State* (Cambridge: Harvard University Press, 1981).

_____ "Taking the State Seriously," in Weiner, Myron, and Samuel P. Huntington (eds.), *Understanding Political Development* (Boston: Little, Brown, 1987), 353–90.

O'Balance, Edgar, *Malaya: The Communist Insurgent War* (Hamden, CT: Archon Books, 1966).

_____ *Terror in Ireland* (Novato, CA: Presidio Press, 1981).

Osgood, Robert E., *Limited War Revisited* (Boulder, CO: Westview Press, 1979).

Owen, John M., "How Liberalism Produces Democratic Peace," *International Security*, 19:2 (1994), 87–125.

Pakenham, Thomas, *The Scramble for Africa, 1876–1912* (London: Weidenfeld and Nicolson, 1991).

Pape, Robert A. Jr., "Coercive Air Power in the Vietnam War," *International Security*, 15:2 (1990), 103–146.

Paret, Peter, *French Revolutionary Warfare from Indochina to Algeria* (London: Pall Mall, 1964).

Pick, Daniel, *War Machine: The Rationalization of Slaughter in the Modern Age* (New Haven: Yale University Press, 1993).

Price, Roger, *A Social History of Nineteenth-Century France* (London: Hutchinson, 1987).

Poggi, Gianfranco, *The Development of the Modern State* (London: Hutchinson, 1978).

Porter, Bruce D., *War and the Rise of the State* (NY: The Free Press, 1994).

Powers, Thomas, *The War at Home: Vietnam and the American People 1964–1968* (Boston: G.H. Hall, 1984).

Pryor, Frederic, "The New Class: Analysis of the Concept, the Hypothesis and the Idea as a Research Tool," *American Journal of Economics and Sociology*, 40:4 (1981), 367–79.

Punch, Keith F., *Introduction to Social Research* (London: Sage, 1998).

Putnam, Robert D., "Diplomacy and Domestic Politics: The Logic of Two-Level Games," *International Organization*, 42:3 (1988), 427–60.

Ray, James Lee, and Ayse Vural, "Power Disparities and Paradoxical Conflict Outcomes," *International Interactions*, 12:4 (1986), 315–42.

Rice, Edward E., *Wars of the Third Kind: Conflict in Underdeveloped Countries* (Berkeley: University of California Press, 1988).

Ringer, Fritz, *Fields of Knowledge: French Academic Culture in Comparative Perspective, 1890–1920* (Cambridge: Cambridge University Press, 1992).

Risse-Kappen, Thomas, "Did Peace Through Strength End the Cold War? Lessons from INF," *International Security*, 16:1 (1991), 162–88.

_____ "Public Opinion, Domestic Structure, and Foreign Policy in Liberal Democracies," *World Politics* 43:4 (1991), 479–512.

Robertson, Geoffrey, *Crimes against Humanity* (London: Allen Lane-The Penguin Press, 1999).

Rokkan, Stein, *Citizens, Elections, Parties* (NY: David McKay, 1970).

Rosecrance, Richard N., *The Rise of the Trading State* (NY: Basic Books, 1986).

Rosen, Stephen, "Vietnam and the American Theory of Limited War," *International Security*, 7:2 (1982), 83–113.

Rosen, Steven , "War Powers and the Willingness to Suffer," in Russett, Bruce (ed.), *Peace, War, and Numbers* (Beverly Hills: Sage, 1972), 167–83.

Rosenau, James N., "Global Changes and Theoretical Challenges: Toward a Post International Politics for the 1990's," in Czempiel and Rosenau, *Global Changes and Theoretical Challenges*, 1–20.

Rosenberg, Milton J., Sydney Verba, and Philip E. Converse, *Vietnam and the Silent Majority* (NY: Harper and Row, 1970).

Russett, Bruce, *Controlling the Sword* (Cambridge: Harvard University Press, 1990).

_____ *Grasping the Democratic Peace* (Princeton: Princeton University Press, 1993).

Salmon, Edward T., *A History of the Roman World* (London: Methuen, 1957).

Schalk, David L., *War and the Ivory Tower: Algeria and Vietnam* (NY: Oxford University Press, 1991).

Schulte Nordholt, Jan W. (trans. Herbert H. Rowen), *Woodrow Wilson: A Life for World Peace* (Berkeley: University of California Press, 1991).

Schumpeter, Joseph A., *Imperialism and Social Classes* (NY: Augustus Kelley, 1951).

Schweller, Randall L., "Domestic Structure and Preventive War: Are Democracies More Pacific?," *World Politics*, 44:2 (1992), 235–69.

Seligman, Adam B., *The Idea of Civil Society* (NY: The Free Press, 1992).

Sharp, Jane M. O., "Appeasement, Intervention and the Future of Europe," in Freedman, *Military Intervention*, 34–55.

Shirer, William L., *The Rise and Fall of the Third Reich* (NY: Simon & Schuster, 1960).

Simon, Herbert A., "Notes on the Observation and Measurement of Political Power," *Journal of Politics*, 15:4 (1953), 500–16.

Singer, Max, and Aaron Wildavsky, *The Real World Order: Zones of Peace/Zones of Order* (Chatham, NJ: Chatham House, 1993).

Skinner, Quentin, *The Foundations of Modern Political Thought*, Vol. 2 (NY: Cambridge University Press, 1978).

Skocpol, Theda, "Bringing the State Back In: Strategies of Analysis in Current Research," in Evans et al., *Bringing the State Back In*, 3–37.

Sluglett, Peter, "The Kurds," in CARDRI, *Saddam's Iraq* (London: Zed Books, 1990), 177–202.

Small, Melvin, *Johnson, Nixon, and the Doves* (New Brunswick, NJ: Rutgers University Press, 1988).

Small, Melvin, and J. David Singer, *Resort to Arms* (Beverly Hills: Sage, 1982).

Small, Melvin, and William D. Hoover (eds.), *Give Peace a Chance: Exploring the Vietnam Antiwar Movement* (Syracuse: Syracuse University Press, 1992).

Smelser, Neil J., "The Methodology of Comparative Analysis," in Warwick, Donald P., and Samuel Osherson (eds.), *Comparative Research Methods* (Englewood Cliffs, NJ: Prentice-Hall, 1973), 42–86.

Spies, S. B., *Methods of Barbarism?* (Cape Town: Human & Rousseau, 1977).

Stake, Robert E., "Case Studies," in Denzin and Lincoln, *Handbook of Qualitative Research*, 236–47.

Steiberg, H. S. (ed.) *The Statesman's Year-Book* (NY: St. Martin's Press, 1958).

Stoecker, Helmuth (ed.) (trans. Bernd Zöllner), *German Imperialism in Africa* (London: C. Hurst, 1986).

—— "German East Africa 1885–1906," in Stoecker, *German Imperialism*, 93–113.

Strang, David, "Global Patterns of Decolonization, 1500–1987," *International Studies Quarterly*, 35 (1991), 429–54.

Sullivan, Anthony T., *Thomas Robert Bugeaud* (Hamden, CT: Archon Books, 1983).

Summers, Anne, "Militarism in Britain before the Great War," *History Workshop*, No. 2 (1976), 104–123.

Summers, Harry G. Jr., "A War Is a War Is a War Is a War," in Thompson, *Low-Intensity Conflict*, 27–49.

Tannenbaum, Edward R., *European Civilization since the Middle Ages* (NY: John Wiley & Sons, 1971).

Taylor, A.J.P., *The Course of German History* (London: Methuen, 1985).

—— *English History 1914–1945*, Harmondsworth (Middlesex, UK: Pelican, 1970).

Thompson, Janice E. and Stephan D. Krasner, "Global Transactions and the Consolidation of Sovereignty," in Czempiel and Rosenau, *Global Changes and Theoretical Challenges*, 195–220.

Thompson, Loren B. (ed.), *Low-Intensity Conflict* (Lexington, MA: Lexington Books, 1989).

—— "Low-Intensity Conflict: An Overview," in Thompson, *Low-Intensity Conflict*, 1–25.

Thompson, Robert G., *Defeating Communist Insurgency: The Lessons of Malaya and Vietnam* (NY: Praeger, 1966).

_____ "Squaring the Error," *Foreign Affairs*, 46:3 (1968), 442–53.

Thompson, W. Scott, and Donaldson D. Frizzell (eds.), *The Lessons of Vietnam* (London: Macdonald and Jane's, 1977).

Thucydides (trans. Rex Warner), *History of the Peloponnesian War* (London: Penguin Books, 1954).

Tibawi, A. L., *A Modern History of Syria* (NY: St. Martin's Press, 1969).

Tilly, Charles, *Coercion, Capital, and European States, A.D. 990–1990* (Cambridge, MA: Basil Blackwell, 1990).

_____ (ed.), *The Formation of National States in Western Europe* (Princeton: Princeton University Press, 1985).

Tse-Tung, Mao, *On Guerrilla Warfare* (NY: Praeger, 1961).

_____ *Selected Military Writings* (Peking: Foreign Languages Press, 1963).

_____ *Strategic Problems of China's Revolutionary War* (Peking: Foreign Languages Press, 1954).

Vagts, Alfred, *A History of Militarism* (NY: Norton, 1937).

Vought, Donald, "American Culture and American Arms: The Case of Vietnam," in Hunt, Richard A., and Richard H Schultz Jr. (eds.), *Lessons from Unconventional War* (NY: Pergamon Press, 1982), 158–190.

Waltz, Kenneth N., *Theory of International Politics* (Reading: Addison-Wesley Publishing, 1979).

Watt, Donald Cameron, "Restraints on War in the Air Before 1945," in Howard, *Restraints on War*, 57–77.

Weber, Eugen, *Peasants into Frenchmen* (Stanford: Stanford University Press, 1976).

Wells, Donald A., *War Crimes and Laws of War* (NY: University Press of America, 1984).

Wells, Tom, *The War Within* (Berkeley: University of California Press, 1994).

Westheider, James E., *Fighting on Two Fronts* (NY: New York University Press, 1997).

Wohl, Robert, *The Generation of 1914* (Cambridge, MA: Harvard University Press, 1979).

World Almanac and Book of Facts (NY: Press Publishing Co., 1969).

Yadin, Yigael, *Bar Kokhba* (London: Weidenfeld and Nicolson, 1971).

Zacher, Mark W., and Richard A. Matthew, "Liberal International Theory: Common Threads Divergent Strands," in Kegley, Charles W. (ed.), *Controversies in International Relations Theory* (NY: St. Martin's Press, 1995), 107–50.

Zaher, U., "Political Developments in Iraq 1963–1980," in CARDRI, *Saddam's Iraq* (London: Zed Books, 1990), 30–53.

Zaroulis, Nancy, and Gerald Sullivan, *Who Spoke Up? American Protest against the War in Vietnam, 1963–1975* (NY: Doubleday, 1984).

Zeman, Z. A. B., *Nazi Propaganda* (London: Oxford University Press, 1964).

MEDIA SOURCES

ABC: Nightline
CBS: 60 Minutes
The Guardian
MacNeil-Lehrer News Hour
PBS: Frontline
Wall Street Journal

FRANCE IN ALGERIA

Ageron, Charles-Robert, "L'Algérie, dernière chance de la puissance française," *Relations Internationales*, 57 (1989), 113–139.

—— "L'Opinion française à travers les sondages," in Rioux, *La guerre d'Algérie*, 25–44.

—— "Les français devant la guerre civile algérienne," in Rioux, *La guerre d'Algérie*, 53–62.

Alleg, Henri et al., *La guerre d'Algérie*, Vol. 1–3 (Paris: Temps actuels, 1981).

Ambler, John S., *The French Army in Politics 1945–1962* (Columbus: Ohio State University Press, 1966).

Argoud, Antoine, *La décadence, l'imposture et la tragédie* (Paris: Fayard, 1974).

Aron, Raymond, *L'Algérie et la république* (Paris: Plon, 1958).

—— *La tragédie algérienne* (Paris: Plon, 1957).

Asselain, Jean-Charles, "'Boulet colonial' et redressement économique (1958–1962)," in Rioux, *La guerre d'Algérie*, 289–303.

Aussaresses, Paul, *Service spéciaux Algerie 1955–1957: Mon témoignage sur la torture* (Perrin, 2001).

Becker, Jean-Jacques, "L'intérêt bien compris du parti communiste français," in Rioux, *La guerre d'Algérie*, 235–44.

Berstein, Serge, "La peau de chagrin de 'l'Algérie française'," in Rioux, *La guerre d'Algérie*, 202–17.

Bonnaud, Robert, "Le refus," in Rioux and Sirinelli, *La guerre d'Algérie et les intellectuels*, 345–52.

Bossuat, Gérard, "Guy Mollet: la puissance française autrement," *Relations Internationales*, 57 (1989), 25–48.

Cogan, Charles G., *Old Allies, Guarded Friends* (Westport, CT: Praeger, 1994).

—— "France, the United States and the Invisible Algeria Outcome," *Journal of Strategic Studies*, 25:2 (2002), 138–58.

d'Abzac-Epezy, Claude, "La société militaire, de l'ingérence à l'ignorance," in Rioux, *La guerre d'Algérie*, 245–56.

Delarue, Jacques, "La police en paravent et au rempart," in Rioux, *La guerre d'Algérie*, 257–68.

Delmas, Jean, "A la recherche des signes de la puissance: L'armée entre l'Algérie et la bombe A, 1956–1962," *Relations Internationales*, 57 (1989), 77–87.

Domenach, Jean-Marie, "Commentaires sur l'article de David L. Schalk," *The Tocqueville Review*, 8 (1986/87), 93–95.

—— "Democratic Paralysis in France," *Foreign Affairs*, 37:1 (1958), 31–44.

—— "The French Army in Politics," *Foreign Affairs*, 39:2 (1961), 185–195.

—— "Un souvenir très triste," in Rioux and Sirinelli, *La guerre d'Algérie et les intellectuels*, 353–57.

Estival, Bernard, "The French Navy and the Algeria War," *Journal of Strategic Studies*, 25:2 (2002), 79–94.

Éveno, Patrick and Jean Planchais, *La guerre d'Algérie: Dossier et témoignages* (Paris: Le Monde/La Découverte, 1989).

Frémont, Armand, "Le contingent: témoignage et réflexion," in Rioux, *La Guerre d'Algérie*, 79–85.

Fouilloux, Étienne, "Chrétiens et juifs: comme les autre?" in Rioux, *La Guerre d'Algérie*, 109–15.
_____ "Intellectuels catholiques et guerre d'Algérie (1954–1962)," in Rioux and Sirinelli, *La guerre d'Algérie et les intellectuels*, 79–114.
Furniss, Edgar S., *France: Troubled Ally* (NY: Harper and Brothers, 1960).
Gaulle, Charles de (trans. Terence Kilmartin), *Memoirs of Hope: Renewal and Endeavor* (NY: Simon & Schuster, 1971).
Girardet, Raoul, *L'idée coloniale en France* (Paris: La Table Ronde, 1972).
Grall, Xavier, *La génération du Djebel* (Paris: Cerf, 1962).
Granjon, Marie-Christine, "Raymond Aron, Jean Paul Sartre et le conflict algérien," in Rioux and Sirinelli, *La guerre d'Algérie et les intellectuels*, 115–38.
Hamon, Hervé, and Patrick Rotman, *Les porteurs de valises: la résistance française à la guerre d'Algérie* (Paris: Albin Michel, 1979).
Heggoy, Alf Andrew, *Insurgency and Counterinsurgency in Algeria* (Bloomington: Indiana University Press, 1972).
Hoffmann, Stanley, *Decline or Renewal* (NY: The Viking Press, 1974).
Horne, Alistair, *A Savage War of Peace: Algeria 1954–1962* (NY: The Viking Press, 1978).
Irving, R. E. M., *The First Indochina War* (London: Croom Helm, 1975).
Jouhaud, Edmond, *Ce que je n'ai pas dit* (Paris: Fayard, 1977).
Julliard, Jacques, "La réparation des clercs," in Rioux and Sirinelli, *La guerre d'Algérie et les intellectuels*, 387–95.
_____ "Une base de masse pour l'anticolonialisme," in Rioux and Sirinelli, *La guerre d'Algérie et les intellectuels*, 359–64.
Kelly, George A., "The French Army Re-enters Politics," *Political Science Quarterly*, 76:3 (1961), 367–92.
L'année politique (Paris: Presses universitaires de France), Vols. 1955–1962.
La guerre d'Algérie par les documents (Vincennes: Service Historique de l'Armée de Terre, 1990).
La Morandais, Alain Maillard de, "De la colonisation à la torture," Ph.D. thesis (Université de Paris-Sorbonne, 1983).
Lacouture, Jean (trans. George Holoch), *Pierre Mendès-France* (NY: Holmes and Meier, 1984).
_____ (trans. Alan Sheridan), *De Gaulle: The Ruler 1945–1970*, Vol 2 (London: Harvill, 1991).
Lancaster, Donald, *The Emancipation of French Indochina* (London: Oxford University Press, 1961).
Liauzu, Claude, "Intellectuels du Tiers Monde et intellectuels français: Les années algériennes des Éditions Maspero," in Rioux and Sirinelli, *La guerre d'Algérie et les intellectuels*, 155–74.
Lustick, Ian, *State Building Failure in British Ireland and French Algeria* (Berkeley: IIS, University of California Press, 1985).
Maran, Rita, *Torture: The Role of Ideology in the French-Algerian War* (NY: Praeger, 1989).
Marseille, Jacques, "La guerre a-t-elle eu lieu? Mythes et réalités du fardeau algérien," in Rioux, *La Guerre d'Algérie*, 281–88.
Martin, Michel L., "Conscription and the Decline of the Mass Army in France, 1960–1975," *Armed Forces and Society*, 3:3 (1977), 355–406.

Massu, Jacques, *La vraie bataille d'Alger* (Paris: Plon, 1971).

Mélandri, Pierre, "La France et le 'jeu double' des États-Unis," in Rioux, *La Guerre d'Algérie*, 429–50.

Monchablon, Alain, "Syndicalisme étudiant et génération algérienne," in Rioux and Sirinelli, *La guerre d'Algérie et les intellectuels*, 175–189.

Montagnon, Pierre, *La guerre d'Algérie: Genèse et engrenage d'une tragédie* (Paris: Pygmalion/Gérard Watelet, 1984).

Morse, Edward L., *Foreign Policy and Interdependence in Gaullist France* (Princeton: Princeton University Press, 1973).

Nozière, André, *Algérie: Les chrétiens dans la guerre* (Paris: CANA, 1979).

Paret, Peter, *French Revolutionary Warfare from Indochina to Algeria* (Princeton: The Center of International Studies, 1964).

Pervillé, Guy, "Bilan de la guerre d'Algérie," in *Études sur la France de 1939 à nos jours* (Paris: Seuil, 1985).

Planchais, Jean, *Une histoire politique de l'Armée*, Vol. 2, 1940–1962 (Paris: Seuil, 1967).

Porch, Douglas, *The French Foreign Legion* (NY: Harper/Collins, 1991).

Rioux, Jean-Pierre (ed.), *La guerre d'Algérie et les Français* (Paris: Fayard, 1990).

—— (trans. Godfrey Rogers), *The Fourth Republic, 1944–1958* (NY: Cambridge University Press, 1987).

Rioux, Jean-Pierre, and Sirinelli, Jean-François (eds.), *La guerre d'Algérie et les intellectuels français* (Bruxelles: Éditions Complexe, 1991).

Rotman, Patrick, and Yves Rouseau, "La résistance française a la guerre d'Algérie," Doctorat de Troisième Cycle en Sciences Politiques (Université de Paris 8, 1981).

Rovan, Joseph, "Témoignage sur Edmond Michelet, garde des Sceaux," in Rioux, *La guerre d'Algérie*, 276–78.

Rudelle, Odile, "Gaullisme et crise d'identité républicaine," in Rioux, *La Guerre d'Algérie*, 180–201.

Ruz, Nathalie, "La force du 'cartiérisme,'" in Rioux, *La guerre d'Algérie*, 328–36.

Sadoun, Marc, "Les socialistes entre principes, pouvoir et mémoire," in Rioux, *La guerre d'Algérie*, 225–34.

Schalk, David L., "Péché organisé par mon pays: Catholic antiwar engagement in France, 1954–62," *The Tocqueville Review*, 8 (1986/87), 71–92.

Simon, Pierre-Henri, *Contre la torture* (Paris: Seuil, 1957).

Simonin, Anne, "Les Éditions de Minuit et les Éditions du Seuil: Deux stratégies éditoriales face à la guerre d'Algérie," in Rioux and Sirinelli, *La guerre d'Algérie et les intellectuels*, 219–45.

Sirinelli, Jean-François, "Guerre d'Algérie, guerre des pétitions," in Rioux and Sirinelli, *La guerre d'Algérie et les intellectuels*, 265–306.

—— "Les intellectuels dans la mêlée," in Rioux, *La Guerre d'Algérie*, 116–130.

Smith, Tony, *The French Stake in Algeria, 1945–1962* (Ithaca: Cornell University Press, 1978).

Sondages: Revue Française de l'Opinion Publique (Paris: Institut Français de l'Opinion Publique [IFOP]), Vols. 1954–1963.

Sorum, Paul C., *Intellectuals and Decolonization in France* (Chapel Hill: University of North Carolina Press, 1977).

Soustelle, Jacques, "France Looks at her Alliances," *Foreign Affairs*, 35:1 (1956), 116–130.

___ "The Wealth of the Sahara," *Foreign Affairs*, 37:4 (1959), 626–36.

Talbott, John, *The War Without a Name: France in Algeria, 1954–1962* (NY: Alfred A. Knopf, 1980).

Tartakowsky, Danielle, "Les manifestations de rue," in Rioux, *La guerre d'Algérie*, 131–43.

Tint, Herbert, *French Foreign Policy since the Second World War* (London: Weidenfeld and Nicholson, 1972).

Trinquier, Roger, *Le Temps Perdu* (Paris: Éditions Albin Michel, 1978).

Tripier, Philippe, *Autopsie de la guerre d'Algérie* (Paris, Éditions France-Empire, 1972).

Vidal-Naquet, Pierre, *Face à la raison d'état* (Paris: Éditions La Découverte, 1989).

___ *La raison d'état* (Paris: Les Éditions de Minuit, 1962).

Wall, Irwin M., "De Gaulle, the 'Anglo-Saxons' and the Algerian War," *Journal of Strategic Studies*, 25:2 (2002), 118–137.

Winock, Michel, "Pacifisme et attentisme," in Rioux, *La Guerre d'Algérie*, 15–21.

ISRAEL IN LEBANON

Aharoni, Yair, *The Israeli Economy* (NY: Routledge, 1991).

Arian, Asher, Ilan Talmud, and Tamar Hermann, *National Security and Public Opinion in Israel*, JCSS study no. 9 (Jerusalem: The Jerusalem Post, distributed by Westview Press, 1988).

Bank of Israel, *National Budget for 1985* (Jerusalem, 1985).

Bar-On, Mordechai, *Shalom Achshav* [Peace Now] (Tel Aviv: Hakibbutz Hameuhad, 1985).

Barzilai, Gad, "Democracy at War: Attitudes, Reactions, and Political Participation of the Israeli Public in Processes of Decision Making," Ph.D. dissertation (Jerusalem, The Hebrew University, 1987).

___ "A Jewish Democracy at War: Attitudes of Secular Jewish Political Parties in Israel toward the Question of War (1949–1988)," *Comparative Strategy*, 9 (1990), 179–194.

Ben-Porath, Yoram (ed.), *The Israeli Economy* (Cambridge: Harvard University Press, 1986).

___ "Introduction," in Ben-Porath, *The Israeli Economy*, 1–23.

Benziman, Uzi, *Lo Otzer Be'Adom* [*Would not Stop at Red Light*] (Tel Aviv: Adam Publishers, 1985).

Berglas, Eitan, "Defense and the Economy," in Ben-Porath, *The Israeli Economy*, 173–91.

The Commission of Inquiry into the Events at the Refugee Camps in Beirut, Final Report (Jerusalem: MFA, 1983).

Cordesman, Anthony H., *The Arab-Israeli Military Balance and the Art of Operations* (Lanham, MD: University Press of America, 1987).

Dorman, Menachem (ed.), *Be'tzel Ha'milhama: Sichot Be'yad Tabenkin* [In the Shadow of War: Discussions in the Tabenkin Memorial] (Tel Aviv: Hakibbutz Hameuchad, 1983).

Dupuy, Trevor and Paul Martell, *Flawed Victory* (Fairfax, VA: Hero Books, 1986).

Edelist, Ran, and Ron Maiberg, *Malon Palestina* [Palestine Hotel] (Modan, Israel, 1986).

Eitan, Refael, with Dov Goldstein, *Sippur Shel Hayal* [A Soldier's Story] (Tel Aviv: Maariv, 1985).

Elam, Yigal, *Memalei Ha'pkudot* [The Orders' Executors] (Jerusalem: Keter, 1990).

Evron, Yair, *War and Intervention in Lebanon* (Baltimore: Johns Hopkins University Press, 1987).

Feldman, Shai, and Heda Rechnitz-Kijner, *Deception, Consensus and War: Israel in Lebanon*, JCSS paper no. 27 (Tel-Aviv: Tel Aviv University, distributed by Westview, 1984).

Gabbay, Yosef, "Israel's Fiscal Policy, 1948–1982," in Sanbar, *Economic and Social Policy in Israel*, 85–112.

Gabriel, Richard A., *Operation Peace for Galilee* (NY: Hill and Wang, 1984).

Hann-Hastings, Elizabeth, and Philip K. Hastings (eds.), *Index to International Public Opinion 1981–82, 1982–83, 1983–84, 1984–85* (Westport, CT: Greenwood Press, 1983).

Heller, Mark A., Dov Tamari, and Ze'ev Eytan, *The Middle East Military Balance – 1983* (Tel-Aviv: JCSS, 1983).

Hermann Tamar, "From the Peace Covenant to Peace Now: The Pragmatic Pacifism of the Israeli Peace Camp," Ph.D. dissertation (Tel-Aviv: Tel-Aviv University, 1989).

Horowitz, Dan, "Israel's War in Lebanon: New Patterns of Strategic Thinking and Civilian Military Relations," in Lissak, Moshe (ed.), *Israeli Society and its Defense Establishment* (London: Frank Cass, 1984), 83–102.

Jansen, Michael, *The Battle of Beirut* (Boston: South End Press, 1982).

The Journalists' Yearbook, 1983 (Tel Aviv: Agudat Ha'itonayim, 1983).

The Journalists' Yearbook, 1985 (Tel Aviv: Agudat Ha'itonayim, 1985).

Khalidi, Rashid, *Under Siege: PLO Decision-Making during the 1982 War* (NY: Columbia University Press, 1986).

Kochav, David, "The Influence of Defence Expenditure on the Israeli Economy," in Sanbar, *Economic and Social Policy in Israel*, 25–45.

Naor, Arye, *Memshala Be'milhama* [Cabinet at War] (Lahav, 1986).

Peleg, Ilan, *Begin's Foreign Policy, 1977–1983* (NY: Greenwood Press, 1987).

Rabin, Ytzhak, *Ha'milhama Be'levanon* [The War in Lebanon] (Tel Aviv: Am Oved, 1983).

Rubin Barry, "The Reagan Administration and the Middle East," in Oye, Kenneth A., Robert J. Lieber, Donald Rothchild (eds.) *Eagle Resurgent?* (Boston: Little Brown, 1987), 431–457.

Sanbar, Moshe (ed.), *Economic and Social Policy in Israel* (NY: University Press of America, 1990).

—— "The Political Economy of Israel 1948–1982," in Sanbar, *Economic and Social Policy in Israel*, 1–23.

Schief, Zeev, and Ehud Yaari, *Milhemet Sholal* [Deceiving War] (Jerusalem: Schocken, 1984).

Sharon, Ariel, with David Chanoff, *Warrior: The Autobiography of Ariel Sharon* (NY: Simon & Schuster, 1989).

Tamir, Avraham, *A Soldier in Search of Peace* [in Hebrew] (Tel Aviv: Edanim, 1988).
Yaniv, Avner, *Dilemmas of Security* (NY: Oxford University Press, 1987).

ISRAELI NEWSPAPERS AND JOURNALS
Be'eretz Israel
Ha'aretz
Koteret Rashit
Maariv
Migvan
Monitin
Yediot Aharonot

Index

British military
    conduct in Boer War, 61–2
    intervention in Russian-Ottoman
        conflict, 58
    mortality rates, 75
Bronfman, Roman (Knesset Member),
    255
Bruller, Jean (Vercors), 116, 138
brutality
    by American army in Philippines,
        62
    Britain and, 62–3, 72–3
    against civilians, 40–1
    domestic dimension of, 60–3
    French army/France and, 94–5,
        105–6, 113, 240
    in empire-building, 36
    escalation of, 48
    insurgent populations as victims of,
        72–3
    Israeli military and, 185–6, 253,
        254
    levels of, 24, 25, 27
    as means of cost management,
        42–6
    military, 47, 71–3
    pacification and, 62
    pragmatic, 43
    tolerance to brutal engagement and,
        19–20, 20f
    Vietnam war, 41
Bugeaud, Thomas Robert, 61, 61n42
Bülow, Bernhard von, 60, 62
Bundy, McGeorge, 235
Bundy, William, 242
Burg, Yosef, 218–19
Buron, Robert, 147
Bush, George, Sr., 247

Callwell, Charles E., 72–3
Camp David accords, 192, 252
Capitant, René, 124
Cartier, Raymond, 109
Cartierist ideas, 110
case studies, 26, 26nn60–1, 28nn64–6
    French, 28–30, 30n68
    hard, 29, 29n67
    Israeli, 28–30

casualties/fatalities
    Algerian War, 87, 87nn11–13, 104,
        240, 240t
    Boer War, 45
    Israeli war in Lebanon, 157, 170,
        171, 175–6, 177–89, 178t, 180,
        181, 183–4, 185–6, 190, 218,
        240t
    levels of, 24
    rate of, 8n15
    tolerance to, 19–20, 20f
    Vietnam War, 232n2, 240t
Chaban-Delmas, Jacques, 147
Challe, Maurice, 38, 85
Charbonnier, Lieutenant, 142
Charles X, 50
Chevallier, Jacques, 116, 117, 134, 135
China
    Japanese conduct in, 45, 45n48
    national destruction of Tibet by,
        37
Churchill, Winston, 63
Clausewitz, Carl von, 12, 12nn36–7
coercion
    diplomacy and, 11, 11n31, 12
    extraction cycle, 49, 49n4
    isolation based on, 38, 38n16, 45
Cohen, Eliezer, 196
Cohen, Haim 221
colonialism/imperialism, 71, 72
    British, 8
    failure of Western, 7, 7n9
    French, 83
    loss/demise of European, 4–5n4
    Portuguese, 79
    wars of, 44, 60–3
comparative study, positive/negative
    techniques of, 27, 27n62
compartmentalization, 66–7, 73, 74,
    78–9
concentration camps, 39, 64
conflicts
    asymmetric, 4, 7, 7n12, 10, 229,
        250, 251
    international, 4, 4n3, 16
    Third World, 48, 246
    underdogs winning, 4–5
confounding variables, 28, 28n66

"economic strain" arguments, 6
education
    anti-war ideas and, 76–7
    in Austria 68n4
    compulsory elementary, 68–70,
        68nn4–5
    critical mass and, 76
    in democracy, 59
    in Denmark, 68n4
    in England, 68–71, 69t
    in France, 58, 68–71, 68n3, 68n5,
        69t, 139–40
    in Germany, 68–9, 69t
    in Prussia, 70
    in Sweden, 68n4
    secondary, 70–1
    state-controlled, 51
Ehrlich, Simcha, 218
Einan, Menahem, 203
Eitan, Rafael, 159, 160–1, 168, 171–2,
        175, 187, 187n29
    battle policy of Sharon, Ariel, and,
        185
    consequences of instrumental
        dependence and, 172–4
    departure from office by, 224
    on Israeli media, 199–200, 200n14
    purging of Geva, Colonel Eli, by
        187–8
    refugee camps massacre and,
        189–90, 215–16, 221–2,
        222t
    reserve paratroop brigade not
        mobilized by, 173–5, 220
    unable to stop army dissent, 188,
        188n32
Eliav, Aryeh, 198
Ellsberg, Daniel, 247
empire-building, 72
    decline of, 7–8, 8n13
    extreme brutality in, 36
England. *See* Britain; United Kingdom
        (UK)
eradication
    fighting insurgency by, 42, 46
    of guerilla forces, 46
    methods of, 41
    as "preventive" doctrine, 42

strategy, 34, 41–2, 46, 252
    "surgical," 46
Erez, Yaakov, 215
Erulin, Lieutenant, 142
Esprit, 119, 120, 137, 143
etatist agendas, 64, 80
    v. bourgeois ethos, 57–9
    free press and, 199, 200
    French, 59n36, 90, 92, 199, 200
    Israeli, 161, 163, 192–3, 201
European colonialism/imperialism,
        4–5n4, 7, 7n9, 43
Even, Jackie, 199
expediency-based difference, 20
extermination, 36

Fallaci, Oriana, 205–6
fascism, Algerian War and, 131–2, 133,
        142
fatalities. *See* casualties/fatalities
Faulques, Major, 142
Faure, Edgar, 112, 115, 122
Favrelière, Noel, 112
Feltin, Cardinal (French Army chief
        chaplain), 126
Ferry, Jules, 68
Finer, Samuel, 49n4
FLN. *See* Algerian Front for National
        Liberation (FLN)
Foreign Affairs, 131–2
France. *See also* Algeria; Algerian War
    Algerian community in, 99, 99n1
    Algerian oil and, 89, 89n17, 151
    Algerianization of, 132
    civil law in, 125–6
    collective consciousness of, 99
    colonial history of, 83
    decolonization and, 110
    despotism in, 127–8, 132
    draft and, 103–4, 239
    economic miracle in, 96
    economy (during Algerian War),
        95–8, 96nn40–1, 97t
    education system in, 58, 68–71,
        68n3, 68n5, 69t, 139–40
    empire of, 88–9
    fascist threat in, 131–2, 133,
        142